Basic Sciences in Ophthalmology

Josef Flammer • Maneli Mozaffarieh
Hans Bebie

Basic Sciences in Ophthalmology

Physics and Chemistry

Springer

Josef Flammer, M.D.
Department of Ophthalmology
University of Basel
Basel
Switzerland

Hans Bebie, Ph.D.
Institute for Theoretical Physics
University of Bern
Bern
Switzerland

Maneli Mozaffarieh, M.D.
Department of Ophthalmology
University of Basel
Basel
Switzerland

ISBN 978-3-642-32260-0 ISBN 978-3-642-32261-7 (eBook)
DOI 10.1007/978-3-642-32261-7
Springer Heidelberg New York Dordrecht London

Library of Congress Control Number: 2012951641

Printed on acid-free paper

Springer is part of Springer Science+Business Media (www.springer.com)

Preface

Ophthalmology training is more than just memorizing pieces of information. Particularly important is a comprehensive understanding of the scientific background. This book on "Physics and Chemistry of the Eye" describes the coherence of ophthalmology with physics and chemistry. It is the ambition to provide a better understanding of clinical observations and the way how we treat patients.

Such a physical and chemical background is only conditionally a prerequisite for practising ophthalmology. However, it helps clinicians interpreting phenomena, gives researcher more independency, and increases enthusiasm of curious scientists.

This book is simply an introduction and is not meant to be complete by any means. The mentioned clinical pictures serve merely as examples. For more comprehensive descriptions, please refer to corresponding textbooks.

This first edition may contain weaknesses and mistakes. We encourage readers to give us feedback in order to improve future editions.

For us, writing this book was not just work but also satisfaction. We admire the beauty of the eye and are fascinated the way it functions and are particularly impressed about the interrelations between basic science and medicine. While writing the book, we realized in what sophisticated way fundamental laws of nature enabled the emergence of life.

We hope that some sparks of our enthusiasm may jump to the reader and that this book contributes to the appreciation of ophthalmology both for the benefit of patients and physicians.

For further information and contact: www.glaucomaresearch.ch

Josef Flammer, M.D.
Maneli Mozaffarieh, M.D.
Hans Bebie, Ph.D.

Authors

Josef Flammer, M.D., Professor and Head, Department of Ophthalmology, University of Basel, Switzerland. Special interests: glaucoma, perimetry, pharmacology, microcirculation and molecular biology.

Maneli Mozaffarieh, M.D., Glaucoma Fellow, Department of Ophthalmology, University of Basel, Switzerland. Special interests: glaucoma.

Hans Bebie, Ph.D., Professor Emeritus for Theoretical Physics, University of Bern, Switzerland. Special interests: optics, science of vision.

Acknowledgments

Project manager:
Daniela Hauenstein

Illustrations:
Natasa Cmiljanovic
Rebekka Heeb
Peter Räber

Proofreading and further support:
Vladimir Cmiljanovic
Arthur T. Funkhouser
Katarzyna Konieczka
Nina Müller
Albert Neutzner
Annick Toggenburger
Gertrud Thommen

Additional contributions:
Martina Anderson, Michael Baertschi, Ralf Beuschel, Tatjana Binggeli, Anna Cybulska-Heinrich, Barbara Dubler, Alex Eberle, Arne Fischmann, David Goldblum, Matthias Grieshaber, Farhad Hafezi, Jörg Hagmann, Pascal Hasler, Tatjana Josifova, Simone Koch, Jürg Messerli, Peter Meyer, Ursula Müller, Anna Polunina, Ulrike Schneider, Eberhard Spoerl, Margarita Todorova, Birgit Vorgrimler

Other colleagues who kindly provided us with illustrations are acknowledged in the figure legends (Courtesy of).

Contents

1 What Is Light?

This book was written for ophthalmologists and light will be one of our main topics. We assume that all of us – the authors and the readers alike – marvel about all the various ways that light appears. When we see the pole star in the night sky – because the process of doing it is apparently so self-evident – we are hardly aware that, through this viewing, we are participating in a fantastic process. The light left the star's hot atmosphere in Newton's time and was then underway, solitarily, through almost empty space until it entered our atmosphere and then passed almost unimpeded through several billion molecular shells. Finally, focused on the retina by the cornea and lens, it triggered a state change in rhodopsin molecules that, via a chain of chemical amplifications, led to the hyperpolarization of photoreceptors and fed electrical signals into the brain's neural network. At the end of the process stands the mystery of how the signal then arrives in conscious awareness.

The question "What is light?" has always occupied us. In 1760, in one of his 200 tutorial letters, Euler[1] wrote, "Having spoken of the rays of the sun, which are the focus of all the heat and light that we enjoy, you will undoubtedly ask, What are these rays? This is, beyond question, one of the most important inquiries in physics." Today, physicists assume that the physical nature of light and the laws of its interactions with matter are completely known.[2] The theory is characterized by the simultaneous concepts of light as electromagnetic waves, on the one hand and, on the other hand, as a stream of particles (photons). It seems that any given phenomenon can be easily understood in terms of one of these two models. The wave model easily explains the fundamental limits of visual acuity due to the diffraction of light passing through a pupil, while the photon model forms a suitable vehicle for understanding the absorption processes in a molecule. We shall deal with both of these models and explain how it is that the fact that both are simultaneously true does not represent a contradiction.

In the following sections, we shall be concerned with the question "What is light?" We start with simple assertions from the photon and wave models and then go on to address the highly interesting relationship between them.

1.1 What Did Einstein Have to Say About Blue and Red Light?

When we open a textbook on light and optics, we soon find a picture showing the refraction of light as it passes through a glass prism and how the red, orange, yellow, green, blue, and violet colors

[1] Leonhard Euler (1707–1783), Swiss mathematician. Early in his life, he lost his vision in one eye and, at the time that he wrote the 200 tutorial letters to the princess of Anhalt-Dessau, he was almost totally blind in the other eye due to a cataract. Nevertheless, he continued his immensely creative work. *Letters to a Princess of Germany* (3 Vols., 1768–72).

[2] Based on the quantum theory of light developed between 1930 and 1960. The interested reader may find a discussion of the matter in Sect. 1.4.

J. Flammer et al., *Basic Sciences in Ophthalmology*,
DOI 10.1007/978-3-642-32261-7_1, © Springer-Verlag Berlin Heidelberg 2013

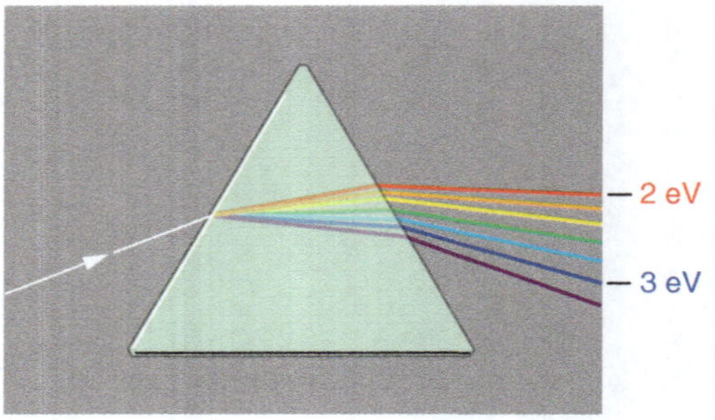

Fig. 1.1 Decomposition of *white light* by a glass prism. If we separate out a small range in the spectrum, we obtain monochromatic light. Photons of *red light* have energies of 2 eV; those of *blue light* have 3 eV

appear in that order. A rainbow has these colors in exactly the same order. With the numbers shown on the right in Fig. 1.1, we note what Einstein postulated in 1905 about the nature of light: it consists of corpuscles (photons), and the energy of red light photons amounts to ca. 2 eV, while that of blue light is ca. 3 eV (electron volts[3]). In other words, light has corpuscular properties, light carries energy, and all photons in the light of a given color have the same energy. This forms the contents of the so-called quantum hypothesis of light. Obviously, a prism can separate the photons of a white beam of light according to their energies. Our eyes also react to these small energy differences: depending on the photon energies, they appear to our minds in their associated colors.

The fact that light is associated with energy is the rather self-evident background to what Einstein said; after all, we let ourselves be warmed by sunlight. However, from the 1,000 W of light power per square meter that enters the atmosphere, only half of it is visible light; the rest consists of infrared and ultraviolet radiation.

Above, we mentioned an initial aspect of light properties without trying to make it plausible. It is well known that Newton already believed that light consisted of particles, but Einstein made this concept more precise. By the way, how big is energy of a few eV? Obviously, it is sufficient to excite rhodopsin and trigger an electrochemical process. A few such processes, taking place in the rods of a fully dark-adapted eye, are enough to create a conscious perception of light. Another possibility of conceiving this is the fact that a few eV of energy suffice to raise a water droplet roughly 1 μm in size by 1 μm.

An impressive demonstration experiment that accentuates light's particle character is elucidated in Fig. 1.2. It is based on the ability of special detectors (e.g., photomultipliers) to react to each photon received with a usable electrical pulse. This can be amplified and fed to a loudspeaker. When a ray of light is so strongly attenuated that the detector absorbs only a few dozen photons per second, a crackling is heard consisting of one click per photon. If the light ray is interrupted, the crackling stops immediately. More precisely, its frequency returns to the so-called dark frequency, triggered by spontaneous thermal effects in the detector.

Light consists of photons – we use this language especially when describing elementary processes that take place in the interactions between light and matter. A single photon excites a single retinal molecule; or an excited atom emits a photon. The absolute threshold of the eye can be expressed in a memorable way in the "photon language." In a completely dark-adapted human eye, a stream of approximately 1,000 photons per second entering the pupil (e.g., at night, from a barely visible star) is sufficient to trigger a consciously perceptible stimulus. The same is true for a brief flash of approximately 100 photons entering the pupil even though only a few photons are absorbed by rhodopsin.

We have not yet discussed the basis behind the quantum hypothesis of light. We are only indicating the questions and problems that Einstein solved at one go:

1. With the quantum hypothesis, Planck's law of the thermal emission of light by hot matter can be theoretically established (this law will be covered in Chap. 3).
2. Without the quantum hypothesis, light would have an infinitely large entropy (inner disorder) – a theoretical argument that was quite prominent for Einstein's deliberations.
3. The quantum hypothesis explains the photoelectric effect; namely, that the energies of electrons,

[3] 1 electron volt (1 eV) is the energy necessary to move an electron with its electrical charge ($1e = 1.6 \cdot 10^{-19}$ Cb, Coulomb) over a voltage difference of 1 V (Volt). $1\ eV = 1.6 \cdot 10^{-19}$ J (Joule). See Appendix for the relations between physical units.

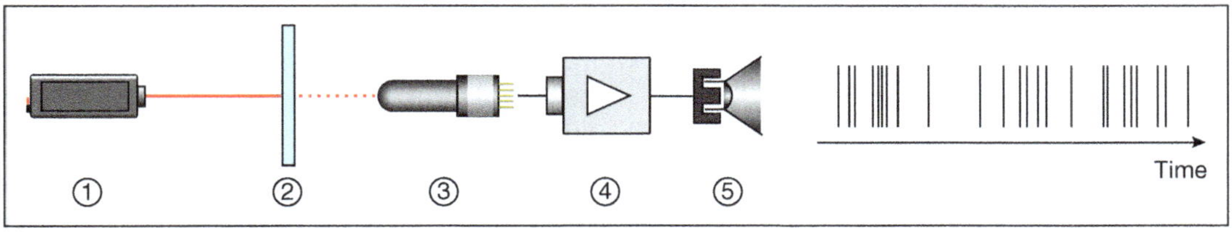

Fig. 1.2 Single photons can be detected individually and made audible as "clicks" in a loudspeaker. Light source (*1*), strongly attenuating filter (*2*), photodetector (photomultiplier, (*3*), electronic amplifier (*4*), loudspeaker (*5*). On the *right*, a typical random series of "clicks" as a function of time is illustrated

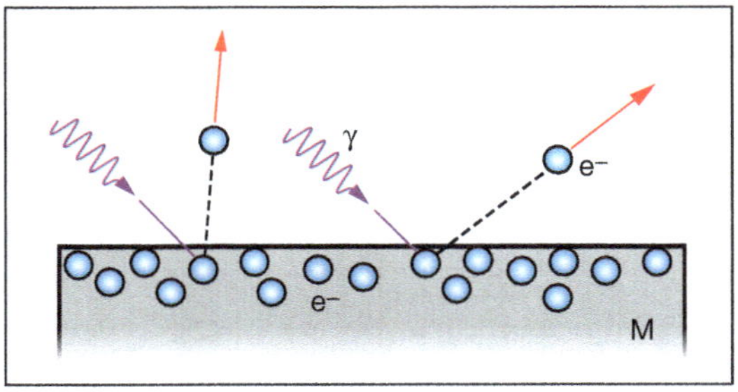

Fig. 1.3 The photoelectric effect. Photons are able to knock electrons out of a metal surface. In doing so, the photon transfers all of its energy to the electron and must transfer at least the energy necessary for it to leave the metal's surface. The size of this so-called work function depends on the metal. For zinc, for example, it amounts to 4.3 eV. The energy that remains after the work function has been overcome is taken by the electron as its kinetic energy. For a material with a work function of 2.5 eV, the effect occurs with blue light but not with red light. An essential feature of the photoelectric effect is that the energies of the electrons do not depend on the light intensity but, rather, only on its color. *M* metal, γ photon, e^- electron

Fig. 1.4 Thomas Young

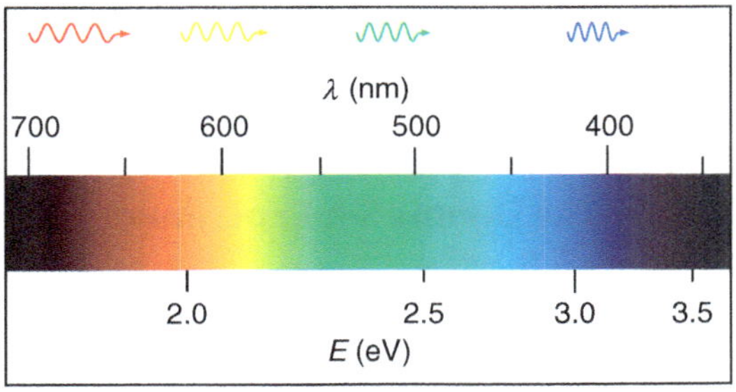

Fig. 1.5 The light spectrum. Each position in the spectrum has an associated wavelength. The wavelengths are indicated above and the photon energies below the colored spectrum

which are ejected from a metal surface when light strikes it, are independent of the light's intensity and depend only on its color (Fig. 1.3).

1.2 Light as a Wave

For a moment, we shall forget about photons and turn to a second, totally different answer to the question concerning the nature of light: light as a wave phenomenon. As evidence, Young[4] developed a very convincing experiment, the double slit experiment, which will be discussed later (Fig. 1.4). For the purposes of orientation, we also assume some facts without substantiating them: red light has a wavelength of ca. 0.7 μm, while that of blue light is ca. 0.45 μm. Figure 1.5 shows more generally the wavelengths of the various colors.

However, when we look around, we see the sun, clouds, blue sky, colored surfaces, light sources, our mirror images – but no waves. Indeed,

[4] Thomas Young (1773–1829), British physician. Originator of the wave theory of light and the three-color theory of vision. He also explained ocular astigmatism.

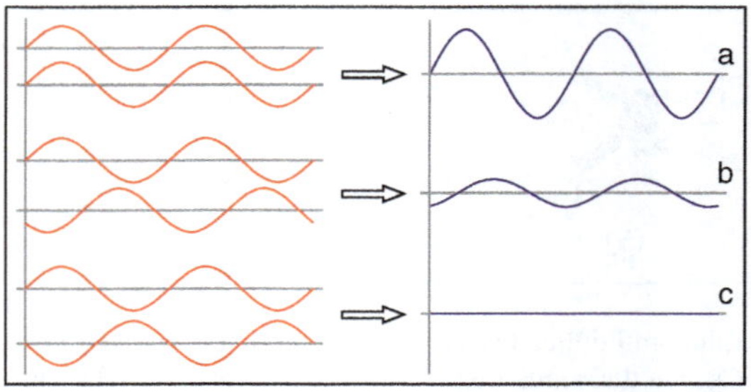

Fig. 1.6 Constructive and destructive interference. (*Left*) Pairs of overlapping waves. (*Right*) The sum of the two components. (*a*) Constructive interference with the same phase with the maximum sum. (*b*) 150° phase shift; the resulting amplitude is weaker. (*c*) Cancellation with a phase difference of 180° (half wavelength) in that the peaks coincide with the valleys

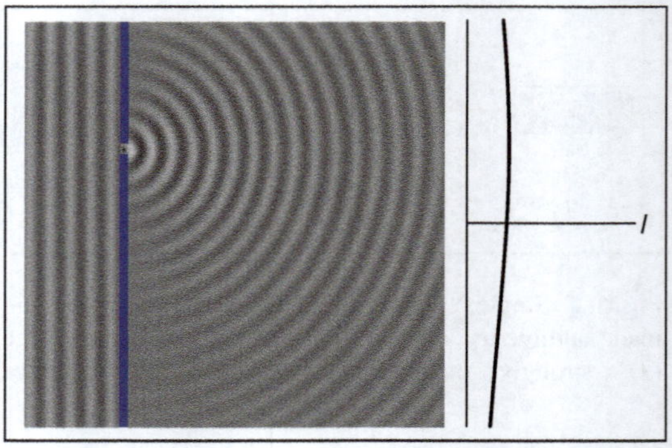

Fig. 1.7 A wave, coming from the left, passes through a single, small opening and spreads out to the right due to diffraction. Averaged over a few vibrational periods, the screen (at the right) becomes uniformly illuminated. *I* intensity on the screen

the wave properties of light are somewhat hidden from the eye due to the very small wavelengths involved (0.4–0.8 μm). The wave properties of light are proven most convincingly, though, with a phenomenon that is observable only with waves: light can cancel out light, just as a peak and a valley of two overlapping water waves neutralize one another. This phenomenon is called interference (Fig. 1.6) and forms the basis for the double slit experiment that we shall now discuss.

1.2.1 The Double Slit Experiment

Young's double slit experiment is deemed decisive evidence of the wave nature of light. Let us start with a single slit. Due to diffraction, as light passes through a single tiny opening, it fans out and produces a uniform illumination of the screen (Fig. 1.7). If a second nearby hole or slit is opened, a pattern of stripes appears on the screen (Fig. 1.8). At certain places, the light coming from the two openings is extinguished. These are precisely the locations where the path differences from the two openings amounts to half a wavelength: a wave crest meets a wave depression.[5]

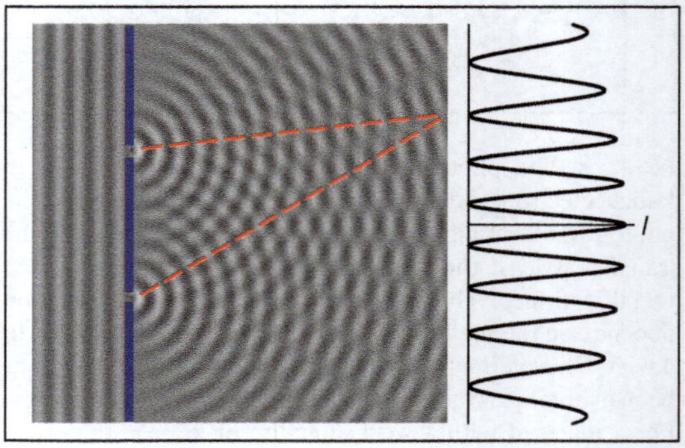

Fig. 1.8 The double slit experiment. Light coming from the left. The screen at the right is illuminated by the light coming from the two openings. An interference pattern appears on the screen. Cancellation (destructive interference) occurs at those locations where the paths from the two openings differ by an odd number (1, 3, 5, …) of half wavelengths. The lengths of the *red paths* differ by 5/2 wavelengths. *I* Intensity on the screen

From the geometric arrangement and the interference pattern, Young could even derive the wavelength. He found the values given above. But what is it that actually vibrates? That was the big question of his time.

1.2.2 A Freehand Interference Experiment

The principle of the interference experiment with light can be observed with the simplest means – with

[5] Young did not carry out his experiment with two openings but split a sun ray with a piece of paper.

the naked eye (Fig. 1.9). The invested time will be repaid by acquiring a more direct relationship with the wave nature of light. Poke two tiny holes into a piece of paper, as close together as possible and, at night, observe a small, bright source of light (e.g., a street lamp at a large distance) through these openings. Through the interference of the light coming from the two openings, a striped pattern arises on the retina that appears to be about the size of the moon. The dark areas arise there where the difference in pathway from the two openings amounts to an odd number of half wavelengths – light waves with a phase difference of 180° extinguish each other. In terms of electrodynamics, at this point, electrical fields with opposite directions meet. The clearest patterns can be seen with monochromatic light. As light sources, the yellow-orange street lamps (sodium vapor lamps, λ=588 nm), for instance, are very well suited. The mechanism is exactly the same as in Young's double slit experiment. For a distance between the holes of 6 mm, the bright stripes on the retina have a separation of roughly 10 μm, corresponding to a visual angle of about 2 min of arc. A model for this demonstration is an apparatus, invented in 1935 by Yves Le Grand, for the interferometric determination of visual acuity (see Sect. 4.2).

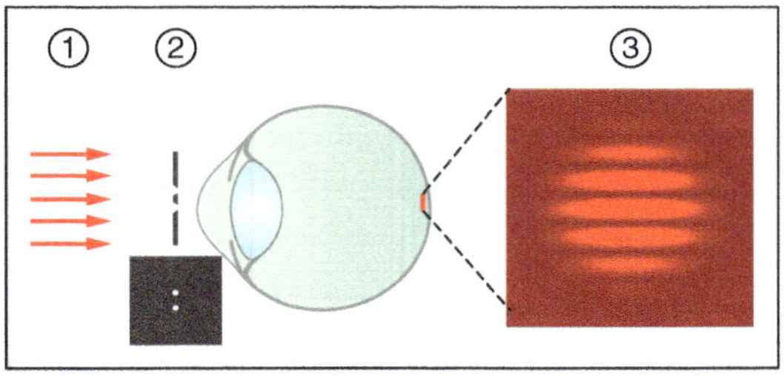

Fig. 1.9 An easy version of Young's double slit experiment: observing light interference with the simplest means. (*1*) Light from a distant street lamp. (*2*) Apertures, consisting of two needle holes (ca. 0.2 mm diameter) close together (ca. 0.6 mm) in a piece of paper. (*3*) Interference pattern on the retina (ca. 0.2 mm diameter, corresponding to 0.5°)

1.2.3 Diffraction

A further manifestation of the wave nature of light is diffraction: When encountering an edge or passing through an aperture, the light's pathway is bent, i.e., deviated from its normal straight line of travel. For example, the play of colors seen when looking at a distant street lamp through an opened umbrella is due to diffraction as the light passes through the periodic arrangement of the fibers of the umbrella textile. Diffraction also occurs when light passes through the pupils of our eyes. This results in a fundamental limitation of visual acuity with small pupils (Sect. 19.1). Due to diffraction, the resolution of a light microscope is also restricted to structures the size of a half wavelength (Sect. 19.4).

Diffraction occurs with every wave phenomenon. With surface waves on water, they can be observed directly: for example, when water waves pass through an opening (Fig. 1.10). For openings

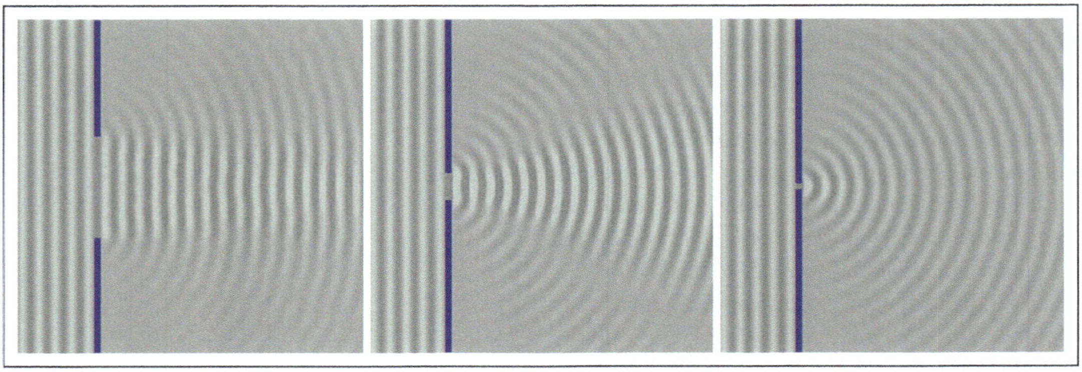

Fig. 1.10 Diffraction of water waves when passing through a harbor entrance. The larger the entrance is in comparison with the wavelength, the less apparent the diffraction will be

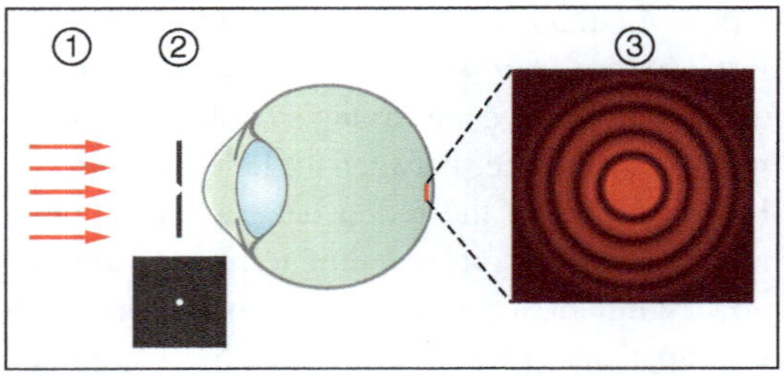

Fig. 1.11 Diffraction rings (*3*) on the retina created by light (*1*) coming from a point-shaped source and through a tiny hole (*2*). As an aperture, a tiny hole is made by sticking a needle point through a piece of paper and this is then held very near the eye

that are much larger than the wavelength, the wave continues without the diffraction being noticeable. However, the narrower the opening is in comparison to the wavelength, the more pronounced the deviation of the wave's direction will be. In the limiting case of an arbitrarily small opening, the wave on the other side spreads out with the same intensity in all directions.

As a variation of the experiment in Sect. 1.2.2, we can try seeing the diffraction image made by a round aperture. This is done by viewing a distant light source (approximating a point source) through a tiny hole (Fig. 1.11).

1.3 Light as an Electromagnetic Phenomenon

Interference and the diffraction of light can be explained by assuming that light is a wave phenomenon without being specific about the precise nature of the vibrations that propagate through empty space in the form of light. We will start with a clear proposition: light is an electromagnetic wave. The waves that travel back and forth between mobile telephone transmitters and cell phones are also electromagnetic waves – the difference lies only in the wavelengths: the physical laws behind them are exactly the same (Fig. 1.12).

As indicated by the section title, a light ray can be understood as a combination of very rapidly oscillating electric and magnetic fields that propagate in empty space at the speed of light. How can we imagine these fields? Briefly put, magnetic fields affect magnetized needles and electric fields exert force on electrically charged particles, e.g., on free electrons or ions. There is an electric field, for example, between the two poles of an electric plug. If the two poles come close enough, the electric field between them is so strong that sparks will be produced in the gap.

A bar magnet produces a magnetic field. It can be perceived by a magnetized needle (such as in a compass) that aligns itself with the direction of the magnetic field. We now consider the situation in which a bar magnet rotates. It generates a magnetic field such that its direction and strength will change at every fixed location. It will oscillate in step with the rotation. Now the laws of electrodynamics take effect: a changing magnetic field engenders an electric field. The rotating magnet also creates an electric field that again oscillates in step with the rotation. Now, another of the basic laws of electrodynamics enters: for its part, a changing electric field once again creates a magnetic field. This mutual creation of changing fields propagates in space with the speed of light.

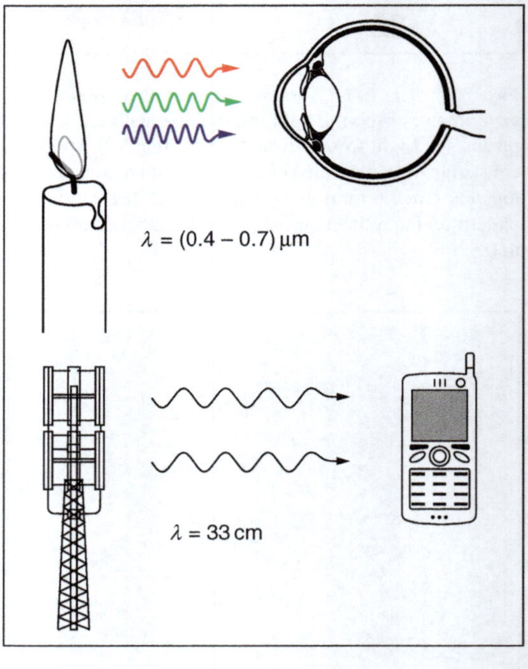

Fig. 1.12 Light as an electromagnetic wave. The difference between light and cell phone waves lies in their respective wavelengths λ (cell phone: λ=33 cm, light: λ=0.4–0.7 µm)

We will not consider the process of the propagation in more detail, but sound and water waves also propagate away from a local disturbance.

With the rotation of the bar magnet, we create an electromagnetic wave that spreads out in all directions at the speed of light. At every fixed location in space, it vibrates in step with the rotation. If the bar magnet were to be rotated with a frequency of 10^8 Hz, radio waves would be produced – and when it is rotated even more rapidly, with a rotation frequency of $5 \cdot 10^{14}$ Hz, yellow light would be seen. Only atoms, though, can achieve such frequencies.

Fig. 1.13 Michael Faraday

Fig. 1.14 James C. Maxwell

Initially, Maxwell's[6] 1864 hypothesis that light consists of electromagnetic waves was purely speculative: at the time, electromagnetic waves were not known but only a possible solution to his equations, resulting from his mathematics. We pay tribute to this event here by concerning ourselves with it a bit further. The empirical foundation was created by the great experimenter Faraday[7] in the first half of the nineteenth century with his research regarding the emergence of electric and magnetic fields from electric charges and currents, as well as the discovery of the laws of induction (changing magnetic fields create electric fields – the basis for transformers). Maxwell succeeded in comprehending all of these phenomena quantitatively with his four equations.[8] In addition, far beyond the laboratory experiments, they exhibited – purely mathematically – a noteworthy solution: electromagnetic waves of any desired wavelength that propagate in a vacuum with a speed of ca. 300,000 km/s – if they were to exist. This speed resulted from the constants measured in the laboratory concerning the relationships between charges, currents, and fields. The agreement with the known velocity of light was spectacular (Figs. 1.13 and 1.14).

In Fig. 1.15, we illustrate an electromagnetic light wave. This is a snapshot of the wave for a moment in time. The whole aggregate would be moving with the velocity of light. This is an especially simple example of a light wave. It has a specific wavelength and does not consist of various colors, and the electric field always vibrates in the same direction. The same is true of the magnetic field. For this reason, we say that such a special wave, like that shown in Fig. 1.15, is linearly polarized. We call the plane in which the magnetic field vibrates the plane of polarization. We shall take up other polarizations in Sect. 1.6. We should not imagine a sunbeam as being so simple; however, it consists of a chaotic overlapping of such waves with all the various wavelengths in the visual range

[6]James Maxwell (1831–1879), Scottish physicist and mathematician. Creator of the fundamental equations of electrodynamics that are still exactly the same today. In a lecture, with three projectors, he demonstrated additive color mixing.

[7]Michael Faraday (1791–1867), English chemist and physicist, investigator of the fundamentals of electricity and electromagnetic fields.

[8]His laws convey, in mathematically exact form, the fact that electric charges and changing magnetic fields create electric fields – electric currents and changing electric fields are sources of magnetic fields. Maxwell's equations are still valid today and are unchanged; they have even survived the "storm" of the special theory of relativity.

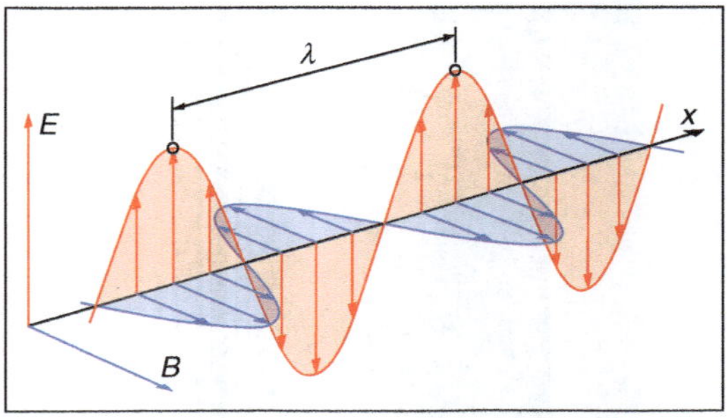

Fig. 1.15 The electric field *E* (*red*) and the magnetic field *B* (*blue*) along a straight line (*x* axis, direction of propagation). For visible light, the wavelength λ amounts from ca. 0.4 μm (*blue light*) to 0.7 μm (*red light*). The figure shows a snapshot. This spatial field structure moves fixedly with the velocity of light in the direction of the *x* axis. Here, the special case of linear polarization is shown: the magnetic field vibrates in a plane (plane of polarization), and the same is true for the electric field

Table 1.1 The electromagnetic spectrum. The energies of the photons are given in electron volts. $1\ \text{eV} = 1.6\cdot 10^{-19}$ J

	Wavelength		
	Min	Max	Energy (eV)
Gamma radiation		0.01 nm	10^5
X-rays	0.01 nm	1 nm	$10^3 \ldots 10^5$
Ultraviolet (UV)	1 nm	0.38 μm	3–1,000
Light	0.38 μm	0.78 μm	1.5–3
Infrared (IR)	0.78 μm	1 mm	0.001–1.5
Microwaves	1 mm	0.1 m	$10^{-5} \ldots 10^{-3}$
Radio waves, communication	0.1 m	1,000 m	$10^{-9} \ldots 10^{-5}$

and nearby frequencies (infrared and ultraviolet) and with all possible polarization directions.

The relationship between wavelength and frequency is not difficult to reason out. When the fields, as shown in Fig. 1.15, move in the direction of the x axis with the velocity of light, a fixed point on the x axis experiences a change in field direction with a frequency $f = c/\lambda$, where $c = 3\cdot 10^8$ m/s is the velocity of light. The wavelength $\lambda = 0.58\cdot 10^{-6}$ m of yellow light yields a frequency of $5\cdot 10^{14}$ Hz. With its range of $\lambda = 0.4 \ldots 0.7$ μm, visible light represents only a narrow region in the spectrum of all electromagnetic waves (Table 1.1 and Fig. 9.8).

1.4 Digression: Are Wave and Particle (Photon) Concepts Compatible?

We will now talk about the relationship between the photon and the wave concepts of light. According to the light-quantum hypothesis, the relation $E = h\cdot c/\lambda$ exists between the energy E of the photons and the wavelength λ of monochromatic light. Here, $c = 3\cdot 10^8$ m/s is the velocity of light and $h = 6.625\cdot 10^{-34}$ Js is Planck's constant. Since $f = c/\lambda$ is the frequency for a wavelength λ, we see the relation often in the form $E = h\cdot f$. For yellow light with $\lambda = 0.58\cdot 10^{-6}$ m, we can easily calculate the other numbers: $f = 5\cdot 10^{14}$ Hz, $E = 3.4\cdot 10^{-19}\ \text{J} = 2.1$ eV.

It is difficult to conceive of light as waves and, at the same time, as a stream of photons. So, what is light? Is it a wave? Or a stream of particles? A simple "both/and" appears helpful and is not wrong. However, we cannot remain satisfied with this statement because a massive problem hides behind it. The necessity of uniting the two points of view brought about a revolution in the physical conception of light and matter. First, we shall phrase the problem.

To do so, we need to return to the double slit experiment (Fig. 1.8) and try to understand it now as resulting from photons rather than from waves. We start with the passage of light through a single opening (Fig. 1.7). Light is diffracted and illuminates the whole screen uniformly. In a pinch, we can also accept a photon concept in which we imagine that the photons are diverted by the edges of the opening. This becomes difficult, though, when we uncover the second opening. As we know, after uncovering the second aperture, the pattern of alternating light and dark stripes emerges. Where the paths taken by the two partial waves differ by an odd number of half wavelengths, the light intensity of the screen disappears. We saw that the intensity distribution as actually observed can be explained by the wave theory without any problem. However, a simple corpuscle idea of light – as a rapid stream of particles, similar to sandblasting – cannot possibly explain the mutual canceling of the two partial beams; the spreading streams from the two openings would simply be added together. In the classical concept, how two streams of particles can cancel each other remains enigmatic. The simple answer of "both/and," thus, has its pitfalls. Nevertheless, both concepts of light indisputably have a justification, depending on the observed phenomenon.

It was the quantum theory (more precisely, the quantum electrodynamic theory of 1928) that came up with the conceptual foundation for understanding the dual nature of light – as both wave and particle. One of the basic ideas states that the wave theory determines nothing more than the probability of detecting a photon at a certain location at a certain time. It is not easy to warm up to this notion – Einstein never believed that such elementary natural events could be based on chance.

A historic experiment[9] showed a way of understanding it. What happens when, in the double slit experiment (Fig. 1.8), the intensity of the light source to the left of the aperture with its two openings is so weak that, only once in a while, maybe once a second, a photon arrives at the aperture? A 1909 experiment showed that the photons at the screen are distributed in precisely the same way as the classical interference pattern. However, when an individual photon goes through one of the two openings, how can it "know" that it should avoid certain places and "favor" others on the screen? Quantum theory maintains that an individual photon behaves in exactly this way: within the framework of the given distribution, it randomly "chooses" a location on the screen for its impact. The sum of many such events, then, crystallizes the distribution that accords with the wave theory. The quantum theory requires us to accept these laws, especially the principle of randomness (unpredictability) in elementary processes, even when these do not coincide with the experiences we have had in the sandbox.

Figure 1.16 shows a modern version of this experiment. The pictures show the locations behind the double slit where the photons impinge, taken with a special CCD camera that is able to register individual photons. When only a few photons are registered, they appear to arrive randomly at the screen. By superposing many pictures, though, it becomes evident that the individual photons "select" the locations of their arrivals with probabilities that accord with the interference fringes of the wave theory. With the so-called statistical interpretation of the quantum theory, the contradictions between the particle and wave concepts are resolved – although it requires an extreme rethinking and acceptance of randomness in individual events of elementary natural happenings. Even this – the so-called statistical interpretation – does not sit easily with us. An example is the question – which we will not pursue any further – of, when an individual photon passes through one of the two openings, how does it "know" about the other opening?

1.5 Light and Color

Our perception of the world we live in is influenced by our sense of color. It is no wonder that we experience this ability again and again as a gift and that we are always fascinated by the richness of the fine, colorful nuances in the moods of a landscape. Here, we have to reduce the sheer inexhaustible subject matter to a few physical aspects.

How do the various spectral combinations of the light that tumbles into our eyes arise? In Chap. 3, we will talk about light sources and how they produce light. Here, we speak briefly about the passive formation of the colors of illuminated objects. When we look around, we see primarily the differing absorption properties of surfaces. The green of a plant leaf comes about because it absorbs the blue and red components of the illuminating sunlight. A red flower absorbs everything except red. The yellow flower absorbs blue, and the remaining green and red is interpreted as yellow. In nature, yellow is often glaringly bright because only relatively little is absorbed – only the blue components that don't contribute much

[9] Taylor GI (1909) Interference fringes with feeble light. Proc. Cambridge Phil. Soc 15:114.

to brightness anyway. Figure 1.17 shows examples of the differing spectra of reflected sunlight.

Less often, colors arise through dispersion (non-uniform refraction depending on color); e.g., in glass fragments or a diamond or from a rainbow. The dependence of light scatter on wavelength bestows on us the blue sky (Sect. 2.7). Nature causes shimmering colors through diffraction at structures – e.g., in the feathers of certain birds or in beetles (Fig. 1.18). We can recognize this in how color reacts to a change of viewing angle. We see the same phenomena in the reflection of light from CD grooves. Colors can also arise due to interference from thin layers, e.g., from a trace of oil or gasoline on water. This occurs when the light reflected from the two interface layers destructively interferes with certain wavelengths. The shimmering colors of certain beetles can also be attributed to this effect.

Our three cone populations with the differing absorption spectra represent the basis for our color perception. The impressive picture in vivo of the mosaic of the cones (Fig. 1.19) was made with the help of adaptive optics (see Sect. 19.3). The hypothesis that our sense of color is based on three receptors with differing reactions to light frequencies was stated by Young[10] at the beginning of the nineteenth century. He went so far as to explain the color blindness of the chemist Dalton as being due to the absence of one of these receptors. The three-color theory was then consolidated

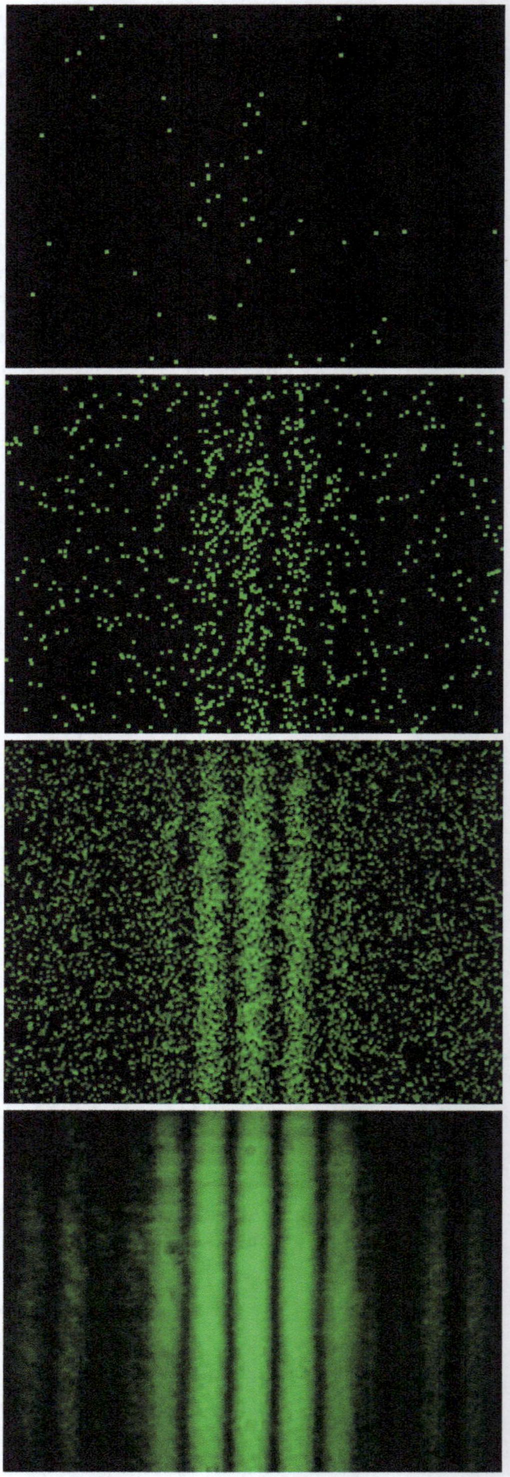

Fig. 1.16 Double slit experiment. Each point indicates the location where an individual photon has impinged on the screen. The individual photons "choose" the random location with probabilities that are determined by the wave concept. Recorded by a single photon imaging camera (image intensifier+CCD camera). The single particle events pile up to yield the familiar smooth diffraction pattern of light waves as more and more frames are superimposed. (Courtesy of A. Weis and T.L. Dimitrova, University of Fribourg, Switzerland)

[10]Mentioned in Sect. 1.2.

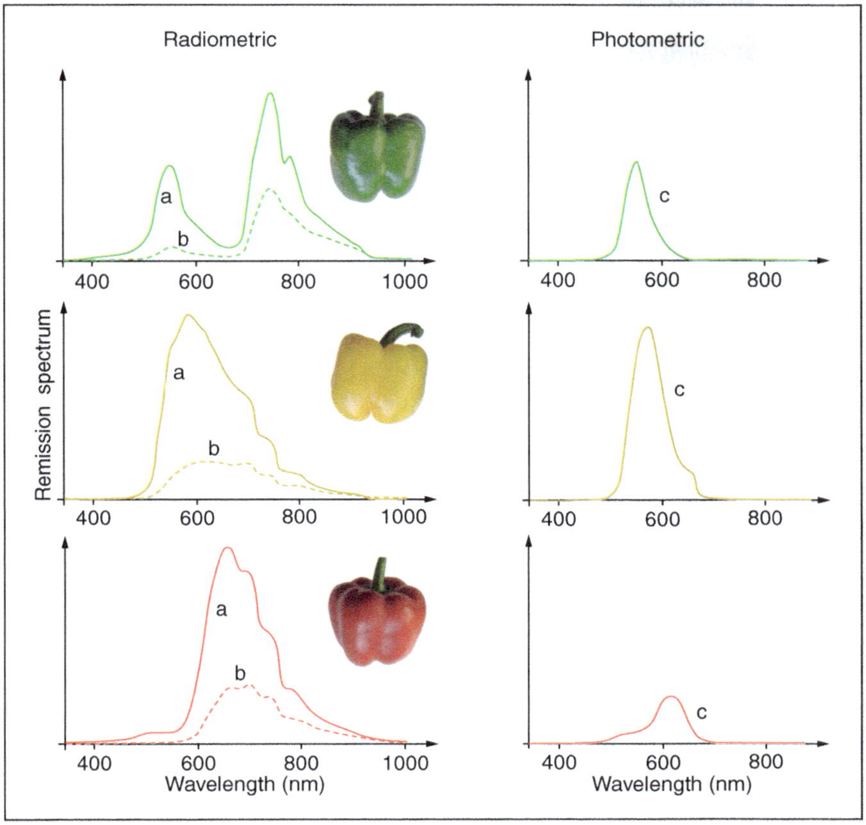

Fig. 1.17 (*Left*) Spectra of the light reflected by green, yellow, and red peppers. (*Solid lines*) In sunlight. (*Broken lines*) In the light of a light bulb (3000 K). The *curves* represent the physical spectra (energy per wavelength interval). Our visual system is able to ignore the differing illuminations. (*Right*) The curves that take into consideration the spectral sensitivity of our eyes. They arise by multiplying the day curves on the left by the V_λ curve (see Fig. 1.22)

and extended by Helmholtz and Maxwell in the middle of the nineteenth century.

In the early phylogenetic stages of our sense of color, only short-wave and long-wave sensors were available for seeing by daylight. Consequently, the perceived color spectrum consisted of a blue-yellow opposition. The corresponding reduction in the range of color perception is indicated in Figs. 1.20 and 1.21. The last developmental stage in the phylogenesis of our sense of color was the differentiation of the long-wave sensitive sensors into ones sensitive to red and green. The protanopia (absence of red cone pigment) and deuteranopia (absence of green cone pigment) represent a regression in two-color vision. Because the sensitivity spectra of the red and green cone pigments are similar (Fig. 1.19), no great difference exists between these two color visions.

However, the differentiation into short and long wave light (blue-yellow opposition) has survived in the retinal coding of the color signals – this is why we experience yellow subjectively as a pure color. The passionate discussions of the time concerning Hering's four-color theory (blue-green-yellow-red) in contrast to the Maxwell-Helmholtz three-color theory (blue-green-red)

Fig. 1.18 Peacock feathers obtain their colors thanks to the diffraction of light from structures

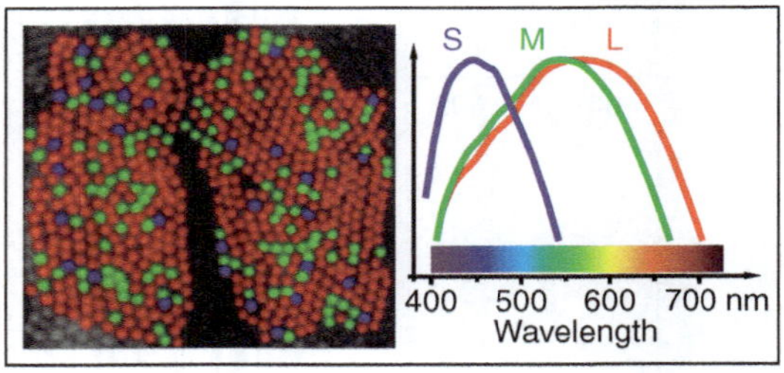

Fig. 1.19 (*Left*) False-color image showing the arrangement of cones in a human retina at a location 10° nasal from the central fovea. The red-, green-, and blue-sensitive cones were identified using bleaching processes and marked in the figure with the associated colorings (Courtesy of A. Roorda and D.R. Williams [Roorda A, Williams DR (1999). The arrangement of the three cone classes in the living human eye. Nature 397:520–522 (With permission)]). (*Right*) The sensitivity spectra of the three cones (arbitrary normalization)

have found their solutions, both in their correlates regarding the construction and organization of the retina.

Our eyes do not have the same sensitivity for all colors. Sensitivity is defined by the ratio of the visual brightness perception to the physical light intensity. Its dependence on wavelength is described by the luminosity function (Fig. 1.22). Toward the ultraviolet and infrared ends of the spectrum, sensitivity falls to zero. For everyday light levels, the sensitivity is given by the internationally defined photopic luminosity function V_λ (cone vision) and by the functio V'_λ (rod vision) for low light levels. These two curves are shown in Fig. 1.22.

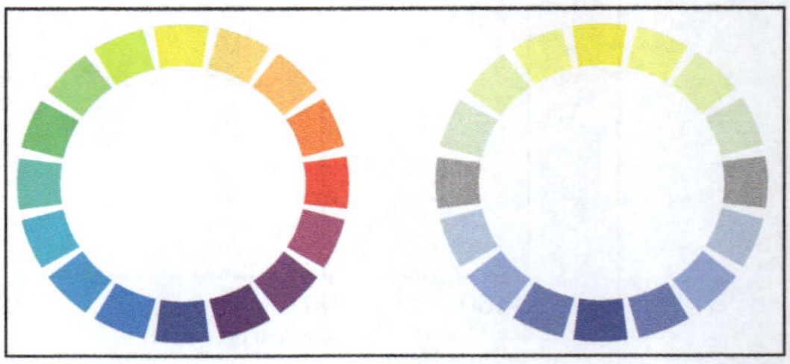

Fig. 1.20 Today's three-color sense and the two-color sense of an earlier stage of development, with the mere distinction between short- and long-wave light. In the development of our color vision, the differentiation of the long-wave light into green and red was the last to form (before approx. 30 million years)

Fig. 1.21 (*Left*) A motif in three-color vision. (*Middle*) Without the differentiation into red and green. The mean of green and red luminosity has been transformed into yellow, which may indicate the kind of loss with red-green dichromacy as compared to three-color vision. No attempt has been made to indicate the difference between protanopia and deuteranopia. (*Right*) With rod monochromacy

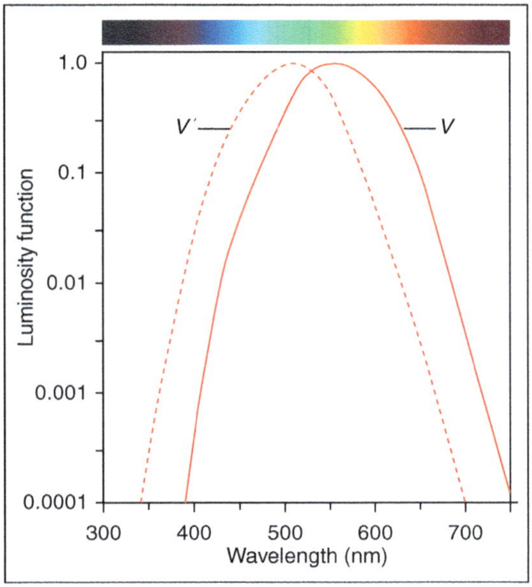

Fig. 1.22 The sensitivity V_λ of cone vision and that of rod vision V'_λ as a function of wavelength λ. Both functions are shown normalized with respect to their maxima. *Abscissa*: wavelength. *Ordinate*: photopic and scotopic luminosity functions. Note that the ordinate is scaled logarithmically

1.6 Polarization

Our eyes have almost no direct access to the polarization of light.[11] Using Polaroid sunglasses, many insights into this phenomenon can be obtained: the brightness of the blue sky changes when the Polaroid lenses are rotated. Reflections, such as those from wet streets, are strongly attenuated. If two Polaroid films are put on top of each other so that their polarization directions are crossed perpendicularly, no light comes through. However, if a few layers of cellophane are put in between the two films, a brilliantly colored picture results (Fig. 1.23). Modern techniques for the projection of 3D films also use polarized light.

The phenomena of polarization originate from the fact that the electric field vibrates perpendicular to the direction that the ray of light travels but, otherwise, it can take on a variety of orientations. Normally, a ray of light is composed of contributions from all possible vibrational electric field orientations. This is the case for sunlight or for the light from an incandescent light bulb. In this case, we speak of unpolarized light. The left half of Fig. 1.24 shows unpolarized light.

Regarding its vibrational orientation, linearly polarized light is more ordered: the electric field vibrates everywhere with the same orientation. This condition is indicated in Fig. 1.24 (on the right). Linearly polarized light arises when unpolarized light passes through a polarizing filter. For example, Polaroid films[12] serve as polarizing filters. They let electric fields of a specific orientation pass and absorb light that has electric fields vibrating perpendicular to that orientation. The orientation of an electric field that is let through is set in the Polaroid film's manufacturing process. Long, parallel molecules that have been made electrically conductive absorb the electric fields that are aligned with them but not the field components perpendicular to them.

We now treat the passage of linearly polarized light through a filter with any given orientation a bit more precisely (Fig. 1.25). The essential idea is the mental separation of the incident light into two components, one of which is parallel and the other perpendicular to the filter's orientation. One component is allowed to pass through while the other is absorbed. This construction explains the amplitudes of the components allowed to pass through in Fig. 1.24.

[11] In Marcel G. J. Minnaert's very beautiful book *Light and Color in the Outdoors*, one finds information on how one can perceive "Haidinger's brush" – as the only weak influence of the polarization of light on our visual perception.

[12] Edwin Land (1909–1991), American inventor and industrialist. As a student, he discovered how to fabricate polarization filters from plastic.

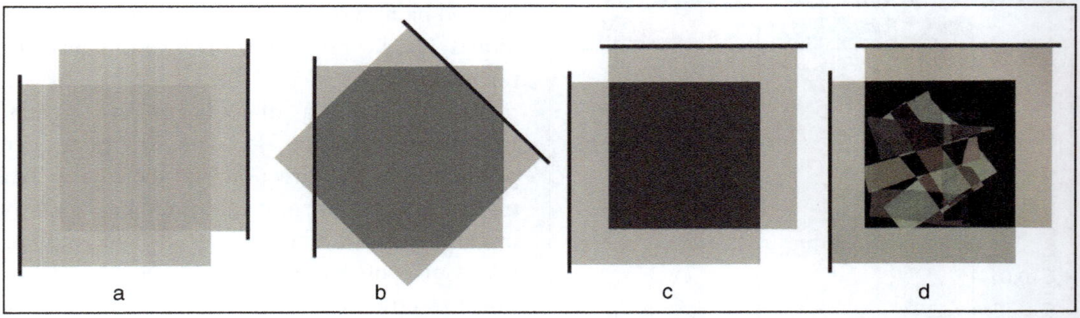

Fig. 1.23 Viewing a white background through two Polaroid films lying one on top of the other and rotated by varying amounts: (**a**) The same angular orientation; no further influence of the second film. (**b**) Turned 45°; reduction of the intensity by half. (**c**) Crossed; the light is completely blocked. (**d**) Crossed but with layers of irregularly shaped cellophane foils between them. The partial transparence is due to the rotation of the direction of polarization by the cellophane foils, depending on wavelength

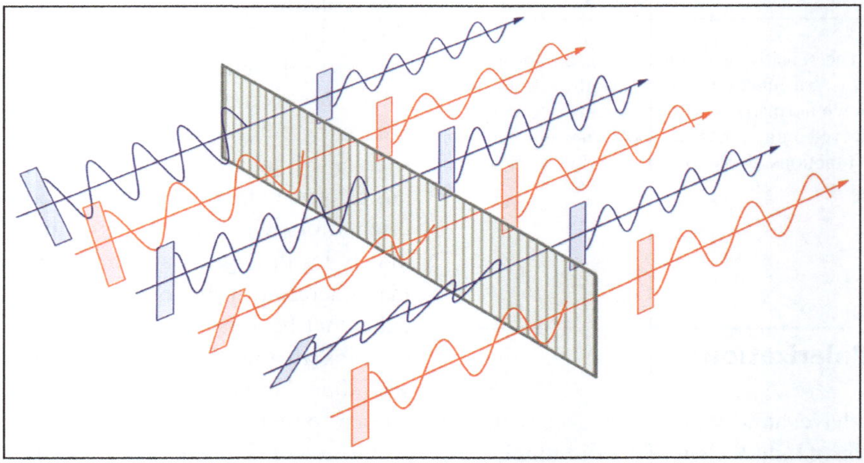

Fig. 1.24 Unpolarized (*left*) and linearly polarized light (*right*). Indicated are the vibrations of the electric fields. Here, unpolarized light passes through a polarizing filter (e.g., Polaroid film) that lets through the vertical components of the electric fields but absorbs the horizontal vibrating components

When reflected off a smooth surface, light becomes partially or completely polarized. Reflected off water, the electric field is mainly polarized horizontally. Polaroid sunglasses block this polarization orientation and attenuate reflections from water and wet streets (Fig. 1.26). By blocking the polarized scattered light from the atmosphere, pictures with improved contrast can be acquired using polarizing filters (Fig. 1.27).

Finally, we will briefly discuss circularly polarized light. In contrast with linearly polarized light, the electric field vectors do not move within a fixed plane; rather, their polarization orientation follows a spiral as the light wave moves forward. Within a distance of one wavelength, electric vectors of this type of light will have made one full turn (360°) about the axis (Fig. 1.28). Left circular and right circular versions exist. Light with this type of

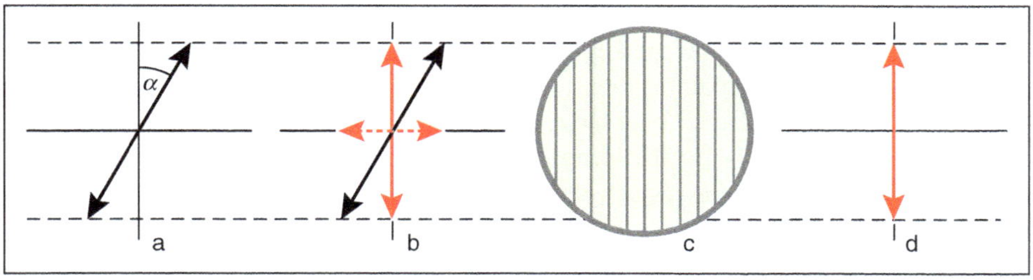

Fig. 1.25 Linearly polarized light passes through a polarizing filter with a vertical transparency orientation. (**a**) Vibration of the arriving electric field, angle α to the transparency orientation of the filter. (**b**) Decomposition into two vibrational orientations: one in the transparency orientation and the other perpendicular to it. (**c**) The filter with a vertical transparency orientation lets one component through (**d**) and absorbs the other

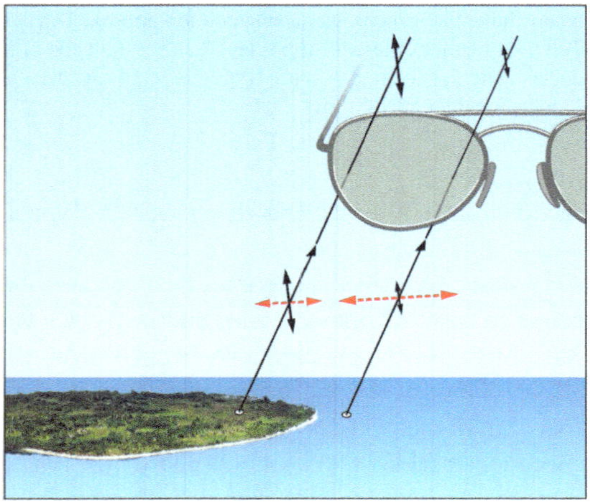

Fig. 1.26 The electric fields of light reflected from water's surface vibrate primarily horizontally. Polaroid sunglasses block this vibrational orientation. On the other hand, light coming from land is made up of all vibrational orientations (unpolarized light)

Fig. 1.27 A suitably oriented polarizing filter blocks part of the polarized scattered light from the sky, as well as light reflected from the water surface (Courtesy of Essilor (Suisse) SA)

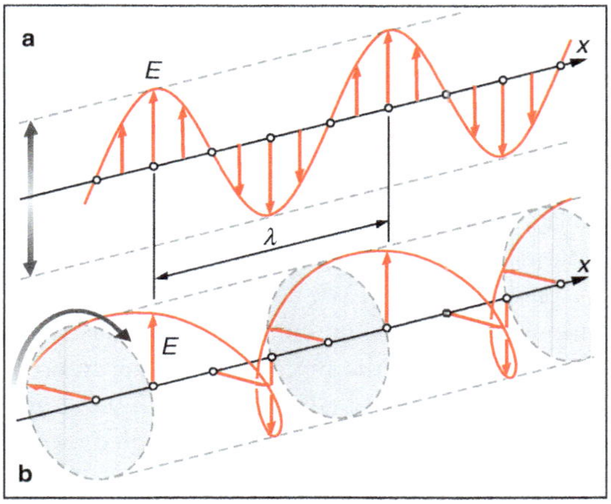

Fig. 1.28 Circular vs. linear polarization. (**a**) Snapshot of linearly polarized light. (*Arrows*) Electric field vector. The field configuration moves with the velocity of light in the direction of the x axis. At any given point in space, the field oscillates with the frequency of light. (**b**) Snapshot of circularly polarized light. (*Arrows*) Electric field vector. The field configuration moves with the velocity of light in the direction of the x axis. At any given point in space, the field rotates with the frequency of light

polarization can also be easily created with a suitable filter. It arises when linearly polarized light passes through a so-called $\lambda/4$ plate. This consists of a birefringent medium of a suitable thickness. Circularly polarized light can be recognized in that it is linearly polarized after passing through a $\lambda/4$ plate. Based on this principle, filters can be manufactured that let either left or right circularly polarized light pass through unattenuated.

Various approaches are available for conveying 3D films. Fundamentally, they must be based on offering the two eyes of the viewer slightly varied images. These technologies make light with differing polarizations available to the two eyes: either two orientations of linearly polarized light or left and right circularly polarized light. The lenses of the polarized eyeglasses select the correct components for each eye. The projection screen must be coated with a metallic layer so that the polarizations of the light sent out by the projector are not lost when they are reflected.

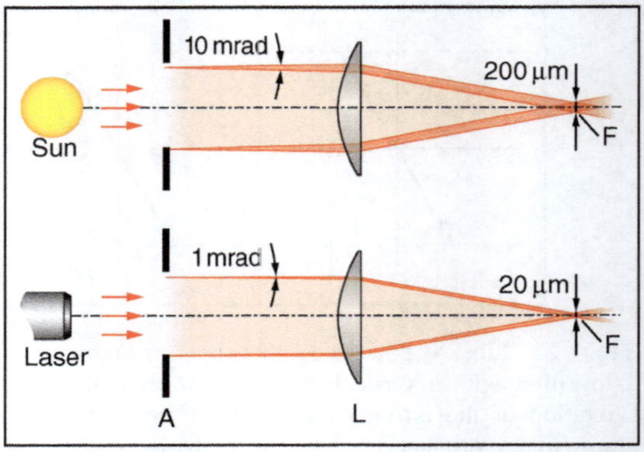

Fig. 1.29 Sunlight cannot be concentrated as well as laser light can. Behind an aperture of 1 mm of diameter, both beams have the same power (laser pointer, approx. 1 mW). The laser beam expands 10 times less than sunlight and produces a smaller focal spot with 100 times more irradiance. *L*: lens, focal length 20 mm

1.7 Laser Light

In 1960, only 2 years after Theodore Maiman was able to get a laser[13] to work, laser light was used for an intervention on a human retina. However, at that time, no one imagined the wealth of applications to come in the following years and decades. Today, in ophthalmology, special surgical instruments and also highly developed imaging systems are based on lasers. We will address these applications, as well as the construction of lasers, in later chapters. At the moment, we wish to bring attention to the properties of laser light. Laser light exhibits several extraordinary characteristics: (1) concentration of the light into a highly directional beam, (2) a very narrow spectrum, (3) coherence, and (4) the possibility of pulsed operation with extremely high momentary powers. In a very memorable image – even if it is not completely precise – we have the impression of a laser beam as being parallel, monochromatic light.

First, we consider the beam of a laser pointer. In a wave picture, it is well described as an electromagnetic wave, as shown in Sect. 1.3. The light is almost monochromatic; i.e., it has a defined wavelength λ and, thus, also a defined frequency $f=c/\lambda$. The electric field oscillates with this frequency at any fixed location. Many types of lasers (but not all) produce a linearly polarized beam, which can be verified using a polarizing filter. In addition, we characterize the beam with its cross-sectional area F as well as the power N. Typical values for a laser pointer are $N=1$ mW and $F=1$ mm^2. Described in terms of corpuscles, the beam consists of photons with an energy $E=h\cdot f$. Since an (almost) monochromatic beam is involved, all the photons have the same energy. The narrow spectrum of many lasers – as a further major difference to thermal light – is not of primary importance in many applications. The wavelength range of a He–Ne laser beam amounts to less than part of 10^{-5} of the wavelength itself (0.6328 µm). In this case, we speak of an exceedingly narrow spectral line. For most applications, it suffices to say that laser light has a specific wavelength, depending on the laser type. Closely associated with this are well-defined absorptions in various media, depending on the wavelength. On the other hand, the sharpness of the spectral line plays a role in laser spectroscopy where we wish to achieve very selective excitations of certain atoms or molecules with light to detect their presence, e.g., in environmental diagnostics. For applications such as this, laser light is almost an ideal instrument.

How does a ray of sunlight differ from the beam of a laser pointer, e.g., behind a cross-sectional area opening of 1 mm^2 (Fig. 1.29)? In terms of power, both beams are practically the same; each is approximately 1 mW. Sunlight

[13] LASER: Acronym for Light Amplification by Stimulated Emission of Radiation.

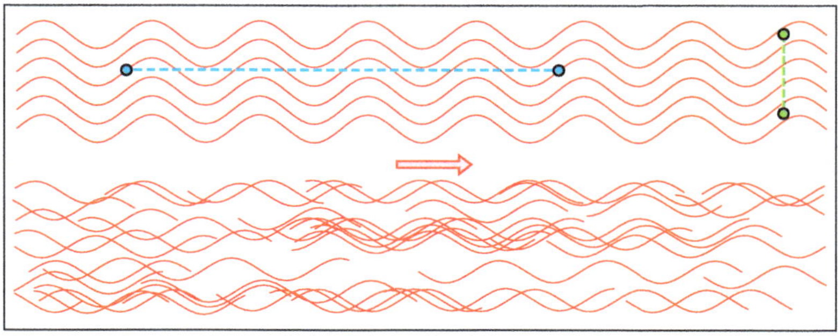

Fig. 1.30 Internal order within a laser beam (*top*) and a thermal beam (*bottom*). (*Top*) Various points within the laser beam oscillate in phase with one another. Spatial coherence: in phase oscillation of points lateral to the direction of the beam (*green points*). Temporal coherence: earlier and later parts of the beam are in phase (*blue points*). (*Bottom*) Electrical fields of thermal light are uncorrelated at various points in space (see text for more precise statements concerning rays of sunlight)

consists of all possible colors. This means that the beam is a combination of components of various wavelengths and frequencies and thereby has photons of a wide range of energies. If we image a sunbeam with a focal length of approximately 20 mm – comparable with the view directly into the sun through an aperture of roughly 1 mm in diameter – a focal spot results that has an irradiance of about 25 mW/mm^2. If, on the other hand, we were to focus the beam of a laser pointer with the same optics, we would have 100 times more irradiance at the focus because the beam divergence of the laser pointer is 10 times less, resulting in a focal spot that is 10 times smaller. The beam divergence of the laser pointer amounts to roughly 1:1,000 (1 mrad), meaning that, at a distance of 10 m, it expands to 1 cm. A sunbeam, on the other hand, has a divergence of 1:100 (10 mrad), corresponding to 0.5°, the size of the sun's disk, and this leads to an expansion of 10 cm at the same distance. For the retina, a glance into a laser is, thus, much more dangerous than a glance at the sun.

Laser light is often described as coherent. This means that the electromagnetic fields oscillate in phase at various points in the beam, whereby the points can be separated transversally as well as along the beam axis. In the terminology of statistics, the coherence of the light at two points means that both fields are correlated in their temporal courses. At two points lying in a cross-section of the laser beam (Fig. 1.30), the fields move in phase with one another – they are spatially coherent. At the two points along the laser beam, the electric fields are also strongly correlated – although they left the laser at different times. This is called temporal coherence within the beam. This is different from a ray of sunlight, in which the spatial coherence is limited to lateral distances of less than 0.1 mm and the temporal coherence for full (unfiltered) sunlight corresponds to a distance along the beam on the order of 1 μm. The picture of the beams in Figs. 1.30 and 1.31 are to be taken as an impression – the quantum chaos of thermal light cannot be depicted faithfully in a figure.

Among typical ophthalmological applications, the coherence of laser light is not of primary importance, except in interferometric measurement methods. Parameters that normally count are those such as beam power, pulse duration, pulse energy, and beam divergence. In this regard, the differences between laser and thermal beams may seem academic. For a deeper understanding of the physical nature of light, though, they are essential. In the following digression, we shall once again consider the topic of the ability to interfere as well as the differing uses of the word "coherence" in classical wave optics and quantum optics.

Fig. 1.31 Birds as symbols for the difference between thermal light (less ordered, *left*) and laser light (coherent, *right*)

1.8 Digression: The Concept of Coherence

In the more general framework of wave optics, coherence has the meaning of the interference ability of light, i.e., the ability of two light waves to mutually (completely or partially) cancel or reinforce when shifted relative to each other. To form the concept, we consider once again the double slit experiment, but now more differentiated in slightly different implementations (Fig. 1.32).

In Fig. 1.32a, a laser beam illuminates two tiny openings, A and B, in an aperture so that the well-known interference pattern appears on the screen behind them. The light fields that come from the two openings cancel each other out at a screen location when the path difference amounts to half a wavelength (or three halves, etc.), such as at point 2. The locations in between are especially bright because constructive interference occurs there (point 1). Interference on the screen presupposes that the two openings, which illuminate the screen as if they were tiny light sources, oscillate in phase. This is guaranteed by the high amount of order in the laser beam. We say that the light fields in the two openings are spatially coherent. The pattern on the screen continues on both sides far away from the middle even though the difference between the two path lengths increases. It is true that the brightness is somewhat less, but the deep modulation remains the same. This is actually surprising because, due to the path length differences, the two contributions had to leave the laser source at different times. Here, the temporal coherence of the laser beam becomes evident: a part of the beam is able to interfere with another part that lags behind it – depending on the type of laser, this distance can amount to meters or even kilometers. These particularities of laser beams become even more pronounced when compared with thermal light.

In Fig. 1.32b, a point-sized incandescent light source illuminates the two openings. In a symmetric arrangement, the two fields in the openings oscillate in step (in phase) with one another. They are, thus, spatially coherent because they left the original point source at the same time. With small thermal light sources, spatial coherence is, therefore, indeed possible. However, can we expect to see an interference pattern on the screen? Certainly, in the middle of the screen, constructive interference with a corresponding increased brightness will appear (point 3). Off to the side, though, only a few variations in brightness are to be expected because the path differences from the two holes mean that light fields that have left the original light source at differing times (points 4, 5) should interfere. The interference pattern is, thus, less distinct because temporal coherence is missing in the illumination. The temporal coherence in a thermal beam can be greatly improved using narrow band filters. For thermal light from a single spectral line, temporal coherence can be present across a distance of a meter along the beam.

Finally, in Fig. 1.32c, two independent thermal light sources illuminate the two openings.

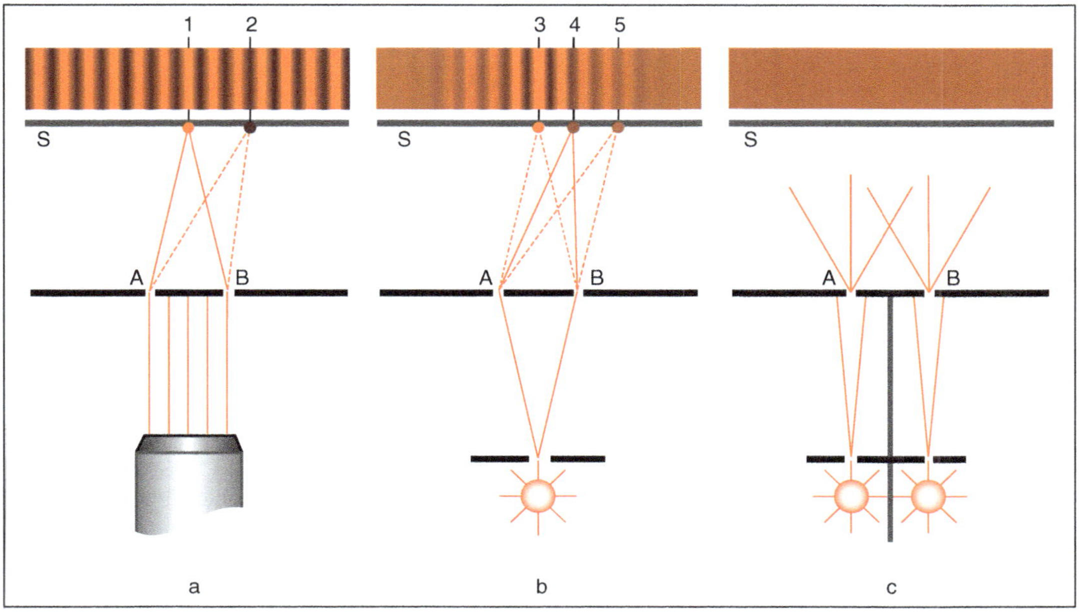

Fig. 1.32 Coherence. The two openings (A, B) in the first screen are considered point light sources that illuminate the second screen (S). (**a**) Monochromatic source. Both point sources A and B oscillate exactly in step; interference is visible on the second screen. (**b**) Incandescent white light. At an off-axis point on the second screen, the beam interferes with a temporally delayed copy of itself. (**c**) Incoherent sources exhibit no interference

Fig. 1.33 Momentary intensity of thermal light (*left*) and laser light (*right*) as a function of time

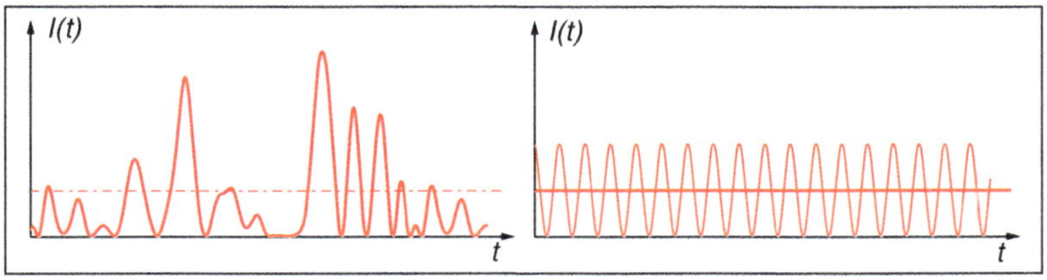

Here, neither spatial nor temporal coherence can be expected. The light coming from the two apertures illuminates the screen uniformly (the figures do not reflect the fact that the intensities away from the center must decrease due to the increasing distance from the openings A and B).

1.8.1 Coherent Light in the Sense of Quantum Optics

The word coherence also has a second meaning: the one where laser light exhibits an inner ordering that differentiates it considerably from the unimaginable chaos present in the beam of thermal light. The associated conceptualizations originate from quantum optics, which was developed in the 1960s as an application of quantum theory to optics. How, then, does this difference manifest itself? One initial manifestation is shown in the fluctuations of the momentary intensity of the light beam. The laser beam exhibits practically constant intensity. Even more amazing are the unavoidable enormous fluctuations of the momentary intensity of a thermal light beam (Fig. 1.33). However, the time in which the intensity noticeably changes is so short that these fluctuations cannot be perceived in normal observations.

This difference also manifests itself in the distribution of the number of photons that arrive

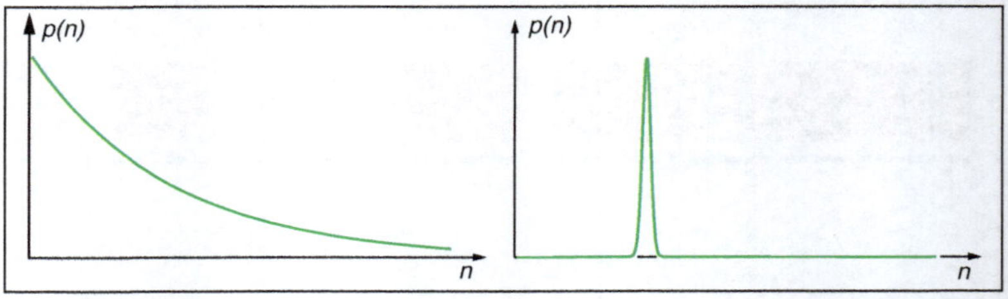

Fig. 1.34 Frequency distribution $p(n)$ of the number n of photons that arrive at a detector at very narrow time intervals, for thermal light (*left*) and laser light (*right*)

at a detector in very short time intervals; in a laser beam, this number fluctuates by very little, while thermal light shows large fluctuations (Fig. 1.34).

An even more basic manifestation of laser beam coherence, in this sense, is seen in the electromagnetic field that comes very close to being the sine-curve shaped wave known from classical electrodynamics, as suggested in Fig. 1.15. The laser is, thus, a demonstration that an electromagnetic wave – like those emitted by radio transmitters – can also be realized at the wavelength of light. This property must be appreciated as distinct from thermal light; there, the electromagnetic field is in a chaotic state that is not in agreement with the classical concept of electromagnetic fields. The force effects of the electric field of a laser beam on an electron are determined at every point in time, while that of thermal light is completely and unpredictably random. The cause does not lie in the broad spectrum of thermal light: even if filters are used that transmit only a very small range of wavelengths, the fundamental difference remains.

A sunbeam cannot cast off the chaos of its creation, even in the case of selecting a very small range of wavelengths, whereas a laser beam already has a much more ordered "ancestry." Once again, as fundamental as this inner property of laser light is for our understanding of the nature of light, it is fully irrelevant for understanding the interactions of laser light with matter in medical or technical applications. There, for the most part, only external properties such as power, power per area, beam divergence, wavelength, and the controllability of the pulses are important. Indeed, it is even difficult to demonstrate this hidden inner quality of laser light in comparison with thermal light.

2 The Interaction Between Light and Matter

What happens when light meets matter? There is always an interaction: light is scattered at a wall's surface, reflected off a surface of water, partially absorbed and partially reflected by a green leaf, refracted when it enters glass, and excites chemical processes in retinal rods and cones, even at very low intensities. The details depend on the structure of the matter and on the wavelength of the light. Additional phenomena are refraction, diffraction, and fluorescence – even the miracle of transparency is fascinating. How is it possible that light passes almost completely unimpeded through a structure like the cornea or through water molecules? In this chapter, we discuss how light is affected by matter. In Chap. 7, we will discuss the special action of light on tissues.

2.1 Phenomenology

Almost all of the processes mentioned above can be illustrated using the eye as an example. Thanks to the refraction of light at the air–corneal interface and at the aqueous humor–lens interfaces, a sharp image is engendered on the retina. The cornea reflects a crossbar or a Placido disk. The aged lens scatters light and reduces the image contrast at the level of the retina. Blood mainly absorbs blue and green light and converts the energy into heat so that red is the dominant color in the light that is scattered back.

The blue iris owes its color to the same process that produces a blue sky: i.e., light scattered by particles that are smaller than the light wavelength. Shorter wavelengths are scattered much more than the longer ones (Rayleigh scattering). The color of a brown iris arises from absorption by a pigment. The white color of the sclera is explained by the almost total scattering of all colors in every direction. In fluorescence angiography, the conversion of light to longer wavelengths is applied. Due to its wave properties, even the diffraction of light is manifest within the eye: the smaller the pupil is, the larger the smallest image of a point source of light at the retina will be. A few of the more important processes are depicted in Figs. 2.1 and 2.2. In the following chapters, we discuss in detail some of these processes and their ocular manifestations.

2.2 Fundamental Physical Processes

We shall occupy ourselves only briefly with the atomic bases of the mentioned processes. The basic principle is always the same with visible, ultraviolet, or infrared light. When light encounters a surface or passes through a medium, inevitable interaction occurs between the light and the electrons of the atoms and molecules of the material. A simplified picture of classical electrodynamics involves the interaction of two fundamental processes: first, the light exerts a force on the electron[1] and, second, as a charged

[1]More precisely, the charged electron experiences an accelerating force in the light's electric field.

J. Flammer et al., *Basic Sciences in Ophthalmology*,
DOI 10.1007/978-3-642-32261-7_2, © Springer-Verlag Berlin Heidelberg 2013

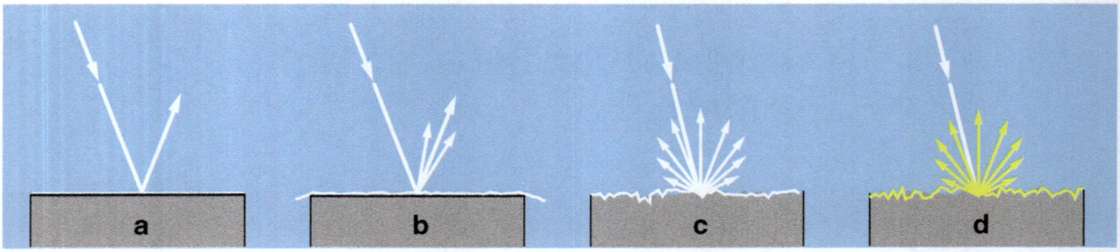

Fig. 2.1 Some of the interactions of white light with surfaces. (**a**) Specular reflection at a smooth surface. (**b**) Lustrous reflection from paper with a slightly rough surface. (**c**) Diffuse reflection from a whitewashed wall; no absorption. (**d**) Diffuse reflection with absorption of the shorter wavelengths at a painted yellow wall

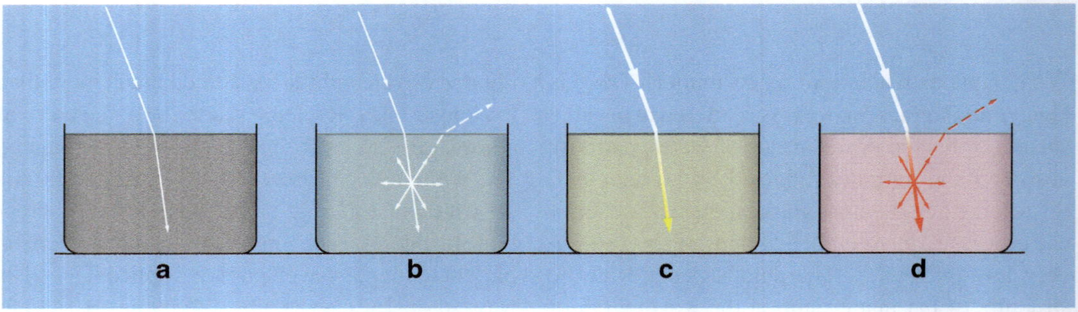

Fig. 2.2 Some interactions of white light with media. Refraction takes place when a ray of light penetrates from above into the medium below (as seen in these diagrams). The media are, for example, gases, fluids, or tissues. (**a**) Refraction. In the denser (lower) medium, light travels more slowly and in a changed direction. (**b**) Scattering, not color-selective (strongly diluted milk). (**c**) Absorption without scattering (clear medium). After blue has been absorbed, the remaining ray of light is yellow. (**d**) Absorption of blue and green, additional scattering of the light (cloudy medium such as blood)

particle, the accelerated electron radiates electromagnetic waves (light).

Scattering of light by a free electron provides an example. When light meets an electron, it is "shaken" at the frequency of the light. As a result, the electron sends out light with the same frequency in any direction. Thus, light scattering takes place. This process represents one of the impediments that solar photons surmount when they must fight their way from where they are produced in the interior of the sun to its outer surface. A second example is that light penetrating through a metallic surface causes the cloud of negative charge – consisting of weakly bound electrons of the metal atoms – to vibrate in phase with the light frequency. This vibrating and charged cloud then produces light of the same frequency, specifically reflected light. A third example is that, inside glass, electrons are also stimulated to vibrate. Instead of reflection, the only consequence in glass is that the light is slowed down somewhat without being absorbed. This slowing down of the light is the basis for refraction (Sect. 2.4).

The electric field of light exerts forces of the same strength on the protons of the atomic nuclei as it does on the electrons. However, due to the much larger mass of the protons and their strong binding within the atom's nucleus, the interaction of visible light with the nucleus is far weaker and is practically negligible in the visible range.

The basic process of the interaction of light with matter can be described more precisely by means of quantum theory: the electron of an atom, a molecule, or an atomic lattice can absorb a photon and use its energy to jump into an energetically higher state (Fig. 2.3). Conversely, an electron can fall into a state of lower energy, with the energy difference being sent out as a photon (Fig. 2.4). Actually, it is usually not just a single electron but,

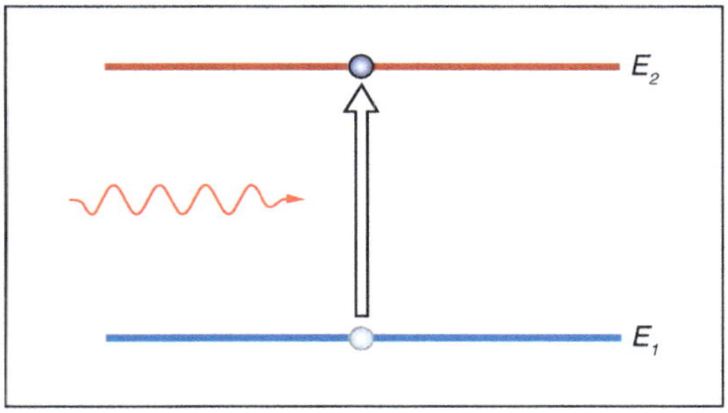

Fig. 2.3 Absorption of a photon. Its energy is transferred to the atom and raises its electron shell onto a higher energetic state. This process is only possible when the photon's energy "fits" a gap in the atom's energy spectrum

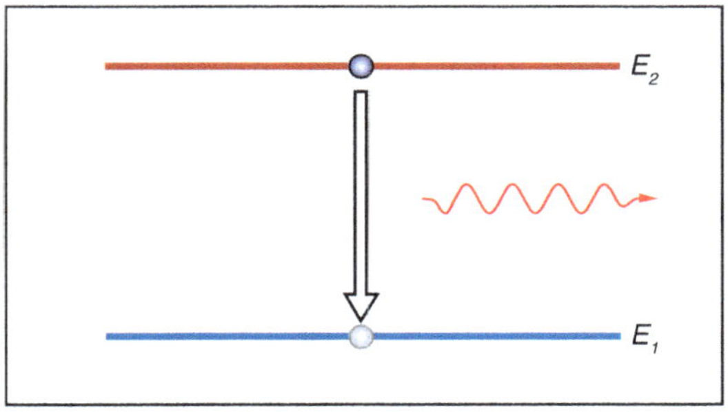

Fig. 2.4 Spontaneous emission of a photon by an excited atom or molecule. Typically, this process occurs spontaneously, often only a few nanoseconds after the absorption of energy. The energy difference between the two atomic states determines the frequency (and, thus, the wavelength) of the departing photon. The direction of flight of the emitted photon is random

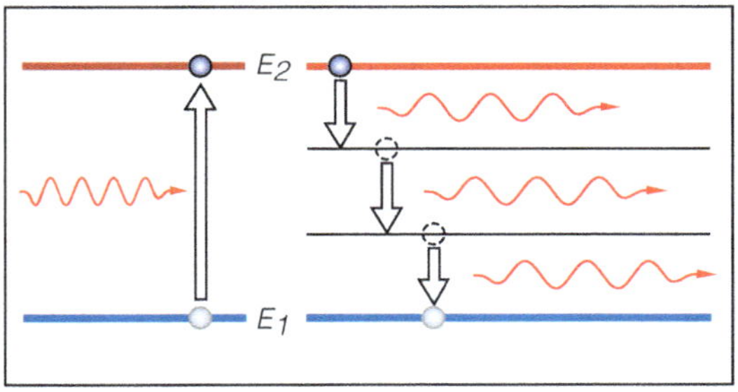

Fig. 2.5 Absorption of a photon and dispersion of the energy into lattice vibrations. The absorbed light energy warms the absorber

rather, the whole shell of an atom or molecule that experiences a change of state in these processes. Besides these two basic processes (absorption and emission), there is a third one: stimulated emission. This will be treated in Sect. 3.4.

Scattering and absorption fit quite simply into this picture of elementary processes. Scattering means that a photon is absorbed and immediately emitted again. The absorbed energy equals the emitted energy and, as a result, it does not change the wavelength of the light. The absorption of light by a black piece of paper or by the pigments of a brown iris follows another scheme: first, the absorption of a photon results in the transition of an atom or molecule to a state of higher energy. This energy is now converted in small portions into vibrations of the material. Heat is generated from the photon's energy (Fig. 2.5). Which one of the aforementioned processes takes place depends on the material, more precisely on its structure and molecular composition.

2.3 Transparency

Keeping in mind that light is scattered when it encounters an obstacle, the existence of transparent media such as glass, water, corneas, crystalline lenses, and air seems quite miraculous. Inside these media, interactions between the light and the materials still occur, but it only leads to the light's traveling more slowly than it would in a vacuum.[2] This slowing down is quantified as the refractive index n: the velocity of light in the medium amounts to $c' = c/n$, where c is the velocity of light in vacuum ($c \approx 300{,}000$ km/s). For example, in water, light travels with a velocity of $c' \approx 225{,}000$ km/s ($n = 1.33$).

Pure water is a classic example of an almost completely transparent medium for visible light. An eye exhibits several portions of tissue that are more or less transparent, such as the cornea, crystalline lens, aqueous humor, and the vitreous body, as well as the inner layers of the retina. A medium is always transparent only to a certain

[2]Why never faster? This is difficult to understand intuitively but follows from Maxwell's electrodynamic equations. The slowing down is the product of a consistent interplay between the electric and magnetic fields of the penetrating light, the vibrations of the electron cloud, and the light generated by these vibrations.

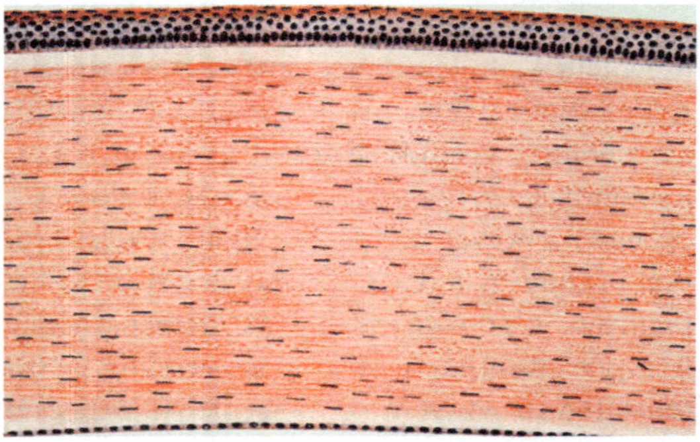

Fig. 2.6 Multi-layer construction of the cornea (Courtesy of E. van der Zypen, University of Bern)

Fig. 2.7 Multi-layer construction of the cornea (Courtesy of H. Elias and J. E. Pauly (1966) *Human Microanatomy*. F.A. Davis Comp., Philadelphia. With permission)

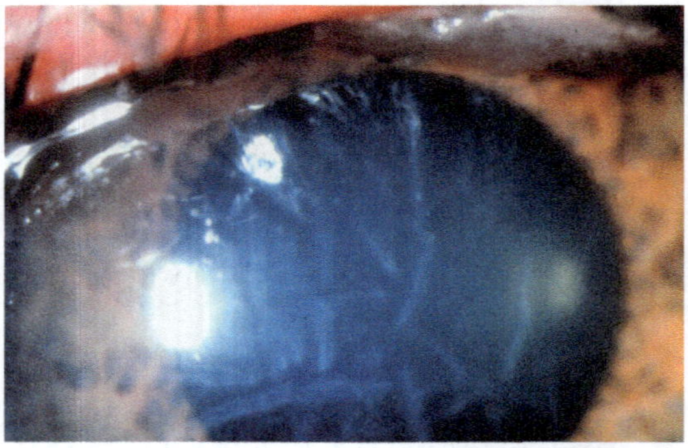

Fig. 2.8 Reduced corneal transparency due to swelling of the stroma

part of the electromagnetic spectrum. For example, water is opaque to radiation in the infrared range (see Sect. 2.8), while the cornea blocks radiation in the ultraviolet range.

It is impressive how nature has been able to construct transparent tissues. The cornea is made up of multiple layers (Figs. 2.6 and 2.7). The largest portion consists of the so-called stroma, which contains relatively few cells but many collagen fibers. For the stroma to be transparent and remain so, a very special arrangement of these collagen fibers must be maintained. The fibers are packed tightly and run from limbus to limbus. The cornea is transparent only as long as the separation between the collagen fibers is less than half a wavelength of the light that passes through. If too much water is present in the stroma (for example, when the pump function of the endothelium fails), the collagen fiber separation increases and the cornea loses its transparency (Fig. 2.8). This can occur, e.g., in cases of corneal decompensation. Here, we have a situation where the incorporation of clear, transparent water leads to clouding of the corneal medium.

The crystalline lens of a healthy person is also transparent. It consists of the capsule, the epithelium, and the lens fibers. The lens fibers run in a meridional fashion from the posterior to the anterior poles (Fig. 2.9). Again, the regular arrangement of these fibers is a prerequisite for the transparency of the lens.

The retina is also transparent, so light can reach the cones and rods unimpeded (Fig. 2.10). However, it can also lose its transparency through water retention (retinal edema). A similar phenomenon can occur at the optic nerve head. The nerve fiber layer continues from the retina into the optic nerve head. The nerve fiber layer is transparent, so, in ophthalmoscopy, the ophthalmologist sees through this layer to deeper layers and, thereby, sees the clear, sharp boundaries of the retina, pigment epithelium, and choroid (Figs. 2.11 and 2.12). The optically sharp delimitation of the optic nerve head is, thus, conditioned by deeper layers. If the nerve fiber layer loses its transparency, either partially or totally, the optic nerve head's boundaries appear blurred. This loss

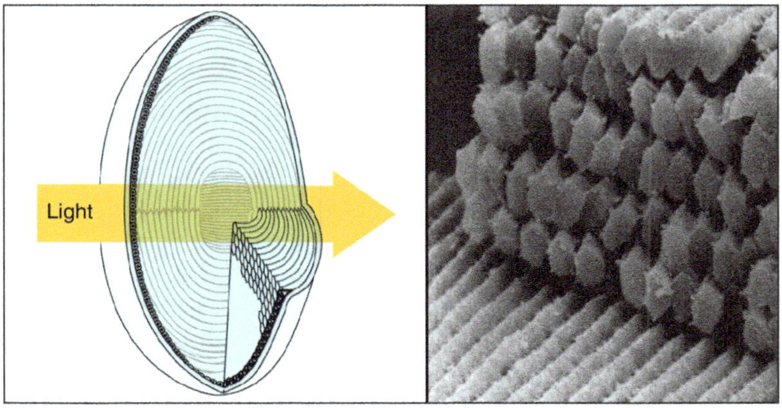

Fig. 2.9 Regular ordering of the lens fibers. *Right*: Scanning electron micrograph showing the orderly arrangement of hexagonal lens fibers. (*Right figure* from *Adler's Physiology of the Eye* (2003) Mosby. With permission. Courtesy of J. Kuszak)

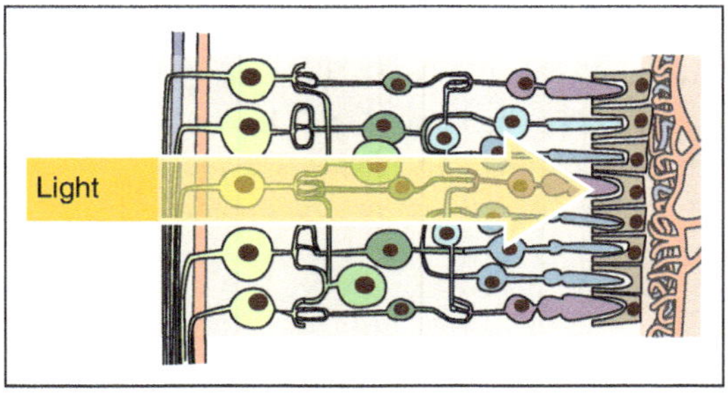

Fig. 2.10 Transparency of the retina

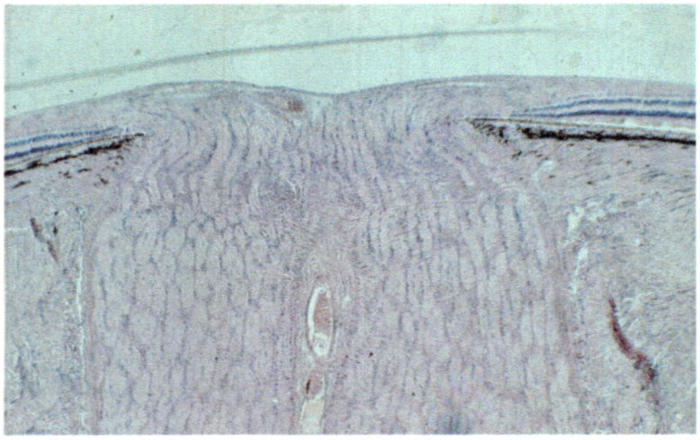

Fig. 2.11 Both the choroid and the pigment epithelium end sharply at the border of the optic nerve head, whereas the superficial nerve fiber layer is continuous (Courtesy of P. Meyer, University of Basel)

of nerve fiber transparency is encountered, for example, in cases of papilloedema in which the axons swell and thereby lose their transparency (Fig. 2.13).

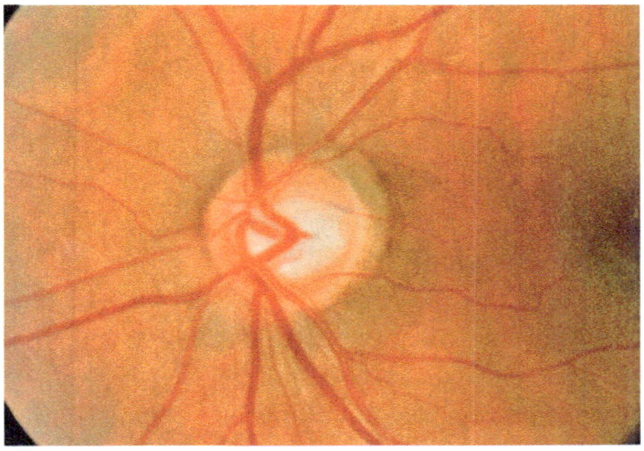

Fig. 2.12 Sharply defined retina, pigment epithelium, and choroid

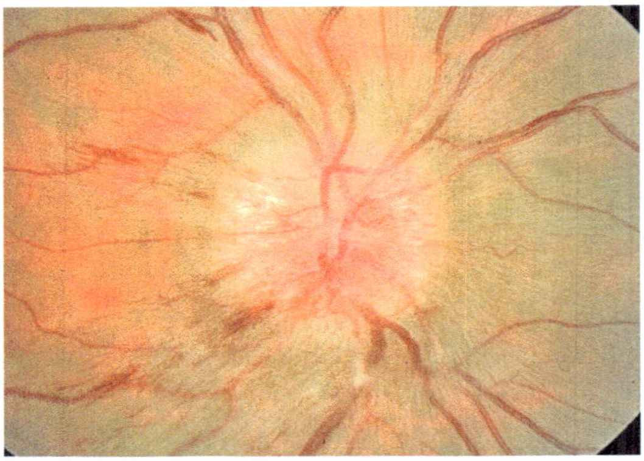

Fig. 2.13 In the case of papilloedema, the nerve fibers lose their transparency. This gives the impression of a blurred-bordered optic nerve head

2.4 Refraction

If a beam of light meets a smooth interface between two transparent media that have different refractive indices, both reflection and refraction occur (Fig. 2.14). The reflection is symmetric with respect to the surface normal,[3] and the percentage of the light reflected increases with an increasing angle α. In Sect. 2.5, we will discuss reflection in more detail. The refraction of light is the basis for the optical imaging through the cornea, crystalline lens, eyeglasses, and optical instruments (e.g., magnifying glasses, microscopes, and refractive telescopes).

2.4.1 The Law of Refraction

The incident ray of light onto a surface, the refracted and reflected rays, and the surface normal all lie in the same plane (Fig. 2.14). The amount of light refracted depends on the ratio of the refractive indices of the two media. The relationship between the two angles α and β is specified by the law of refraction (Fig. 2.14). Depending on which culture one comes from, it is ascribed to either Snell[4] or Descartes,[5] both of whom rediscovered it at roughly the same time.[6]

Refraction is a consequence of the differing speeds of light in two media (Fig. 2.15). To understand this, we first note that the frequency of the light vibrations remains the same in both media (the vibrations on both sides of the interface are in step). Therefore, inside the medium with the slower light speed, the wavelength is smaller since the light moves one wavelength further during one period. Figure 2.15 shows that the continuous transition at the interface is possible only with a change in direction.

Lenses that are thinner in the center are called diverging lenses ("minus lenses"). Their optical powers are given as negative values. Lenses that are thicker in the center than the periphery are called collecting lenses ("plus lenses"). They bundle parallel light into a focus. The reciprocal value of the focal length (in meters) is called the optical

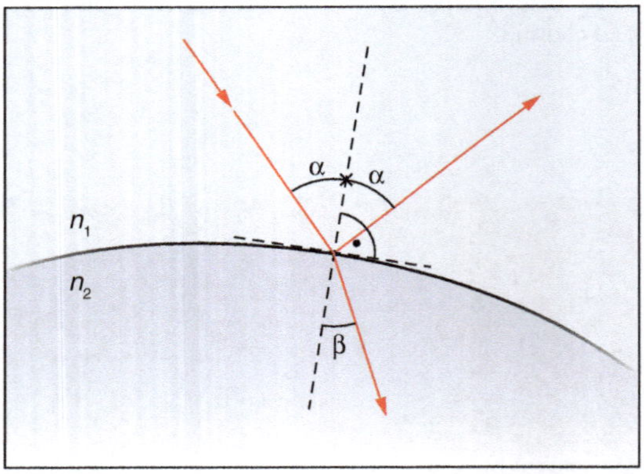

Fig. 2.14 Refraction at the interface of two media. The primary ray is partially reflected and partially refracted. α and β are the angles of the rays with respect to the surface normal. The law of refraction determines the angle β when α and the refractive indices n_1 and n_2 are given: $\sin\alpha / \sin\beta = n_2 / n_1$

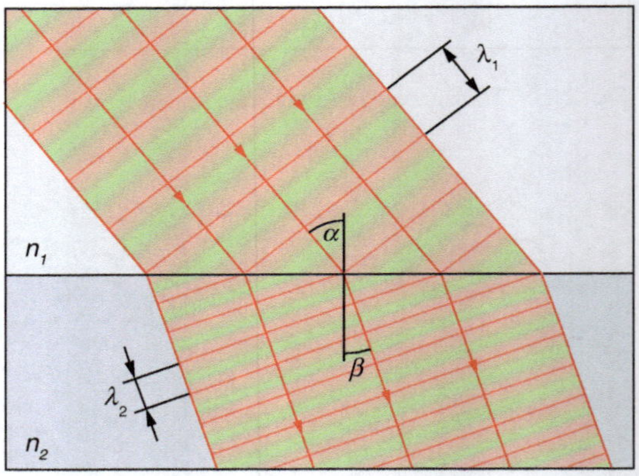

Fig. 2.15 Wave image of refraction. The differing light speeds in the two media give rise to differing wavelengths. The continuous transition of the phases at the interface is possible only with a change in direction. In a medium with an index of refraction n, the wavelength is n times shorter than in a vacuum. $\lambda_1 / \lambda_2 = n_2 / n_1$

[3]The surface normal is perpendicular to the surface.

[4]Willebrord van Roijen Snell (1580–1626), Dutch astronomer and mathematician.

[5]René Descartes, mentioned in Chap. 1.

[6]The earliest known discoverer was Ibn Sahl (940–1000), Persian mathematician and physicist in Baghdad. In 984, he wrote a tract concerning magnifying mirrors and glasses.

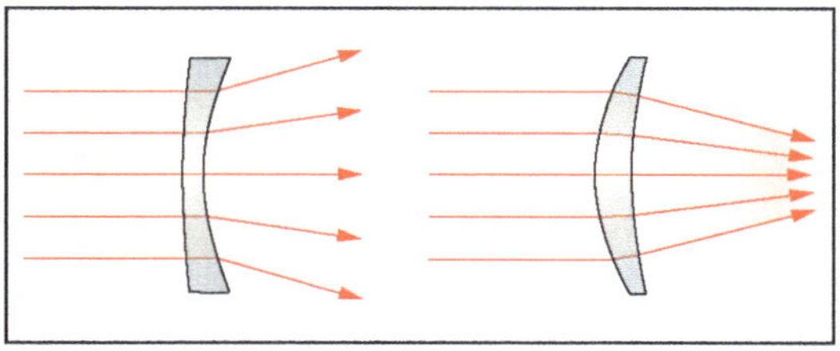

Fig. 2.16 *Left*: diverging lenses are thinner in the middle than on the edge. *Right*: collecting lenses are thicker in the middle than on the edge. Bending a lens inward or outward has little influence on its optical power

power and is measured in diopters, D (to give an example, a focal length of 0.25 m corresponds to a refractive power of 4 D). Bending a lens inward or outward slightly does not influence its refractive power (Fig. 2.16).

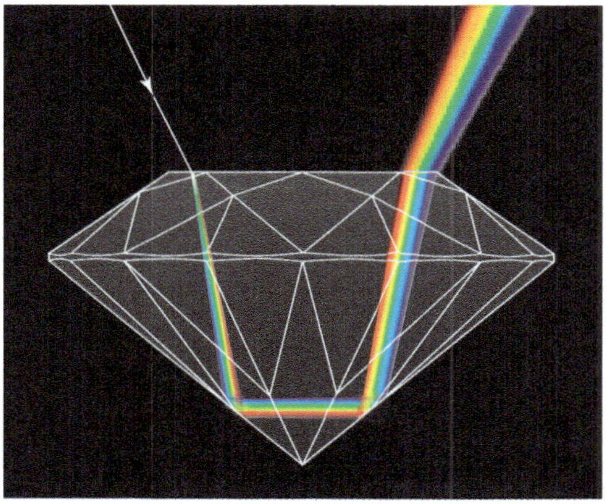

Fig. 2.17 Dispersion in a diamond. The refraction depends on the color: blue light is more strongly refracted than red light (exaggerated in the figure)

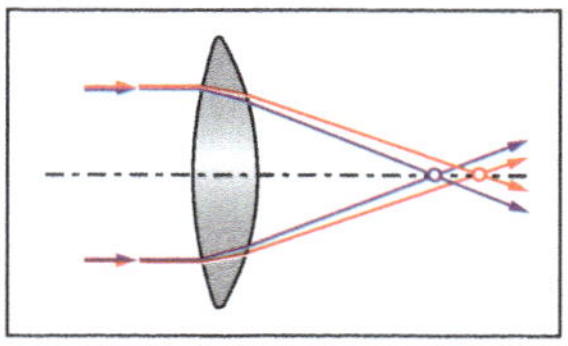

Fig. 2.18 Chromatic aberration in a simple lens: the focal length for blue light is shorter than for red light

2.4.2 Dispersion

The refractive index of a transparent medium is slightly dependent on the wavelength and increases with shorter wavelengths.[7] This gives rise to dispersion during refraction, i.e., to a breaking up of white light into various colors, as we mainly know from prisms or crystals (Fig. 2.17). The colors of the rainbow are also based on the dispersion in water droplets. In imaging systems, the corresponding color error is referred to as "chromatic aberration," which can be observed as colored edges toward the periphery of the field of view of some binoculars. The error occurs because the lens edges split up the light into its spectral components like a prism. In other words, a simple lens has a slightly different focal length for each color (Fig. 2.18). Since the light that passes through the edges of the lens contributes most to aberration, this error is minimized with aperture stops. The basic idea for correcting color errors in imaging systems is outlined in Sect. 19.2.

Newton was interested in the chromatic aberration of the human eye. Its focal plane for blue light lies approximately 1 mm in front of that for red light. It is amazing that we perceive the chromatic error of our eyes only under rare conditions, even though the difference between the refractive power for red and blue light amounts to ca. 1.5 D (Fig. 2.19). One reason could be that the innermost

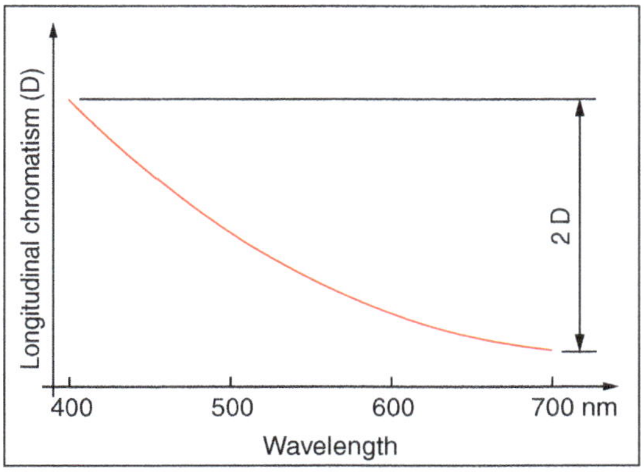

Fig. 2.19 The refractive power of the human eye depends on the color of the light. *Abscissa*: wavelength. *Ordinate*: variations of the refractive power. The total refractive power of the eye amounts to ca. 58 D

[7]From 700 to 400 nm, the refractive power of water increases by ~4 %.

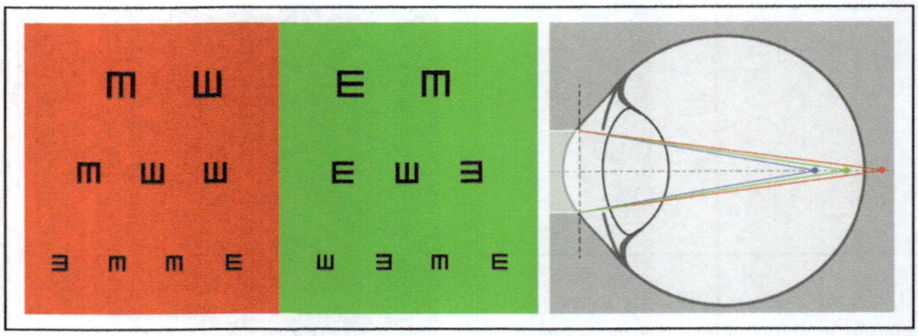

Fig. 2.20 Red-green balance of the refractive correction of an eye. *Left*: the correction is considered ideal when the symbols on the red and green backgrounds are seen to be equally sharp. *Right*: the red focus lies, then, just behind the retina and the green one just in front. The distance of the blue, green, and red foci from the retina is exaggerated by a factor of 10. The distance between the red and green focus amounts to about 0.3 mm

part of the central visual field is insensitive to blue since the blue-sensitive rods are completely absent in a circular area of approximately 20 min of arc. We can make use of the eye's chromatic aberration in red-green balancing: a refractive correction is considered optimal when the edges of black letters are seen to be equally sharp against both red and green backgrounds. Green and red foci are, then, a little bit in front and a little bit behind the retina, respectively. When the perception on the red background seems sharper, the spectacle correction is slightly too much on the hyperopic ("plus") side (Fig. 2.20).

2.5 Specular Reflection

The oldest encounter of humans with the phenomenon of reflection is seen when looking at a quiet surface of water. Reflection occurs at every smooth interface between media of differing optical density, i.e., other refractive indices (Fig. 2.14). Reflecting surfaces can be smooth or uneven. A smooth surface yields specular reflections: we can see sharp images of the mirrored objects. An uneven surface, such as a painted wall, our skin, or normal paper, leads to diffuse reflections; we treat this in more detail in the following chapter. Reflection is also the reason that moist eyes are shiny (Fig. 2.22). Eyes that are not moist – for example, those of a deceased person – seem to be dull (Fig. 2.21).

The fraction of the light that is reflected depends on the angle of incidence, the ratio of the refractive indices of the two media, as well as from the state of polarization of the incident light (see Sect. 1.6), but it does not depend on the color in most situations. For a perpendicular incidence from air to glass (or for a perpendicular exit out of glass into air), approximately 4 % of the light is reflected. For the transition from air to water, this is approximately 2 %. In a window with double-glazing, we see four reflected images of a candle flame.

A flat surface reflects an image that is true to scale. Spherical surfaces reflect objects either enlarged or reduced in size. Deviations from ideal forms are evident in the images they reflect. This can be used in diagnostics. If we observe the shape of a cross reflected by a cornea, we can draw conclusions about the shape of the corneal surface (Fig. 2.22). If, for example, a corneal erosion is present, the image of the cross will show a corresponding step. Modern keratometers that measure the corneal outer surface by processing a video recording of the reflected image of a pattern of concentric rings will be dealt with in Sect. 4.1.9.

So-called total internal reflection occurs when a ray of light tries to exit an interface into a

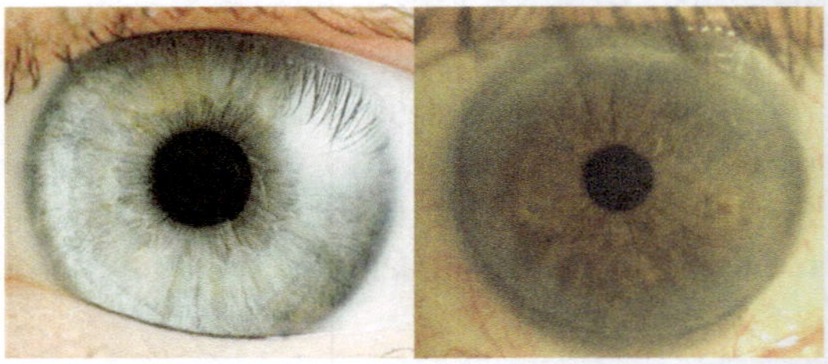

Fig. 2.21 *Left*: moist healthy eye. *Right*: dull eye of a patient with lagophthalmos

medium with a lower index of refraction and the angle of incidence with the surface normal exceeds a certain critical angle (Fig. 2.23). With total reflection, no light energy is lost. Light-conducting glass fibers – having cores with higher and claddings with lower refractive indices – transmit signals with large bandwidths over large distances. The light is "trapped" within the fiber (Fig. 2.24). Another application consists of image transmission via a fiber optic imaging cable that is constructed from a large number of individual fibers. An example is an endoscope (Fig. 2.25). Due to total reflection at the cornea, the anterior chamber angle cannot be observed directly. A contact lens is utilized to eliminate the total reflection of the cornea (Fig. 2.26).

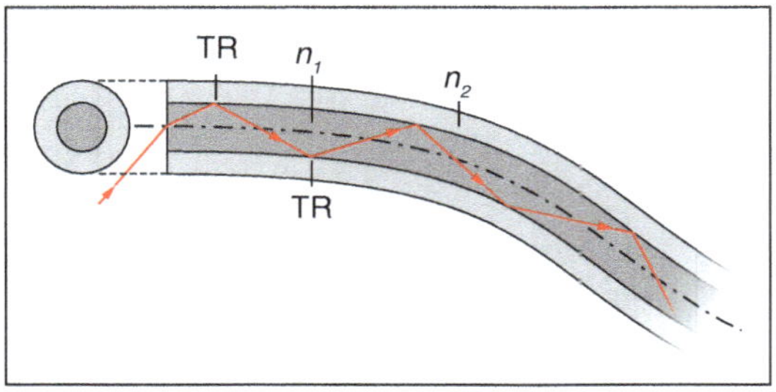

Fig. 2.24 Light conductor. The core has a higher index of refraction (n_1) than the cladding (n_2). The light travels along the core and, due to total reflection (TR), it cannot leave it

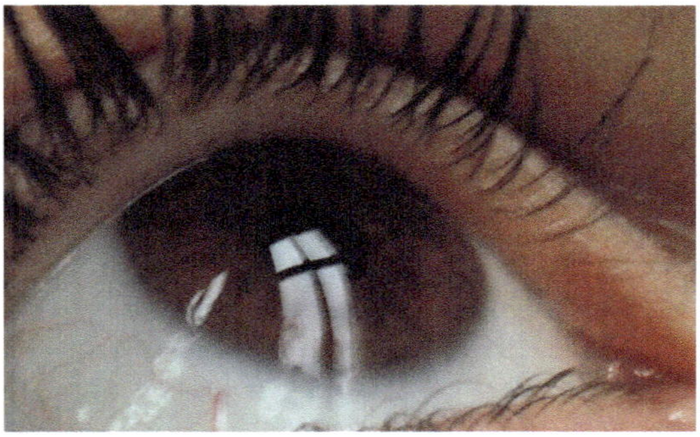

Fig. 2.22 Specular reflection: the tear film covering the cornea reflects a crossbar

Fig. 2.25 Rastered imaging through an image-conducting fiber optic cable. Each fiber produces a point of light. The diameter of the fibers limits the resolution. This principle is applied in the endoscope

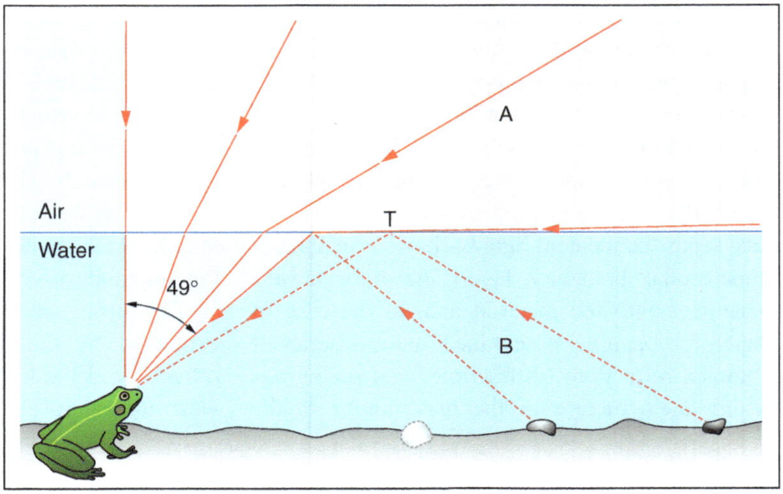

Fig. 2.23 Total reflection. In a part of its visual field, the frog sees a mirror image of the pond bottom (*B*) and, in another part, the outer world (*A*). For the transition from water into air (or vacuum) the critical angle is 49°. In accordance with the law of refraction, it corresponds to an angle of 90° in the optically thinner medium (air)

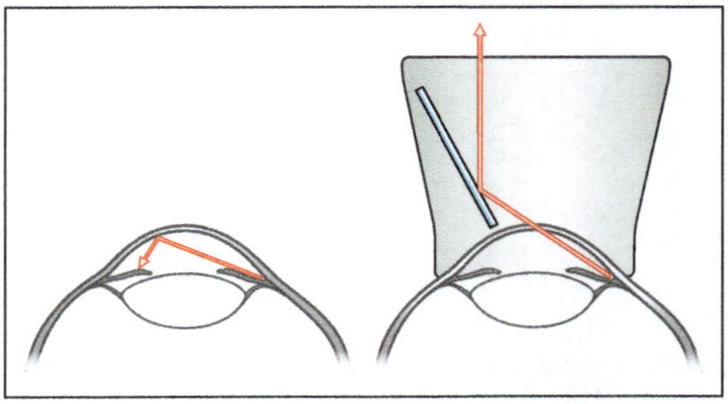

Fig. 2.26 Due to total reflection at the cornea–air interface (*left*), the anterior chamber angle is not visible without using a contact lens (*right*)

2.6 Diffuse Reflection at Surfaces

Rough surfaces, such as a piece of paper, reflect light back in all directions. This also occurs when sunlight strikes the wall of a house or the green leaf of a plant. Thanks to the diffuse character of the reflection, we see the illuminated object from every angle (Fig. 2.1c, d). This general picture will now be made more precise.

The most obvious is the phenomenon of the color of the reflecting surface. The wall of a house, being illuminated by the sun, appears white when its paint reflects all wavelengths of the incident light completely (Fig. 2.1c). The yellow color of a sunflower arises through the absorption of blue: together, the remaining green and red produce the perception of yellow. If a surface partially absorbs all the spectral portions of the light uniformly (50 % of it, for example), it appears to be gray, that is, without any color.

An example of a completely white surface is provided by snow. Its white color has a simple explanation: the tiny ice crystals reflect the light without any absorption. For the same reason, crystal sugar appears completely white. Strictly speaking, these phenomena are not surface effects; rather, the elementary processes with which a surface sends the incident light back occur in a three-dimensional boundary layer: the light always penetrates into the material and, in the case of paper, for example, is sent back out in a series of complex refractions, diffractions, and scatterings within the structure of the material. From the transparency observed in very thin paper, we can recognize that the layer in which the aforementioned processes occur must have a certain thickness.

The directional distribution of diffusely reflected light can exhibit differing characteristics depending on the material and surface texture. An extreme case is the mirror that reflects the incident light in only one direction. The other extreme case is the so-called Lambertian reflector: it reflects the light in all directions, independent of the angle of incidence, such that a surface appears equally bright when viewed from any direction (Fig. 2.1c, d). Between these extremes is shiny paper that returns light mainly in a relatively narrow set of directions (Fig. 2.1b) or shiny surfaces that mirror a part of the light and diffusely reflect another part. A water surface, moved slightly by the wind, can exhibit both diffuse and specular reflections at the same time. An interesting phenomenon is that dry sand is bright, while wet sand is dark.

2.7 Light Scattering in Media

An outer surface does not always scatter back all the light that penetrates it. Instead, the scattering can also take place deeper inside the interior of the medium. Among innumerable examples, the blue of the sky is the most well known: the air molecules scatter sunlight and mainly the shorter wavelengths (Fig. 2.27). Without the scattering, the sky would appear black to our eyes and light would come into our eyes only when looking directly at the sun (Fig. 2.28). Other examples include, say, the visibility of a laser beam from the side when it passes through smoke or strongly diluted milk. A glass of beer absorbs light of short wavelengths and scatters light with longer wavelengths in all directions.

The pure white color of clouds can be explained by the fact that the reflection and refraction of the light in water droplets happens with almost no absorption. The gray or black

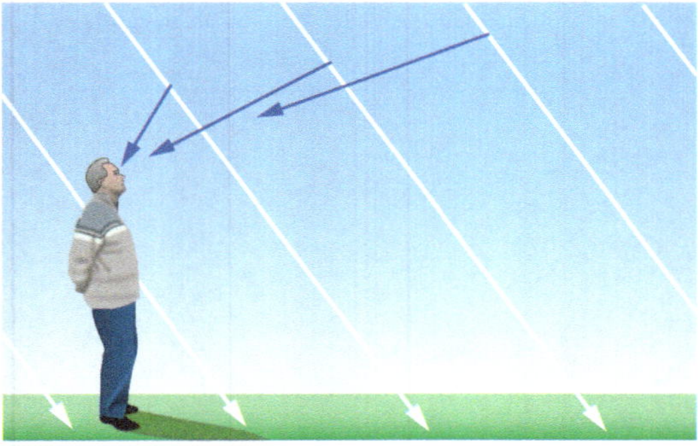

Fig. 2.27 The observer sees a blue sky because the atmosphere's molecules scatter the blue wavelengths of sunlight more than the red ones

Fig. 2.28 The astronauts on the moon saw the earth against a black sky (Courtesy of NASA)

appearance of heavy clouds derives from the fact that the light coming from above is mainly scattered and reflected back upward, while only a small part passes through.

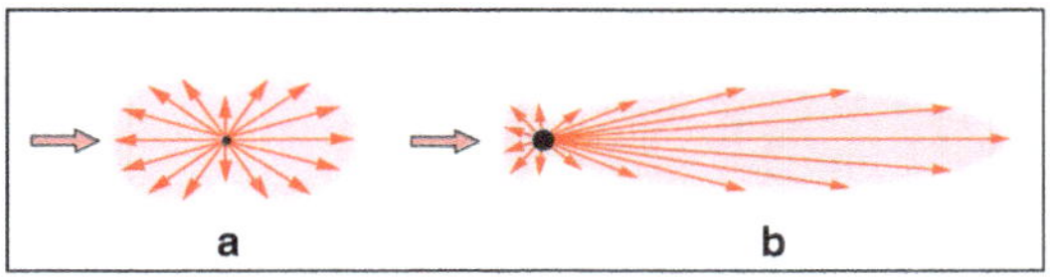

Fig. 2.29 Angular distribution of the scattered light. (**a**) Scattering due to particles with diameters far below the light wavelength (Rayleigh scattering). (**b**) Scattering due to particle diameters of a few light wavelengths (Mie scattering). Light arrives from the left. With large particles, the scattering takes place mainly in the forward direction

The scattering of light by particles depends on the particle size. The scattering due to particles that are considerably smaller than the light wavelength is known as Rayleigh scattering.[8] The best known example is the scattering of sunlight by the molecules of the atmosphere. The blue portion of the sunlight is approximately six times more strongly scattered than the red portion (Fig. 2.27). Rayleigh scattering occurs in all directions. The blue iris with its missing pigmentation of the stroma represents a further example of Rayleigh scattering. It occurs due to scattering on structures of the iris that are considerably smaller than the light wavelength.

For larger particles, e.g. from atmospheric pollution, with diameters on the order of light wavelengths or larger, the scattering takes place mainly in the forward direction, and it is less color-dependent (Mie scattering,[9] Fig. 2.29). The scattered light loses the blue dominance of the Rayleigh scattering and becomes increasingly whiter with the increasing diameter of the scattering particles. A nice manifestation of the forward direction of the scattering of sunlight on atmospheric particles is the whitish appearance of the sky near the sun: The scattering due to water droplets and atmospheric pollution (aerosols, salt

[8]John W. Strutt, Baron Rayleigh, 1842–1919. English physicist. Nobel Prize 1904.

[9]Gustav Mie, German physicist, 1868–1957.

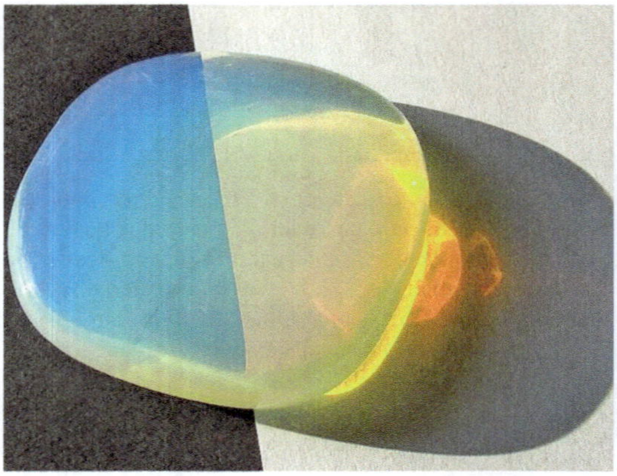

Fig. 2.30 Opal glass illuminated from the left. With a black background, it appears bluish (scattered light). The light that passes through it is orange-yellow because the blue light is partially scattered away. The black shadow appears to the right since no light arrives there due to refraction. (Courtesy of D. Zawischa, University of Hannover, http://www.itp.uni-hannover.de/~zawischa/ITP/streuung.html#tyndalleffekt)

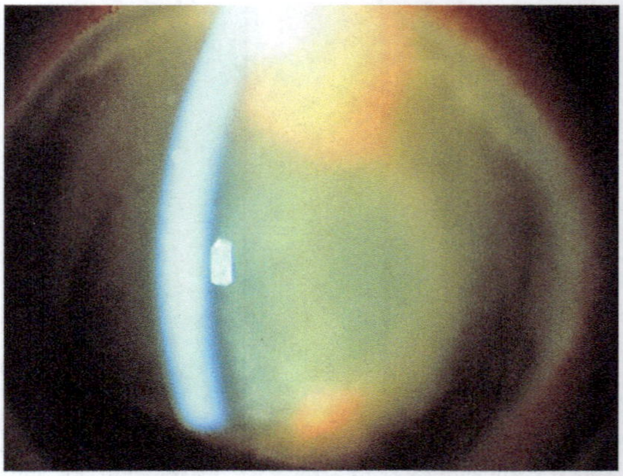

Fig. 2.31 Tyndall effect due to exudation of proteins

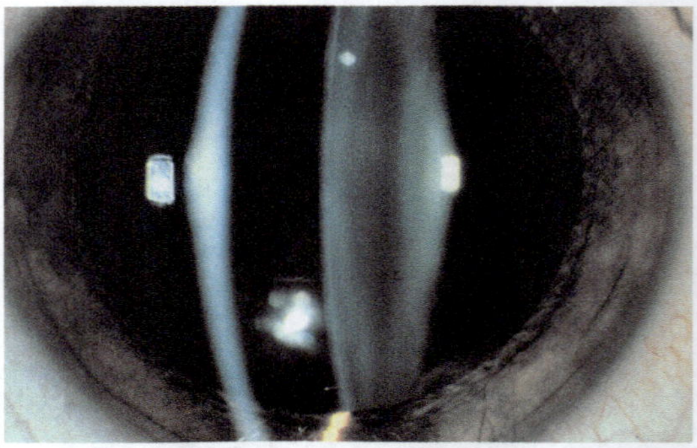

Fig. 2.32 Backscattered light from the cornea and lens as observed with the slit lamp

particles) often overpowers the blue Rayleigh scattering.

Light scattering due to submicroscopic particles in an apparently homogenous medium is known as the Tyndall effect.[10] An especially beautiful example – opal glass – is shown in Fig. 2.30. In ophthalmology, the glow caused by the incident slit lamp light, such as in the anterior chamber, is denoted as a positive Tyndall effect. It indicates that protein molecules are present in the aqueous humor and this, in turn, is a manifestation of a disturbed blood-aqueous barrier, mostly in connection with inflammation (Fig. 2.31).

The so-called transparent media of the eye are practically never fully transparent but scatter light to some extent. Observation of the cornea and lens with the slit lamp is based on the light scattered by these media[11] (Fig. 2.32). If a photograph of the anterior part of the eye is analyzed densitometrically, we can obtain a measure of the scattering, for example, of a cataract (Figs. 2.33 and 2.34). The forward scattered light disturbs patients, while the physician perceives only the light that is backscattered. Their ratio depends on the type of cataract. The visual impairment by light scattered by the lens is well known (Fig. 2.35).

[10]John Tyndall, Irish physicist, 1820–1893.

[11]In photography of the anterior chamber, the Scheimpflug principle is frequently applied. In 1907, by tipping the image plane, Theodor Scheimpflug (Austrian naval officer and photographer, 1865–1911) was able to achieve sharply focused images of planes that were not perpendicular to the direction of view.

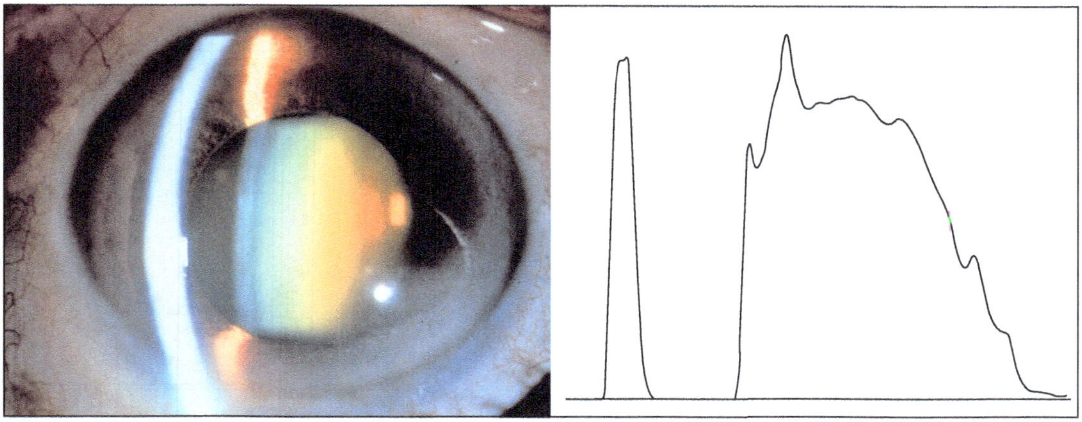

Fig. 2.33 Backscattered light from a nuclear cataract as seen with the slit lamp (*left*) and as shown in a densitometric profile (*right*)

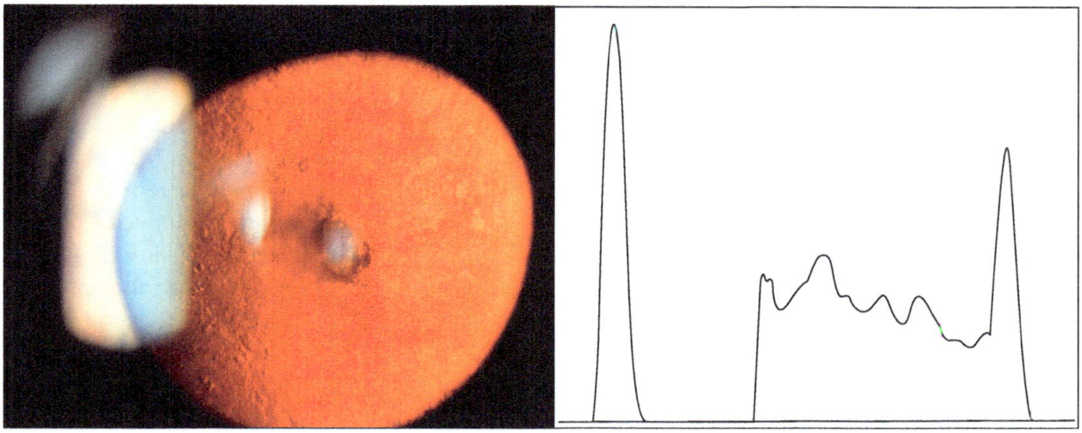

Fig. 2.34 A posterior subcapsular cataract produces more forward light scatter but only little backscatter. *Left*: observation with the slit lamp. *Right*: the corresponding densitometric profile (Note the difference between the profiles in Figs. 2.33 and 2.34)

Fig. 2.35 Simulation of light scatter caused by lens opacity

2.8 Absorption

Scattering at an exterior surface can be modified by the absorption of a portion of the incident light. A black surface swallows up all incident light. A piece of paper appears red when it absorbs the blue and green components of the light. If, for all colors, the same amount is absorbed, the surface appears gray. Absorption can also occur in the interior of a material and be also connected with light scattering. Red wine lets a part of white light pass through, absorbs the blue and green components, and scatters the red component sideways and back. In Sect. 1.5, we showed examples of body colors with their associated absorption spectra.

In the language of atomic physics, absorption takes place in the following way: the energy of the "swallowed" photon is transformed into the excitation energy of a cloud of electrons. Instead of this energy being re-emitted immediately as a single photon with the original energy, as occurs in a scattering process, it can be converted into vibrational energy of the material (i.e., heat) or into several energetically weaker infrared photons (Fig. 2.5).

Molecules that are intensively absorbing are called pigments. Through differing absorption spectra, brightness and color contrasts arise. We distinguish between inorganic and organic pigments. Inorganic pigments are crystals, polycrystalline powder, aggregates, and agglomerates. They come in the form of oils, lacquers, etc. They were used in cave paintings as early as 30,000 years ago (Fig. 2.36). Organic pigments are present in

Fig. 2.36 Cave painting

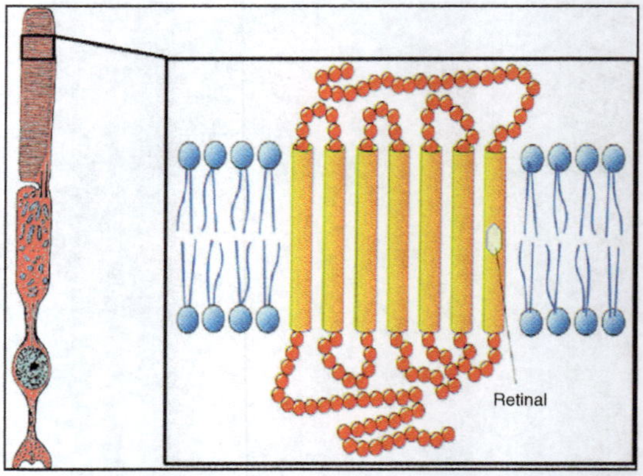

Fig. 2.37 Absorption of light by retinal. The same molecule (retinal) stands at the beginning of the cascade of processes leading from the arrival of a photon down to an electrical pulse in all three cones (red-, green-, and blue-sensitive). The differing sensitivity spectra go back to the embedding of cis-retinal in the protein opsin, the amino acid sequences of which differ slightly from those of the rods and also between the cone types. (From Flammer J (2006) *Glaucoma*. Hogrefe & Huber, Bern. With permission)

almost all creatures. Hemoglobin, for example, gives blood its red color, chlorophyll turns leaves green, etc. Organic pigments share many double bonds; i.e., they are multi-unsaturated. Thus, an electron cloud is formed that can absorb light.

Black and colored pigments are also applied in so-called xerography. The Greek word "xeros" means "dry." In 1937, the printer Chester Carlson developed a printing process that does not require the use of liquid chemicals. Today, it is still the basic principle underlying laser printers and copy machines. In this process, a thermoplastic powder – the toner – is applied to locations that were not exposed to light on a roller; these are marked by differing photoelectric charges. The toner is then transferred onto the paper and the toner image is fixed to the paper with heat and pressure.

The absorption of light by chlorophyll is the basis for the energy acquisition of plants from sunlight. The absorption of light by the retinal molecule in our retina is the physical basis for vision (Fig. 2.37). The resulting conformation change of retinal leads to phototransduction in the retina. The retina is not completely transparent. Depending on the wavelength, light is more or less strongly absorbed in the differing layers.

This is important when regarding the photocoagulation of the retina (see Sect. 7.1).

The absorption of light by water or pigments strongly depends on wavelength (Fig. 2.38). For this reason, for the specific heating of certain ocular media, lasers with the correspondingly best-suited wavelength are utilized. For example, wavelengths around 500 nm are especially suited for absorption by blood at the fundus of the eye. More information on this topic can be found in Chap. 7. Table 2.1 shows the absorption of light by water in the various parts of the spectrum.

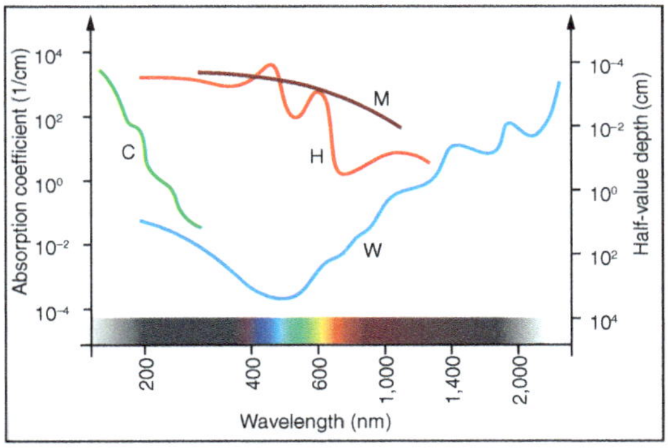

Fig. 2.38 Absorption coefficients of various materials. *W* water, *H* hemoglobin (in the physiological concentration of 150 g/l). *M* melanin, *C* collagen. *Left ordinate*: in units of cm^{-1}. *Right ordinate*: half is absorbed by the time it has traveled this far. Water is very transparent in the visual range and strongly absorbing in the infrared

Table 2.1 Absorption of light by pure water (approximate values for orientation). Half-value depth is the depth at which half of the light has been absorbed

Wavelength (nm)	Absorption coefficient (cm^{-1})	Half-value depth (cm)
200	0.08	10
500	0.0003	2,000
1,000	0.1	10
1,500	1	1
3,000	10,000	0.0001

2.9 Fluorescence

In daylight, fluorescent materials are markedly bright. They are utilized, for example, in laundry detergents. This visual effect often originates from blue or UV light, which illuminates a material and, in turn, excites the emission of yellow or orange light to which our eyes are especially sensitive. The conversion always happens toward the longer wavelength and, thus, in the direction of decreasing photon energy because some energy is converted into molecular vibrations (heat). This behavior (absorption by short-wave light, emission of longer-wave light) is termed "fluorescence" (Fig. 2.39). The name stems from the mineral fluorite. This phenomenon can occur in organic materials and also in minerals. If we irradiate minerals with ultraviolet light, we can ascertain that individual mineral samples shine more or less brightly in various colors.

Fluorescent materials often produce light weakly. The emitted light is normally overwhelmed by the considerably more intense light of the illumination. For this reason, in both microscopy and ophthalmology, filters are employed. An initial filter verifies that, e.g., only blue light illuminates the object (excitation filter).

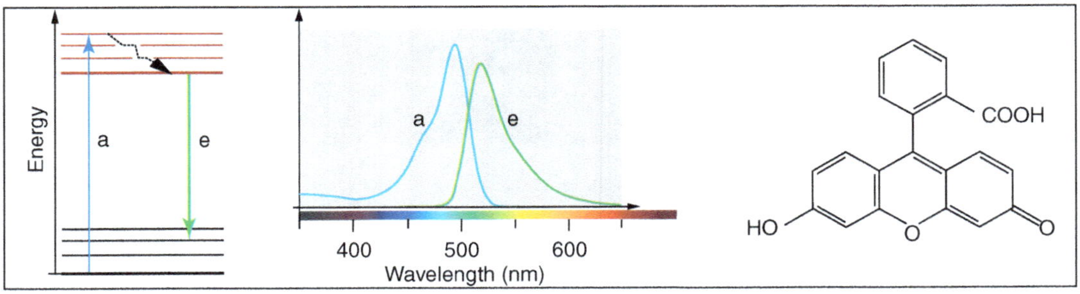

Fig. 2.39 Fluorescein. *Left*: term scheme. *Middle*: absorption spectrum (*a*) and emission spectrum (*e*). Absorption maximum at 485 nm, emission maximum at 514 nm. *Right*: formula of fluorescein

In the observation path, a second built-in filter, the so-called band stop filter, cancels out the excitation wavelength and lets only the fluorescent light through (e.g., green). In this way, the fluorescing molecules are markedly more visible than the rest of the specimen. In ophthalmology, fluorescein is mainly used, which leads to the emission of green light when stimulated with blue light (Fig. 2.40).

Fluorescence is also an important element in the so-called Goldmann tonometry (Fig. 2.41). In 1957, Goldmann,[12] at that time head of the Department of Ophthalmology of the University of Bern, described the principle of his applanation tonometry. A force that can be adjusted is transmitted to the measurement unit by means of a movable transfer arm mounted in a plane perpendicular to the eye. The force is adjusted so that a defined corneal area is flattened by the measurement unit's surface. A round adhesion meniscus forms in the tear film between the anterior surface of the tonometer unit ("tonometer tip") and the corneal epithelium. Thanks to previously instilled fluorescein, it is clearly visible in cobalt blue light (Fig. 2.42). Through a prism placed in the transparent measurement unit, the fluorescent fluid meniscus is divided horizontally into upper and lower half-rings. The prismatic shift corresponds to the diameter of the desired applanation. The pressing force of the tonometer tip is now increased until the insides of the half-rings just touch each other (Fig. 2.42). The surface area of the flattened cornea has now been attained and the intraocular pressure can be read according to the standards set by Goldmann (Fig. 2.43).

The fluorescein used in ophthalmology dissolves in the tear film but cannot diffuse through the lipophilic epithelium layer of the cornea. When an epithelial defect is present, the

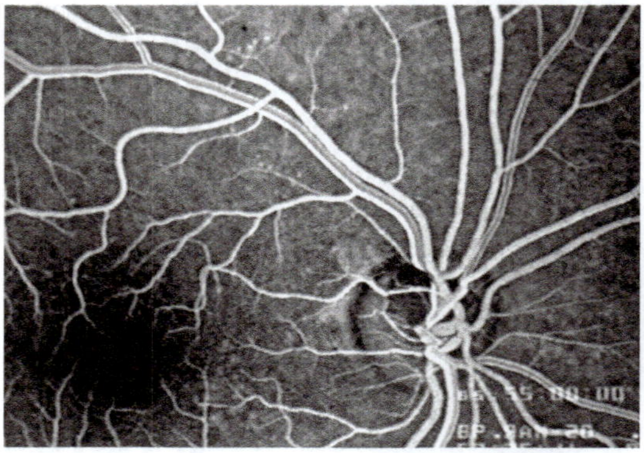

Fig. 2.40 An example of fluorescein angiography of a healthy eye

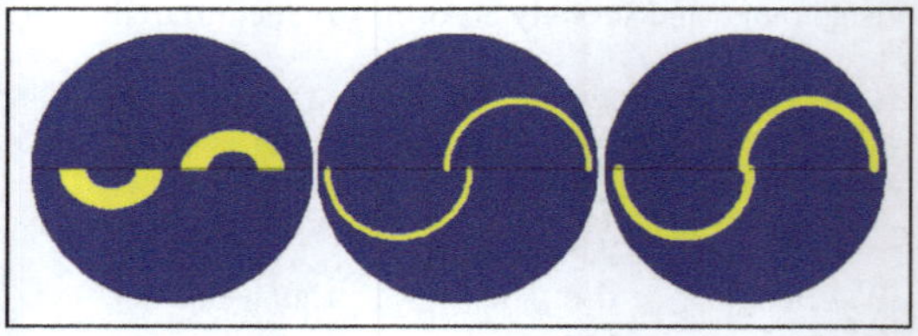

Fig. 2.42 Goldmann tonometry: *Left*: the applied force is too weak. *Middle*: it is excessive. *Right*: the force is correct for the desired applanation surface area

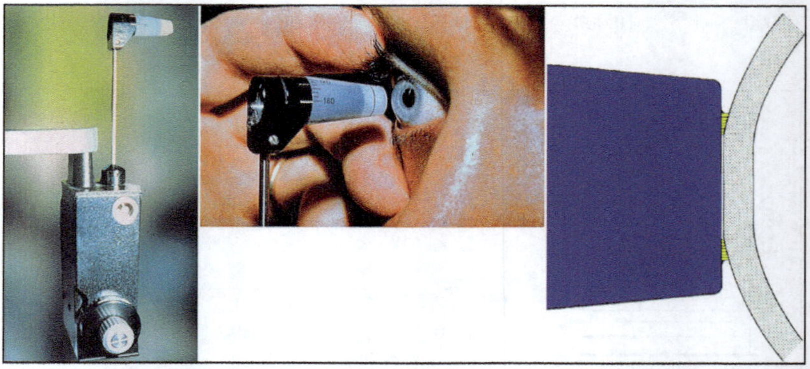

Fig. 2.41 *Left*: entire Goldmann tonometer. *Middle*: tip of the tonometer. *Right*: adhesion meniscus (*yellow*) around the applanated cornea

[12]Hans Goldmann (1899–1991). Swiss ophthalmologist. Famous for his applanation tonometer, contact lenses, perimeter, and contribution to the slit lamp.

fluorescein can diffuse into the hydrophilic corneal stroma (Fig. 2.44). At higher concentrations of fluorescein in the tear film, a small amount can diffuse into the anterior chamber even with an intact cornea. The temporal decay of the fluorescein concentration in the anterior chamber is measured in the so-called fluorophotometry procedure to quantify the turnover of the aqueous humor.

Another fluorescent substance frequently used in ophthalmology is indocyanine green (Figs. 2.45 and 2.46). It is suitable as an indicator and has the property that it does not diffuse out of the capillaries when bound to proteins. Its absorption and fluorescence spectra lie in the infrared (maximum fluorescence at approximately 810 nm in water).

Fig. 2.43 Hans Goldmann

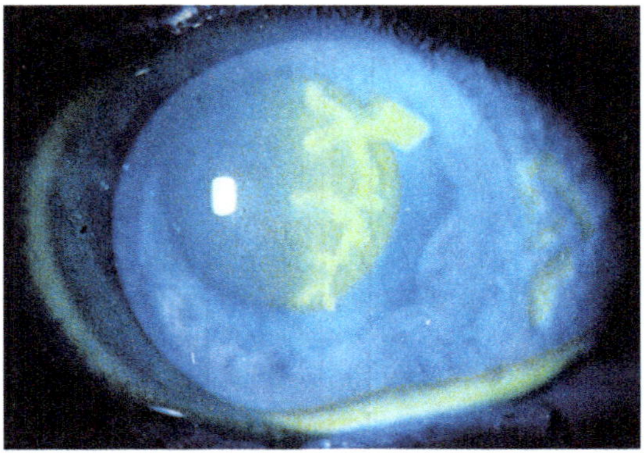

Fig. 2.44 Diffusion of fluorescin into the corneal stroma where the corneal epithelium is damaged by the herpes simplex virus

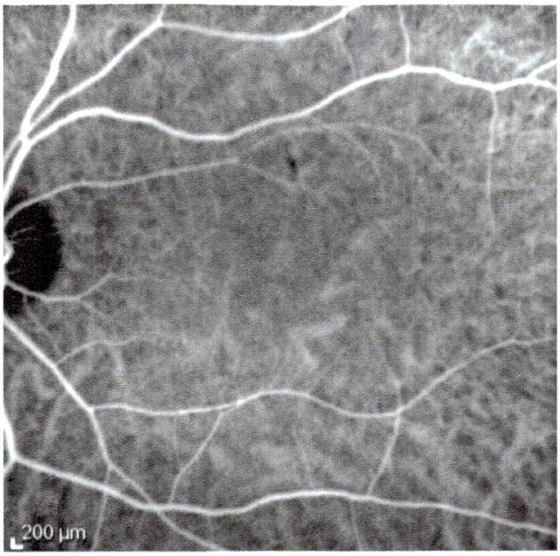

Fig. 2.46 Example of indocyanine green angiography. This patient suffers from a punctuate inner choroidopathy

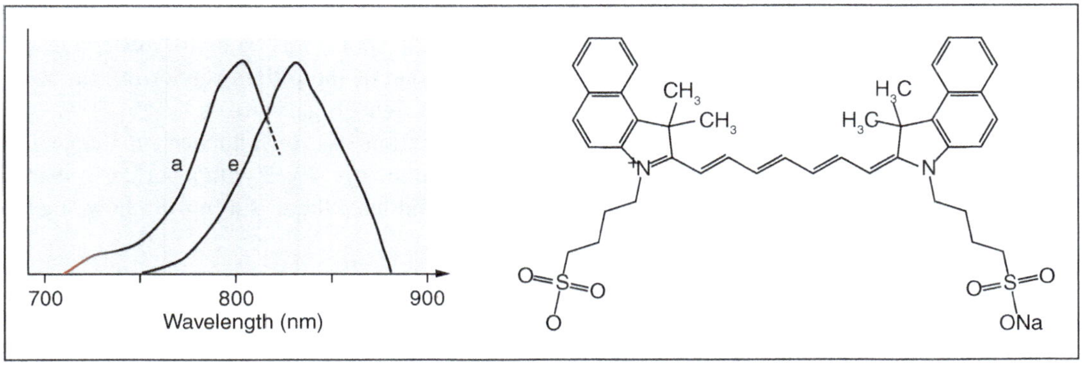

Fig. 2.45 Indocyanine green. *Left*: absorption (*a*) and fluorescence (*e*) spectra. *Right*: chemical formula

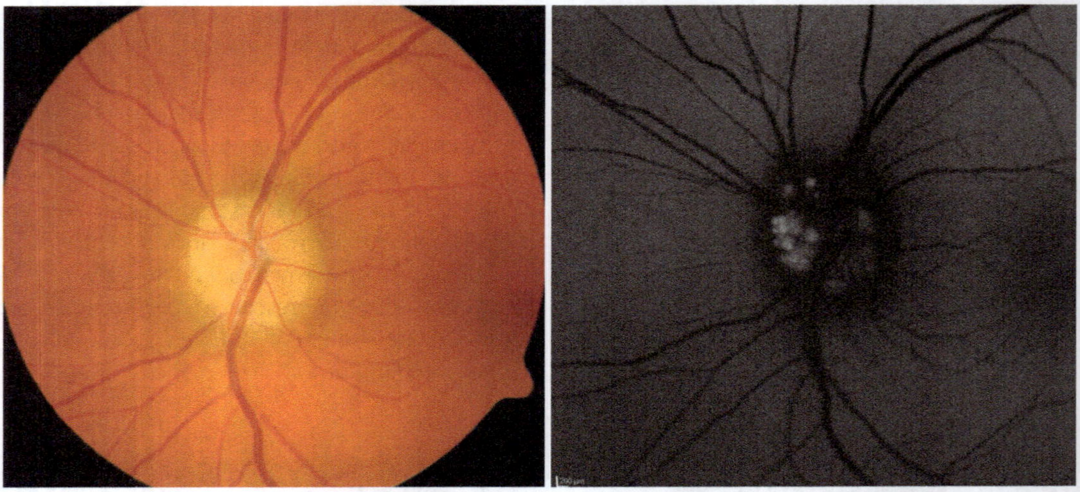

Fig. 2.47 Optic disc drusen. *Left*: regular fundus photo. *Right*: photo taken with appropriate filter to demonstrate autofluorescence

Many structures of the eye also show autofluorescence; i.e., they fluoresce without any fluorescent dye being applied. Examples include the optic nerve head drusen (Fig. 2.47) and the drusen of the retina. These can readily be seen clinically but even better in photos when the appropriate light filters are used. The crystalline lens also has a certain degree of autofluorescence. Changes in retinal autofluorescence are an early sign of retinopathy.

In some minerals, we can observe an additional property after the source of UV light has been turned off: they continue to glow for a few seconds, most often in a color other than the one they fluoresce in. This is called phosphorescence. In everyday life, using various time constants, this effect has found applications on the inside of monitor screens or for marking the ways to exits.

2.10 Diffraction

As mentioned earlier, all forms of waves, including light waves, are diffracted when they encounter an edge or pass through a narrow opening – like water waves when passing through the entrance to a harbor (Fig. 1.9). A diffraction image can also be understood as an effect of interference among the entity of the waves that extend from all the points of the opening. Strictly speaking, diffraction is less the consequence of certain interactions between light and matter and more the expression of an inner property of light: its wave nature.

A curious phenomenon of diffraction was the object of controversy when the wave theory of light was being established. In 1818, Poisson[13] pointed out that, as a consequence of the wave theory, a bright spot must appear in the center of a sphere's shadow because the waves originating from all the edges would arrive there in phase, irrespective of the position of the screen. To him, this seemed so absurd that he believed he had therefore disproved the wave theory. However, a short time afterward, "Poisson's spot" was actually observed and became one of the pillars supporting the wave theory of light (Fig. 2.48).

Diffraction has an influence on image formation in the eye. Even with a perfectly shaped cornea and lens, there is a limit to how small a

[13]Siméon D. Poisson, French mathematician (1781–1840).

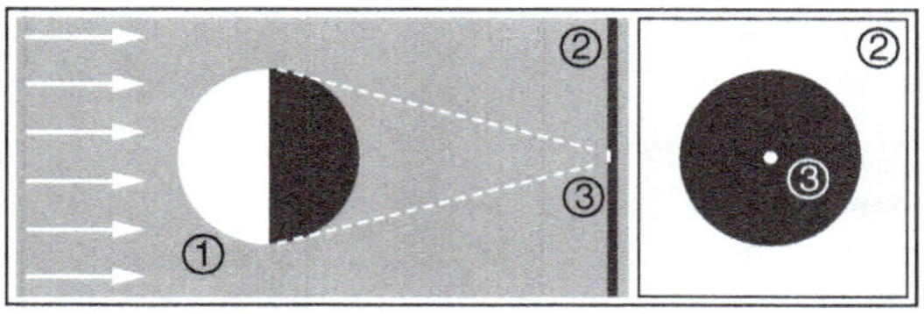

Fig. 2.48 Diffraction of light: Poisson's spot (*3*) on a screen (*2*) behind a sphere (*1*). The spot is observed at any distance behind the sphere

focal point parallel light can be focused on. It will always be a small disk rather than a spot: the so-called Airy disk. For a pupil of 2 mm, the Airy disk at the retina has a diameter of about 12 μm, corresponding to an angle of roughly 2 min of arc. For a pupil of 1 mm, the diffraction image at the retina is twice as large and reduces the visual acuity from 1.0 to 0.25 (from 6/6 to 6/24). We will return to the Airy disk in Sect. 19.1.

We can experience a nice demonstration of diffraction when looking through an opened umbrella or fine curtain material at a light source that is far away. The colored pattern we see is the result of the diffraction caused by the periodic structure of these textiles.

3 Light Sources

In one of his Christmas lectures,[1] Michael Faraday spoke about a classic source of light: a hot material (here, the flame of a candle) emitting light, known as thermal light. We find this process in a light bulb as well as in the radiation produced by the sun and stars or in the Olympic torch. In nature, other sources of light are rather rare, such as lightning, auroras, or the bioluminescence of a firefly. On the other hand, technology has made different (non-thermal) light sources available, e.g., fluorescent lamps, lasers, and light emitting diodes.

Almost without exception, all light sources share a basic process in which electrons (or a system of electrons) return to a state of lower energy and release the energy by emitting photons. Beforehand, the electrons have to be brought into a state of higher energy. It is possible for this so-called pumping mechanism to occur in a number of ways: through the absorption of photons, through collisions of high-velocity particles in a very hot environment, or in semiconductors via electrical current. The light emitted by each source has a characteristic spectrum.

[1]His six lectures on "The Chemical History of a Candle" (1860) are available online. In 1825, Michael Faraday inaugurated the Christmas lectures for young people at the Royal Institution in London. Apart from a few, the delivery of which was prevented by WWII, the lectures have been running ever since.

3.1 Thermal Light

Unaided, we can detect that a hotplate on a stove is overly heated in two ways: one via the red glow perceived by our eyes and also via the thermal radiation that receptors in our skin perceive as warmth. In both perceptions, electromagnetic radiation emitted by the hotplate in relation to its temperature is involved. Only a small part of the energy lies in the visible portion of the spectrum and is perceived by the eye (Table 3.1). While our eyes react to only a limited part of the spectrum, our sensory system for heat reacts to all the absorbed radiation energy, independent of the wavelength. If the hotplate cools down, the total radiated energy diminishes and the visible portion is reduced even more rapidly (Table 3.1).

The surface of the sun, the hotplate, human skin, and an iceberg, thus, all have in common that they spontaneously radiate electromagnetic waves in accordance with their temperature. Even an iceberg at night radiates an ample amount of energy per square meter. The origin of this radiation lies in the thermal movements of the molecules in every material that only cease at a temperature of absolute zero. Between thermal radiation and light, there is no fundamental difference except the wavelength.

Light that is radiated spontaneously from a hot material is termed thermal light. This describes the light in the visible part of the spectrum. Stars, glowing iron, and the tungsten

J. Flammer et al., *Basic Sciences in Ophthalmology*,
DOI 10.1007/978-3-642-32261-7_3, © Springer-Verlag Berlin Heidelberg 2013

Table 3.1 Electromagnetic radiation emitted spontaneously from a hot material as a function of temperature. The radiant emittance describes the emitted power per area. With decreasing temperature, the visible part of the radiation decreases very rapidly. (The figures are upper limits, not reached by all materials)

Temperature		Emittance	Power fraction in visible part	
°C	°F	kW/m²	(0.4–0.7 µm)	
5,600	10,112	70,000	0.3	Sun
700	1,292	50	10^{-6}	Glowing hotplate
400	752	12	3×10^{-10}	Hot hotplate
100	212	1	3×10^{-20}	Teapot
0	32	0.3	10^{-28}	Iceberg

wire of a light bulb give off thermal light. The radiation of these emitters is never limited to the visible range but, depending on the temperature, also contains ultraviolet and infrared portions. Figure 3.1 shows the spectra for several temperatures. Some simple laws are applicable: (a) with increasing temperature, the power per radiating area increases rapidly[2]; (b) the wavelength of the spectral maximum is inversely proportional to the absolute temperature, meaning that the lower the temperature is, the further the radiation will be in the infrared, i.e. further away from the visible range. For the radiation of a body at room temperature, the spectral wavelength maximum lies at roughly 10 µm.

A naked human body radiates several hundred watts, although almost totally in the infrared (maximum at ca. 10 µm). In an environment that is at the same temperature as our bodies, we receive approximately the same amount of radiation that we give off. In a cold environment, this balance no longer holds and we tend to freeze. By the way, the evaporation of perspired water (sweating) also takes away heat just as a cool blast of wind does.

The curves in Fig. 3.1 are derived from Planck's equation, which exactly specifies the behavior of the thermal light spectrum for any given temperature. Planck[3] developed it in 1900 based on purely theoretical considerations. Strictly speaking, the curves represent the spectral intensity that cannot be exceeded by any emitter. However, it is an empirical fact that many radiating bodies operate very close to this limit, such as the sun or the filaments of incandescent light bulbs (Fig. 3.2).

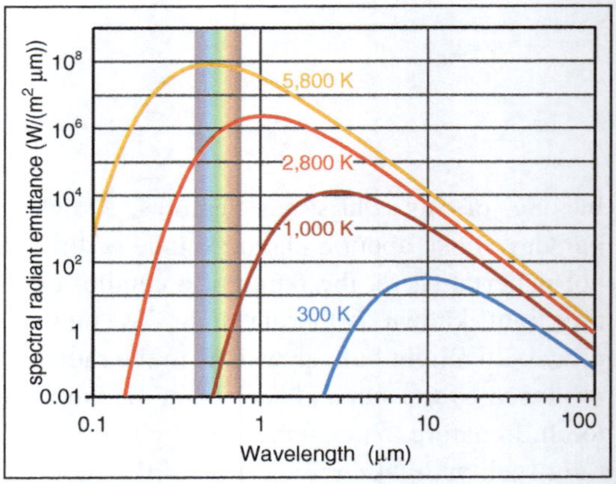

Fig. 3.1 Thermal radiation (power per radiating area) in the ideal case (the possible maxima). For the sun's surface temperature (5,860 K), the maximum lies in the visible range; for a human's surface temperature, it lies far in the infrared. The visible spectrum (between 0.4 and 0.7 µm) is indicated

Fig. 3.2 Max Planck

Well-known applications of these laws are the heat distribution images of buildings or of somatic regions (Fig. 3.3). The intensities of the recorded radiation as a measure of the temperature of the radiating object are displayed in pseudo-colors. This measurement principle is unproblematic if it is known that the radiation of the body or object involved obeys Planck's curves. In materials such as a wall, earth, water, or human skin, this condition

[2]Proportional to the fourth power of the absolute temperature.

[3]Max Planck, 1858–1947, Nobel Prize for physics for his discovery of energy quanta (1918).

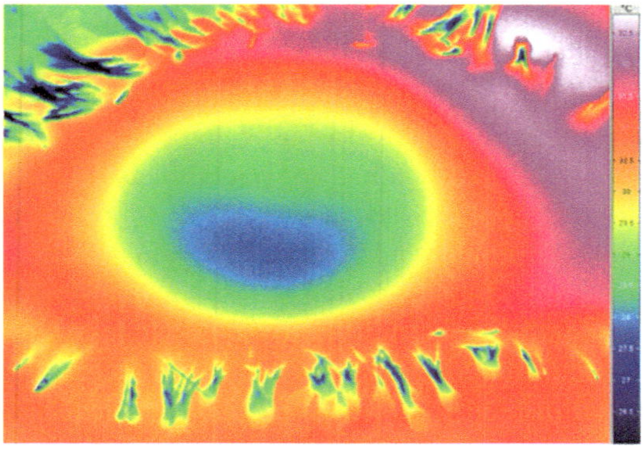

Fig. 3.3 Thermal image of an eye. The camera determines the temperatures from the infrared radiation and displays them in pseudo-colors

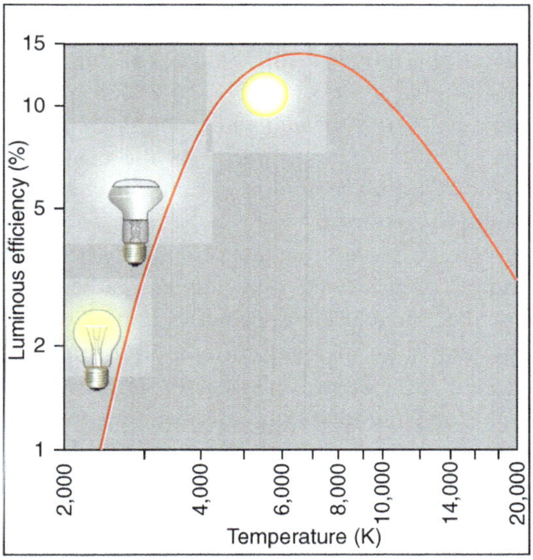

Fig. 3.4 Efficiency of thermal light as a function of the emitter's temperature. *Ordinate*: luminous efficiency in percent. The luminous efficiency is the part of the radiant energy utilized by the eye

is nearly but not precisely fulfilled. Better precision can be obtained with instruments that determine the temperature from the radiation ratios of two different wavelengths.

3.1.1 Luminous Efficiency

The thermal generation of light, e.g., with incandescent light bulbs, has low optical efficiency because only a small portion of the emitted radiation lies in the visible range. This is determined by the fraction of the light that the eye's daylight spectral sensitivity curve V_λ selects from the spectrum produced by the light radiator. Figure 3.4 shows the efficiency as a function of temperature T. For T=2,800 K, it amounts to only 2.5 %; the rest is heat. Even at solar temperatures, where the overlap between the Planck spectrum and the V_λ curve is almost optimal, the efficiency reaches only 14 % (Fig. 3.5). Therefore, it is understandable that fluorescent tubes or light emitting diodes, which give off their energy almost completely in the visible range, rate much better with regard to efficiency than do thermal light radiators (Table 3.2).

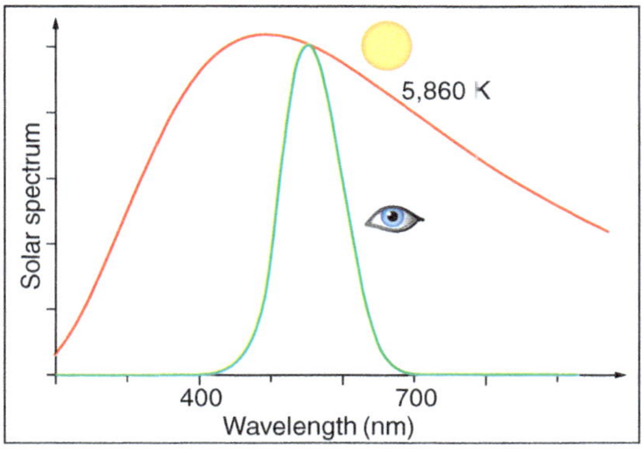

Fig. 3.5 Our eyes utilize 14 % of solar radiation. *Red*: spectrum of the solar radiation (5,860 K). *Green curve*: utilized by the eye

Table 3.2 Conversion of electrical energy into light: comparison of various light sources. The luminous efficiency is the part of the applied electrical energy converted into light and utilized by the eye. It is presented in percentages of the best possible conversion. This is given by light of 560 nm, generated with the complete conversion of the electrical energy into light energy

Light source	Luminous efficiency (%)
Light, wavelength 560 nm	100
Sun (5,860 K)	14
Light bulbs, halogen lamps	2–3
Fluorescent tubes	7–10
Commercial white LED	5–12
Sodium vapor lamp	30

3.2 Fluorescent Tubes

The basic principle of fluorescent lighting is very simple: an electrical voltage between two electrodes causes a gas to glow. More precisely, it

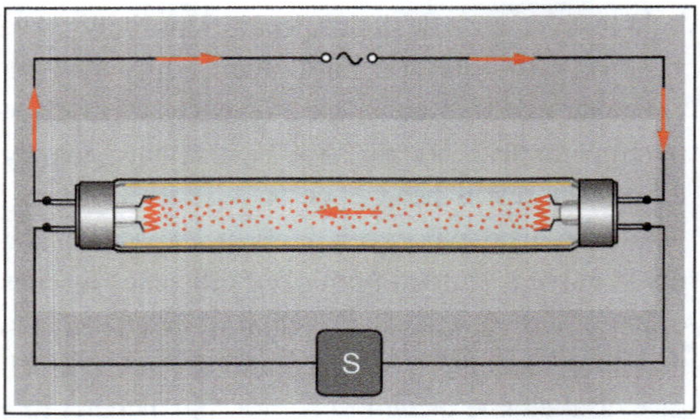

Fig. 3.6 Fluorescent bulb with a fluorescing layer on the inside of the glass tube, the prototype for modern energy-saving bulbs. When turned on, the starter (*S*) generates a high-voltage pulse

is free electrons that are accelerated between the two electrodes and collide with the gas atoms. The principle was discovered in 1858 (Geissler tubes). In the 1880s, fluorescent light bulbs were already being produced in large numbers, although mainly for entertainment purposes. Fluorescent bulbs in the form of the traditional straight tubes (Fig. 3.6) and as energy-saving bulbs have yields that are three to six times those of "normal" incandescent light bulbs. In typical constructions, mercury atoms are the sources of the radiation. However, this radiation lies in the ultraviolet range and is transformed into visible light by the fluorescing interior coating of the glass tube. By means of the coating, various spectra can be generated: cold, warm, and colored. In contrast to thermal sources, fluorescent tubes do not have a continuous spectrum but one consisting of discrete lines.

Sodium vapor, low-pressure bulbs function similarly but need no fluorescing wavelength conversion because sodium, unlike mercury, has visible emission lines. The light is engendered directly in the yellow-orange sodium double lines (589 nm). The light yield can be up to twice that of a fluorescent tube. We encounter these lamps mainly in the form of street lamps.

3.3 Light Emitting Diodes (LEDs)

The light emitting diode is the child of the semiconductor technology of the previous century. Among the semiconducting materials are silicon or germanium crystals. As the name indicates, they can conduct electrical current, but only poorly. In their ability to conduct electricity, they stand between isolators (glass, for example) and conductors (metals). In technological applications, traces of other elements are intentionally added to semiconductors; i.e., they are "doped," which improves their ability to conduct electricity and makes it less temperature-dependent. Semiconductors were first used in rectifiers, which permit electrical current to pass preferentially in one direction, an effect that was explained for the first time by Schottky in 1939.[4] Today, this type of rectifier consists of two semiconductor layers that are doped differently. The next application was the transistor, consisting of three semiconductor layers that allow a weak electrical current to steer a far stronger one. The transistor has massively transformed the technology of the past few decades, primarily as the basic element of highly complex circuits in computer microprocessors.

Numerous components principally have the same composition: two semiconductor layers that have been doped differently: diodes for rectifying electric current, light emitting diodes, laser diodes, photocells for measuring light, and photovoltaic cells for transforming light into usable electrical energy (Table 3.3). If a battery pushes an electric current through this type of a boundary layer, the electrons – as carriers of the current – fall from a higher to a lower energy level (Fig. 3.7). The decisive difference between a rectifying diode and a light emitting diode (LED) is the fact that, in a rectifying diode, the energy is converted into heat; in an LED, it is partially converted into radiated photons. Ultimately, the energy is derived from the battery that actuates the electric circuit. In the photovoltaic cell, the reverse process happens: a photon raises an electron over an energy threshold, thereby "pushing" it into the electric circuit.

[4]Walter Schottky (1886–1976), German physicist, electro-engineer and inventor.

Table 3.3 Concepts found in semiconductor technology: materials and components. With the exception of the transistor, the decisive processes in the components take place in a single interface layer between two semiconductors that have been doped differently. The last column indicates the doped layers (simplified)

Pure semiconductor	For example, pure silicon or germanium crystal	
Doped semiconductor	With a small portion of another element added	
n-doped semiconductor	Doping supplies free (conducting) electrons	n
p-doped semiconductor	Missing electrons behave like positive charges	p
Interface layer	Very thin layer at the boundary between two semiconductors	
Rectifying diode	Electronic switching element	p–n
Light emitting diode	Light source with a narrow spectrum (LED)	p–n
Superluminescent diode	Light source with a broad spectrum, high intensity	p–n
Laser diode	Light source, monochromatic, temporally coherent	p–n
Photodiode	Converts light into electric current for measurement purposes	p–n
Photovoltaic cell	Like a photodiode, used in photovoltaic applications	p–n
Transistor	Electronic amplifier, switching element	p–n–p

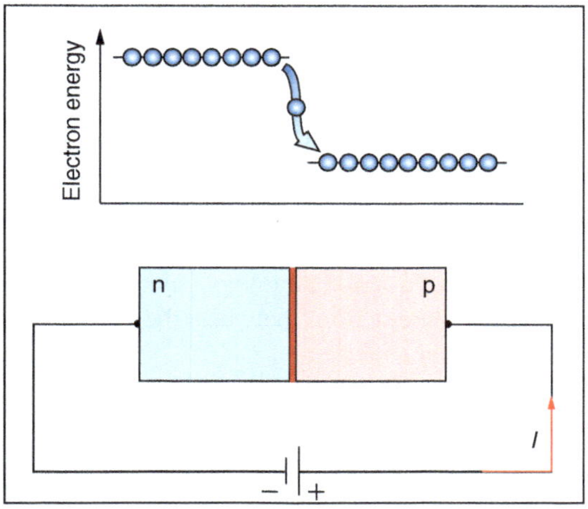

Fig. 3.7 The functioning principle of the LED. Driven by an external voltage source, electrons move from the left to the right through the boundary layer between two semiconductors that have been doped differently. They fall to a lower energy level and give off the energy difference as a photon

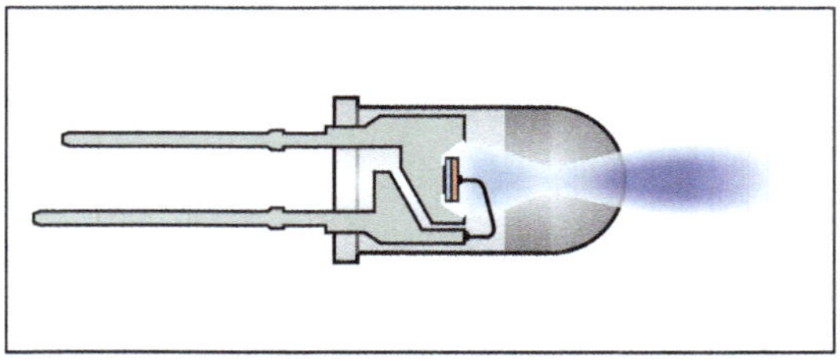

Fig. 3.8 Light emitting diode (LED)

A light emitting diode transforms electric energy directly into light. This takes place via the flow of an electric current through the boundary layer between two differently doped semiconductors. The active element radiates primarily in all directions. The focusing of the light is accomplished by embedding it in a reflector, as well as through the lensing effect of a half-dome exit surface (Fig. 3.8). The LED light exhibits a narrow spectrum that is typically 20 nm wide and, thus, has a very well-defined color. There are LEDs for a variety of colors, depending on which semiconductors are utilized.

By combining luminescent dyes, broad-spectrum light is generated that is spread across the whole visible range (Fig. 3.9, spectrum C). For this, a diode is needed that produces blue or UV light (Fig. 3.9, spectrum A), from which a part excites the dye to emit longer-wavelength light. For the eye, the combination of blue and yellow results in white light. With the doping of various luminescent dyes, a variety of spectra and hues can be produced, including finely tuned white hues, i.e., various color temperatures. The efficiency lies considerably above that of an ordinary light bulb and is comparable with that of a fluorescent tube (Table 3.2).

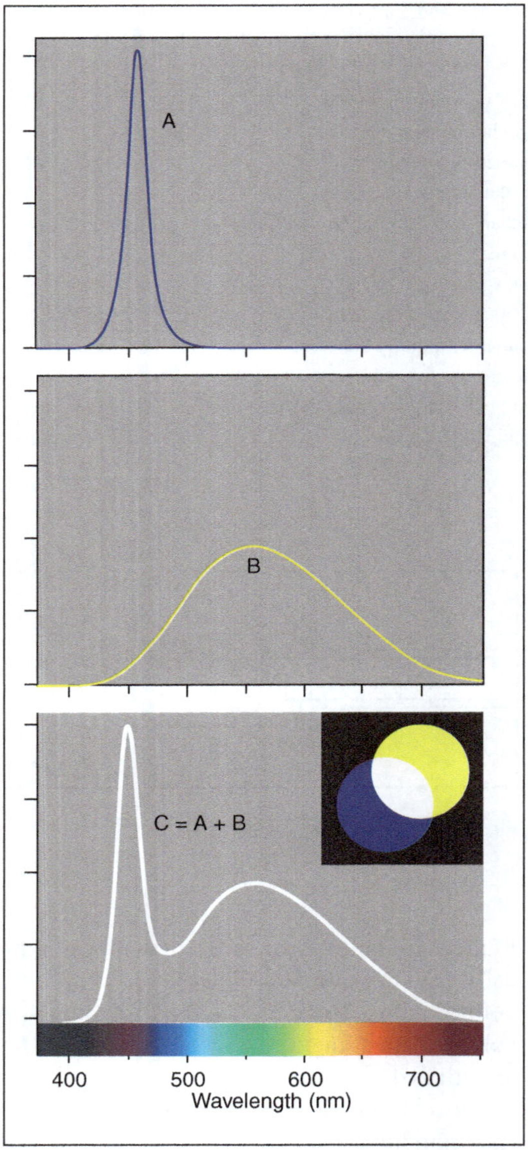

Fig. 3.9 Principle of white LEDs. (*A*) Spectrum of a blue LED. (*B*) Fluorescence spectrum (*yellow*). (*C*) Spectrum of a white LED derived from the combination of the primary blue light with the radiation of the embedded luminescent dyes

3.4 Lasers

Shortly after the laser was born (1961), experimental medical applications began. Not surprisingly, the first deployments involved the optically accessible eye. Indeed, the very first application of the laser in medicine is seen as Campbell's intervention.[5] At the time when the first lasers were being developed, no one was able to imagine the multitude of applications that would arise in the following decades. Today, in ophthalmology, both special surgical instruments as well as highly developed imaging systems are based on the laser. The properties of laser radiation, as discussed in Sect. 1.7, contribute to the multifaceted success of laser applications in ophthalmology. Their success is also attributed to the variety of various laser types that work in the desired wavelength range depending on the application. In addition, pulsed operation gives the possibility to achieve very high power densities and, thus, trigger various interactions between the laser light and the tissue. In a continuous mode, the radiated power is constant and relatively limited, as with a laser pointer. In pulsed operation, the laser internally stores up the engendered light energy and then releases it abruptly, a process that repeats periodically. All laser types are based on the same physical principle, which is discussed below. In Chap. 7, we will consider specific laser types that are applied in ophthalmology, along with the associated interactions between the laser beam and various materials.

3.4.1 How Laser Light Is Created: The Principle

In terms of the lowest common denominator, the laser as light source is based on a two-step process: first – as with every light source – energy is added to the atoms of a suitable medium from the outside, followed by the release of this energy as light in a very special form in a sort of chain reaction. In the case of the very first laser, the ruby laser, the chromium atoms in a ruby crystal were the agents. The focused energy of a flash lamp transferred blue or green light energy to them, bringing their electron shells into an excited state. Normally, excited electrons release their energy

[5]Campbell is mentioned in Sect. 7.1.

within nanoseconds as photons that fly off in random directions, and the associated atoms return to their ground state. With chromium atoms within an aluminum oxide lattice (ruby), this release is somewhat delayed and these atoms are then said to be in a so-called "metastable" state. With some atoms, however, spontaneous emission of their photons occurs.

What happens when one of these photons collides with an excited atom? In 1916, decades before any application, Einstein found the answer: if the energy of that photon corresponds to the energy difference between an excited atom's energy and its energetically lower state, the atom is forced to release an identical "copy" of the stimulating photon. Both photons are identical in energy since both stem from the transitions between the same energy levels and, in addition, have an identical flight direction. Einstein called this process stimulated emission (Fig. 3.10). Now there are two identical photons that can, in turn, trigger the release of further photons from neighboring excited atoms. It is easy to imagine that this process can become a sort of chain reaction when many atoms have been brought into an excited state (or, in lasers that operate continuously, are continuously brought into this state). This chain reaction, based on stimulated emission, is the basic phenomenon of a laser. Stimulated emission is the third basic process of light interaction with matter (after absorption and spontaneous emission, Figs. 2.3 and 2.4).

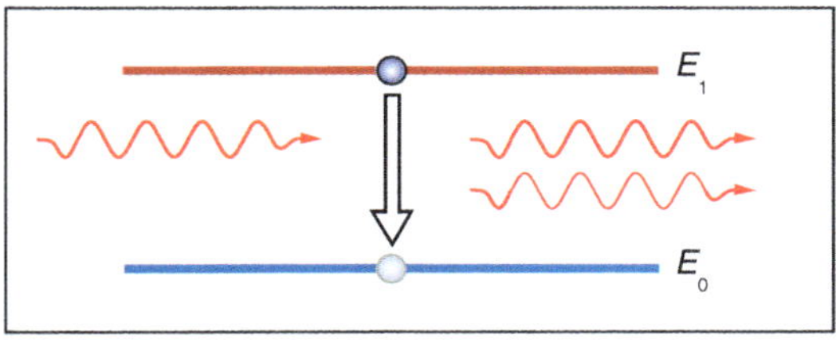

Fig. 3.10 A basic process in the interaction of light with matter: stimulated emission. A photon forces an excited atom to return to a lower energy state immediately by emitting a further, identical photon. As in spontaneous emission (Fig. 2.4), the atom falls into an energetically lower state and gives off the energy difference as a photon. After the stimulated emission, both photons have the same energy, flight direction, and phase

Figure 3.11 illustrates this cascade: among all the excited atoms, one spontaneously releases a photon and returns to its ground state. This photon carries exactly the energy difference between the two atomic states. In another excited atom, this photon stimulates the emission of an identical photon. Now, two photons with the same energy are present. In the next step, these stimulate two further atoms to emit their photons, and so on. Note that all of the photons carry the same energy: the energy difference between the two atomic states. In accordance with the law of stimulated emission, they all have the same flight direction. These photons form the laser light.

To build a functional laser, parallel mirrors at both ends are necessary. The mirrors send the photons back and forth, creating a very intense light field and increasing the feasibility of the stimulation processes. In this way, the lasing process is supported for those photons that move precisely perpendicular to the two parallel mirrors. The laser light that is perpendicular to the two mirrors is, thus, built up as it travels back and forth. The exiting beam from the laser arises

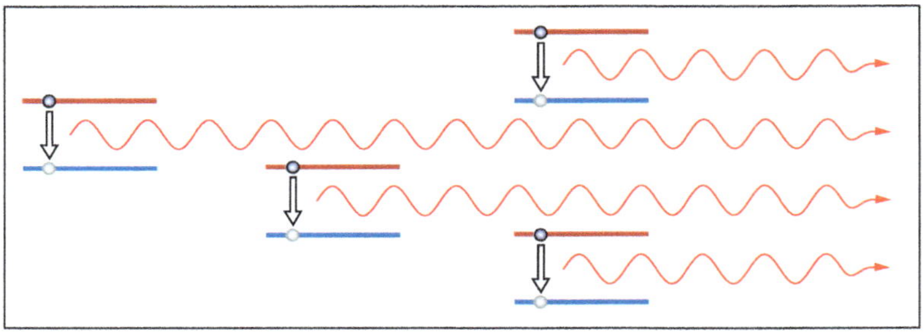

Fig. 3.11 Stimulated emission produces a chain reaction

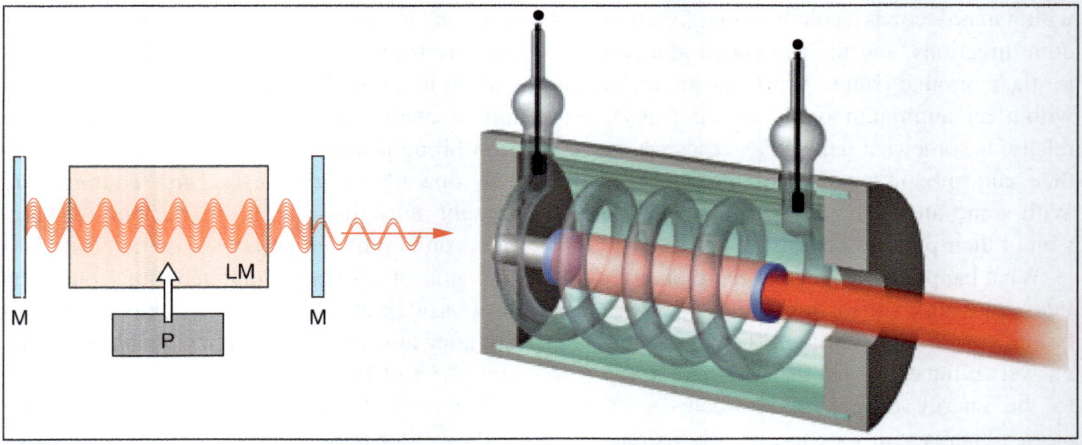

Fig. 3.12 *Left*: The elements of a laser. The two ends of the ruby rod are made into parallel mirrors, the one on the right reflecting incompletely. *P* pumping energy source, *LM* lasing medium, *M* mirror. *Right*: Realization as a ruby laser

via the fact that one of the two end mirrors does not reflect with 100 % efficiency but lets a certain amount of light pass through (outcoupling). The two mirrors form a resonator for the light that travels back and forth. The name indicates that the arrangement selects the particular wavelength that fulfills the resonance requirements, whereby an integral number of half wavelengths must correspond to the length of the resonator. In this way, one or more very narrow spectral lines are engendered. The construction principle of a laser and its essential elements are shown in Fig. 3.12.

In ruby lasers, only the chromium atoms are involved in the lasing process. Ruby is a variety of corundum (Al_2O_3) with a small amount of chromium (typically 0.05 %). Figure 3.13 shows the processes that take place in chromium atoms. The laser light arises via stimulated transitions of the electrons from energy level E_1 to E_0. The energy difference $E_1 - E_0$ of this transition determines the wavelength of the laser light. The supplied energy (pumping) takes place via blue or green light from the flash lamp. This brings the atoms into an excited state; i.e., it raises the electrons to energy levels E_2 or E_3, from which spontaneous transition to the state with energy E_1 occurs.

The lasing process commences only when the pumping power exceeds a certain threshold.

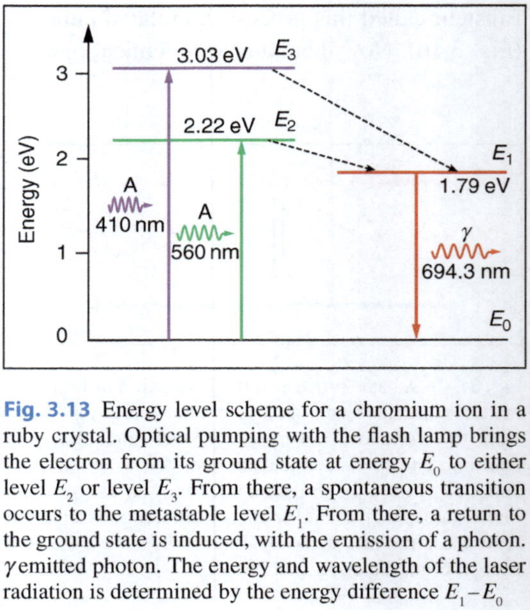

Fig. 3.13 Energy level scheme for a chromium ion in a ruby crystal. Optical pumping with the flash lamp brings the electron from its ground state at energy E_0 to either level E_2 or level E_3. From there, a spontaneous transition occurs to the metastable level E_1. From there, a return to the ground state is induced, with the emission of a photon. γ emitted photon. The energy and wavelength of the laser radiation is determined by the energy difference $E_1 - E_0$

It must be powerful enough to ensure that, despite the competition of the spontaneous emissions, level E_1 is better populated than level E_0 (so-called population inversion). To understand this prerequisite, we have to keep in mind that, for the light that is already present in the lasing medium, two processes can occur: either absorption by atoms

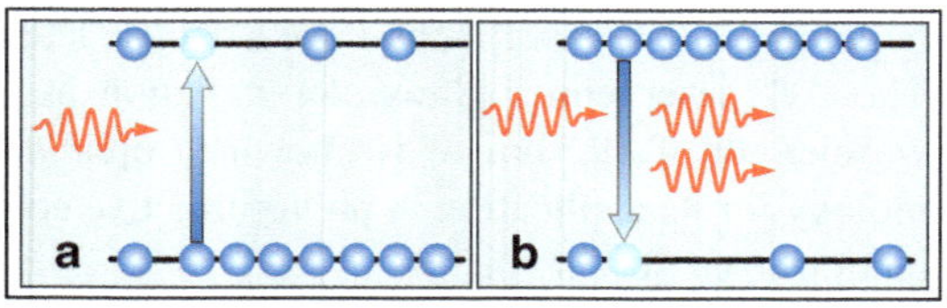

Fig. 3.14 Inversion as the prerequisite that the probability of stimulated emission exceeds the probability of absorption. This assumes sufficient pumping power. (**a**) In thermal equilibrium, the lower level is more populated. The probability is higher that an arriving photon will be absorbed than that it will trigger a stimulated emission. (**b**) Inversion. Powerful pumping leads to a greater population at the higher energy level. The probability that an arriving photon will trigger an induced emission is larger than that it will be absorbed. This means amplification of the light: from one photon, there are now two

at lower energy levels or stimulated emission with atoms at higher energy levels. The relationship of the numbers of these two processes is given by the ratio of the numbers of the atoms in the two possible states (E_0 and E_1). The light that is already present, thus, experiences both amplification (via stimulated emission) and attenuation (via absorption). Amplification prevails when the upper level is filled more than the lower one (Fig. 3.14). This explains why the lasing process takes place only when a population inversion is present. In a thermal equilibrium, e.g., in the hot interior of the sun, the situation is precisely reversed according to the laws of statistical thermodynamics: the upper levels are always less populated. Thus, the sun can never turn into a laser. Thermodynamics,[6] as a basic principle of nature, says this still more directly: out of the maximum disorder of a thermal equilibrium, a state of higher ordering (as is the case with population inversions and laser light) cannot arise spontaneously.

3.4.2 Laser Types

Following its invention, numerous lasers were developed that differ from one another in various ways.

(a) The active medium can be solid, liquid, or gaseous.
(b) The wavelength can lie in the infrared, visual, or ultraviolet range of the spectrum.
(c) For many types of lasers, the pumping takes place optically, whereas, with semiconductor lasers, on the other hand, it occurs directly via electric current.
(d) The energy output can be continuous with low power or in pulsed mode operation with correspondingly short pulses with high power densities, thereby qualitatively influencing the kind of interaction taking place between the light and the target (see Chap. 7).

3.4.3 Semiconductor Laser

Similar to light emission in LEDs, the light from semiconductor laser diodes (LD) arises via the passage of electrical current through the interface layer between variously doped semiconductors (Fig. 3.15). In contrast to LEDs, mirrored

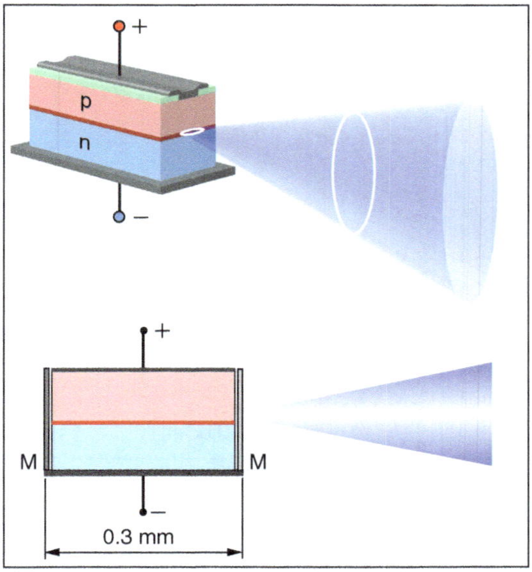

Fig. 3.15 The structure of a semiconductor laser. Electric current flows through the interface layer between two semiconductors (p, n). The lasing process occurs within this interface (shown in *red*). Light is emitted in a cone. The end surfaces are mirrored (*M*)

[6]More precisely, the second law of thermodynamics.

end surfaces form a resonator in which a lasing process takes place, bringing into play the principle of stimulated emission just as with the other laser types. With semiconductor lasers, extreme miniaturization is possible. Typical dimensions are 1 × 1 × 0.1 mm with an exit surface 1 μm high and 5–200 μm wide. Applications today are widespread (CD players, laser printers, laser pointers, etc.).

3.4.4 The Excimer Laser

Argon fluoride (ArF) is a two-atom molecule that is in an energetically excited state but then decays into separated Ar and F molecules after a photon has been sent off (Fig. 3.16). The pump mechanism consists of creating ArF molecules in a mixture of Ar and F gases by means of electrical sparks. This is the principle of the so-called excimer[7] laser. Before the first laser was operating, Houtermans[8] had already pointed out the possibility of this type of laser. Basov succeeded in building the first experimental one a decade later. The ArF laser emits ultraviolet radiation at a wavelength of 193 nm. It is known in ophthalmology for its application in photorefractive corrections (see Sect. 7.3).

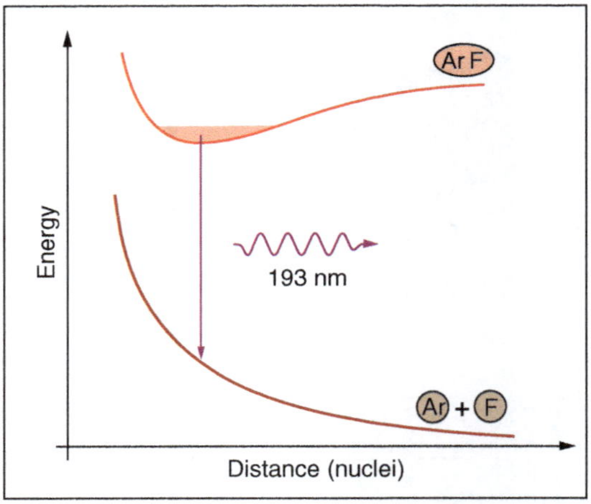

Fig. 3.16 Potential scheme of ArF: energy as function of the distance of the two nuclei. *Upper curve*: unstable bound state of the ArF molecule. *Lower curve*: Ar and F atoms separated spontaneously; total energy lower than in the bound state

3.4.5 Digression: Technical History of Lasers

At the beginning of the 1950s, Einstein's discovery of stimulated emission was technically implemented, not with visible light but with microwaves that have wavelengths 10^4 times larger. This laser predecessor, called the maser,[9] was constructed in 1954 by the American physicist Charles H. Townes, who was searching for an ideal amplifier for radar signals. He used the transition between two energetically neighboring states of the ammonia molecule (vibrations of the nitrogen molecule through the plane of the three H atoms) and could thereby engender coherent microwaves with a wavelength of 12.7 mm. In 1964, Townes received the Nobel Prize for physics, along with the two Soviet physicists, Nicolai Basov and Aleksandr Prokhorov, who had also independently come up with the theoretical foundations for the maser and laser principles.

3.5 Superluminescent Diodes (SLED)

Superluminescent diodes have a structure that is similar to laser diodes, although no lasing process takes place. These components differ from semiconductor lasers in the absence of the two mirroring surfaces (i.e., no resonators) so that the light can exit directly from the interface layer without repeated back-and-forth movements. Through the high population of the excited level, an amplification of the light takes place via stimulated emission but without a true lasing process. In this way, the SLED achieves considerably

[7] The name is derived from "excited dimer."

[8] Fritz Houtermans (1903–1966), German physicist.

[9] MASER, acronym for microwave amplification by stimulated emission of radiation.

Table 3.4 Comparison of LEDs, SLEDs, and LDs. The coherence concept is explained in Chap. 1

	Light emitting diode (LED)	Superluminescent diode (SLED)	Laser diode (LD)
Principle	Spontaneous emission	Amplified spontaneous emission	Stimulated emission
Optical spectrum	Relatively narrow (20 nm)	Relatively narrow (10–100 nm)	Almost monochromatic
Beam direction	All directions (~90°)	Small cone (~10°)	Small cone (~10°)
Light intensity	Low	Large	Extremely large
Temporal coherence	Short	Short	Very long
Spatial coherence	Low	High	High

higher light intensities than do LEDs. The light of a SLED borrows certain properties from LEDs and others from laser diodes. The spectrum resembles that of an LED, with bandwidths of 10–100 nm. This implies short coherence lengths that are suited for interferometric methods (see Sect. 4.2). The light beam has a directional characteristic like that of a laser diode (Fig. 3.15) and concentrates the light to a much narrower directional interval than an LED. Thanks to the extremely small beam cross-section at the exit surface, an efficient coupling into an optical fiber is possible. SLEDs are employed where the monochromatic spectrum of a laser diode causes problems. In Table 3.4, we compare several characteristics of the semiconductor light sources (LEDs, SLEDs, and LDs) we have presented here.

4 Examinations with Light

We now turn to the role of light as the most important "instrumentarium" for ophthalmological examinations. In Sect. 4.1, we discuss methods that are based on geometrical optics, such as, for example, ophthalmoscopes, contact lenses, and slit lamps. The appearance and functions of today's instruments should be quite familiar. We primarily present the optical principles, but we also intersperse a few historical illustrations. Optical coherence tomography (OCT) utilizes the wave nature of light and, for this reason, we shall treat its elements in a separate section (Sect. 4.2). We shall also dedicate a section (Sect. 4.3) to discussing optical Doppler effects, which allows movement to be observed.

4.1 Methods on the Basis of Classical Optics

4.1.1 The Ophthalmoscope (Direct Ophthalmoscopy)

From a purely optical point of view, two emmetropic people could quite possibly see the other's retina by looking through each other's pupils (Fig. 4.1). The reason that this does not work in real life is the lack of illumination. The light that exits the eye of the person sitting opposite is so weak that the pupil appears black. The great contribution of Helmholtz[1] was the fact that he was able to illuminate the interior of a patient's eye in a way that the observer could still look through the patient's pupil. Semitransparent mirrors did not exist back then. To achieve a similar effect, he used a stack of several plates of glass. The reflection of these glass plates made it possible to guide light into the patient's eye while, at the same time, the physician could look through them. He saw an upright and non-reversed image of the retina.

The fact that the pupil appears black has several causes. When light comes to us from all directions, the light intensity at the level of the retina, with a pupil diameter of 4 mm, is roughly 70 times less than that at the level of the iris. In addition, strong absorption also occurs: the fundus absorbs roughly 90 % of red light and roughly 99 % of blue light. Therefore, inside an eye, a darkness is present that is very similar to that inside a room with a small window and brown walls. In light of observations of the fundus – visually or photographically – the blackness of the pupils can also be explained as follows: depending on a photon's color, when it enters the eye, it has a probability of just 10^{-4} to 10^{-3} of coming out again.

In 1851, when Helmholtz invented the ophthalmoscope, he knew about its practical importance (Figs. 4.2 and 4.5). Ten years later, the

[1] Hermann von Helmholtz (1821–1894). German physiologist and physicist. Inventor of the ophthalmoscope. Author of the *Handbuch der Physiologischen Optik* (1856–1867). Promoter of Thomas Young's three-color theory of vision. One of the founders of the general law of energy conservation. He measured the velocities with which excitations reproduce along motor nerves. The German version of the *Handbook of Physiologic Optics* is available online: http://www.archive.org/details/handbuchderphysi00helm.

J. Flammer et al., *Basic Sciences in Ophthalmology*,
DOI 10.1007/978-3-642-32261-7_4, © Springer-Verlag Berlin Heidelberg 2013

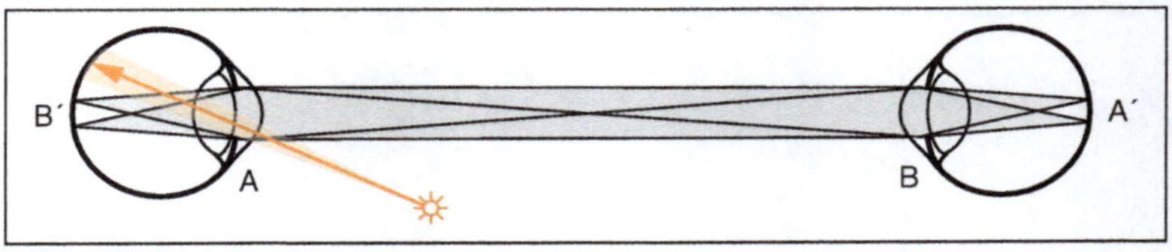

Fig. 4.1 Eyes seeing each other. A′ and B′ are the images of the pupils A and B, respectively (mutual focusing on the observed pupils and ideal imaging are assumed). The light that pupil B receives from pupil A originates exactly from the image B'. Without a light source between the two eyes, the images A′ and B′ remain unilluminated (disregarding possible stray light from other parts of the eyes). *Arrow*: a source of light that does not lie between the two pupils is unable to brighten the images

Fig. 4.2 Helmholtz and his 46-page monograph “Beschreibung eines Augen-Spiegels zur Untersuchung der Netzhaut im lebenden Auge” (“Description of an Eye Mirror for Examining the Retina in Living Eyes”) (1851). It is devoted to optics, handling, and the first physiologic observations. It is available online: www.archive.org

Fig. 4.3 Albrecht von Graefe

ophthalmoscope and the observations that had become possible with it were already the subjects of several dozen publications. With the help of this ophthalmoscope, von Graefe[2] was the first to describe optic nerve head excavations in glaucoma (Fig. 4.3).

The light path of direct ophthalmoscopy is shown in Fig. 4.4. The examiner views the patient's fundus through the patient's optical system, which has the effect of a positive lens and leads to a magnification of the funduscopic image by a factor of approximately 14. To enlarge his

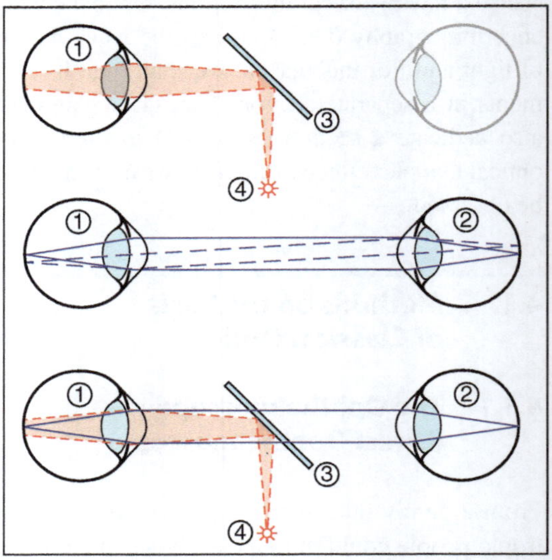

Fig. 4.4 Principle of direct ophthalmoscopy. *Top*: the illumination (*4*) is introduced via a semi-transparent mirror (*3*). With a point light source from a distance near the eye, the field of view can be illuminated. The cornea and lens give rise to reflections. Middle: observation pathways. The patient's retina (*1*) is imaged one to one on the retina of the observer (*2*). A correction lens, used to correct for possible refraction deficits, is not shown (in the case of two emmetropic eyes, no optical corrections are necessary). The field of view is limited by the diameter of the pupil; it gets wider the closer the two eyes are positioned in relation to one another. *Bottom*: both pathways combined

[2] Albrecht von Graefe (1828–1870) is considered the founder of ophthalmology in Germany. He contributed greatly to the dissemination of the ophthalmoscope. He died early from tuberculosis.

viewing angle, the examiner has to come very close to the observed object. If we are used to viewing the retina by means of direct ophthalmoscopy, we are surprised at how small the actual retina is in an anatomic preparation when viewed by the naked eye from a distance of 25 cm. If either the patient or the examiner is ametropic, this is corrected by the addition of a small lens that is brought into the observation path (these lenses are built into the so-called Rekoss disk) (Fig. 4.5).

White light is normally used for examinations of the eye. This enables the examiner to recognize not only shape alterations but also changes in color. For specific questions, certain parts of the color spectrum are preferable. For example: if we would like to see the retinal nerve fiber layer, blue light is superior for two reasons. First, the nerve fiber layer backscatters very little light. Of the amount of light reflected, blue is reflected more than any other color. In addition, most absorption of blue light takes place through the choroid, so minimal light is scattered back from there and it does not outshine the image of the nerve fiber layer. Therefore, corresponding filters are even more important in fundus cameras. For example, a blue interference filter (SE-40) of 495 nm wavelength is used for photographing the nerve fiber layer (Fig. 4.6).

Red-free green light is almost completely absorbed by hemoglobin and, for this reason, it gives better contrast of the blood columns in the blood vessels against the surrounding tissues. Hemorrhages are also easier to see. The dyes of the macula lutea (lutein and zeaxanthin) also strongly absorb this type of light and make it easier for the examiner to find the macula (Fig. 4.7).

The disadvantages of the Helmholtz's ophthalmoscope were reflections, which originated

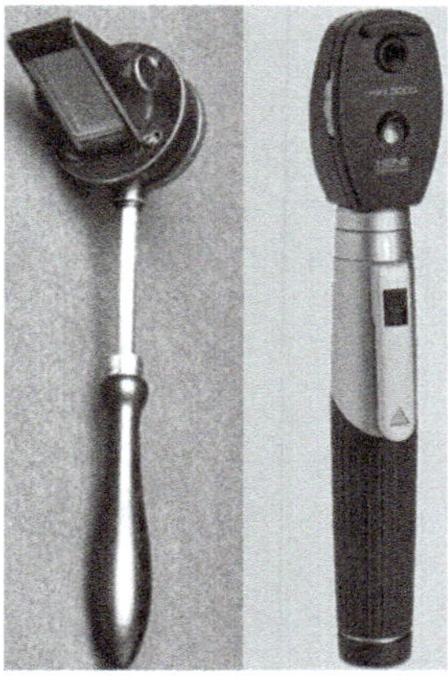

Fig. 4.5 Direct ophthalmoscopes. *Left*: construction according to Helmholtz. *Right*: modern version

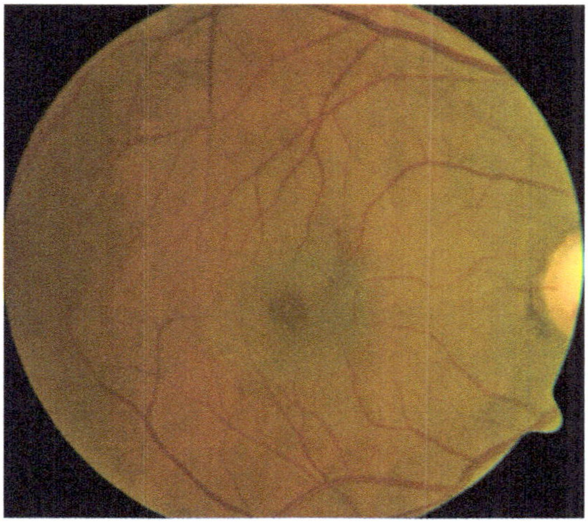

Fig. 4.7 The macula lutea appears slightly yellow due to the pigments lutein and zeaxanthin, which absorb blue light in particular

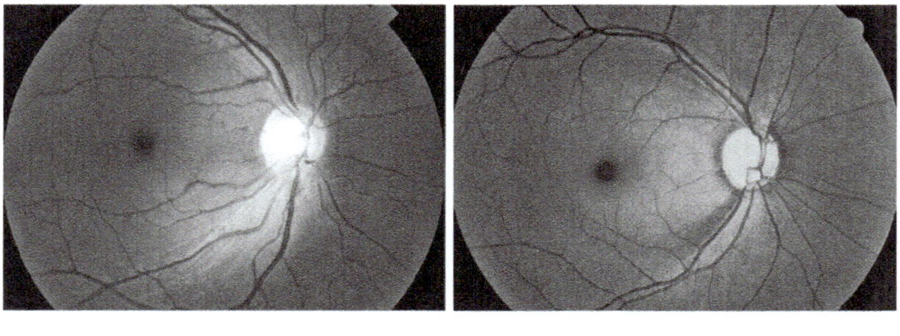

Fig. 4.6 Fundusphotography made through a blue-interference filter. *Left*: normal eye; *right*: nerve fiber bundle defect in a glaucomatous eye

Fig. 4.8 Allvar Gullstrand

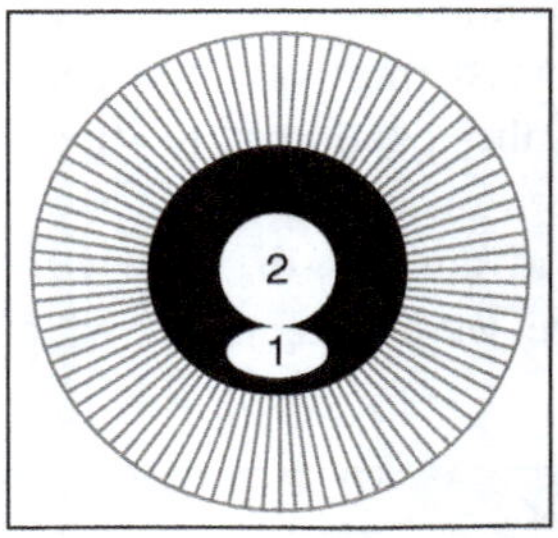

Fig. 4.9 Principle of separating the illumination pathway (*1*) from that of the observation (*2*) in the pupil and lens of the patient. Central ophthalmoscopy (observation through the center of the pupil) results in the best image resolution but requires a wide pupil

mainly from the cornea. Here, Gullstrand[3] brought about a decisive improvement (Fig. 4.8). The illumination path was separated from the observation path so that, on the one hand, the reflections were eliminated, while, on the other hand, the view through an enlarged pupil was still possible (Fig. 4.9).

4.1.2 Indirect Ophthalmoscopy

Indirect ophthalmoscopy allows monocular and binocular (stereoscopic) observation of the patient's retina with a variety of magnifications and fields of view. For indirect ophthalmoscopy, illumination of the retina is also necessary, but this will not be discussed here. A collecting lens in front of the patient's eye creates a real, mirrored, and inverted image of the ocular background that is then observed by the physician (Figs. 4.10 and 4.11). From a distance of 25 cm and with a lens of 70 mm focal length, the ophthalmologist observes an image of the retina that is approximately three times smaller than the image in direct ophthalmoscopy. In return, indirect ophthalmoscopy offers a substantially larger field of view. In terms of enlargement and field of view, direct and indirect ophthalmoscopes complement each other.

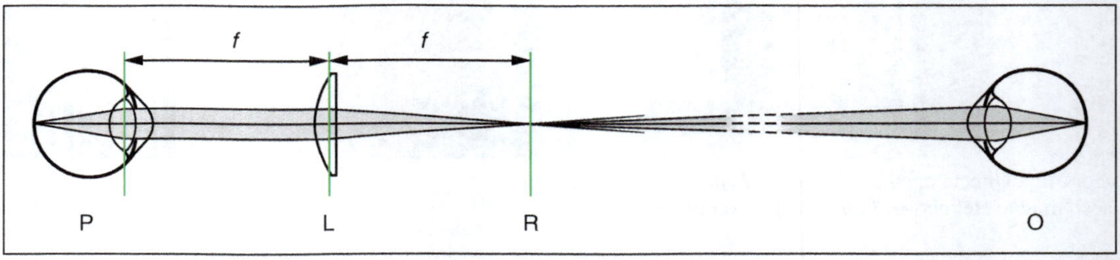

Fig. 4.10 Central beam paths of monocular indirect ophthalmoscopy. *P* patient's eye, *L* strong convex lens, *R* real inverted and mirrored image of the patient's retina, *O* observer. The distance between L and R is given by the focal length of L

[3]Allvar Gullstrand (1862–1930). Swedish physician. 1911 Nobel Prize for physiology or medicine. Researched the dioptrics of the eye as the scientific basis for eyeglass corrections. His most important inventions are the slit lamp and the reflection-free ophthalmoscope.

In 1947, Schepens[4] developed the binocular indirect ophthalmoscope, which also contained the illumination. Figure 4.12 shows that the light beam in the observer's direction is narrow. It becomes narrower as the patient's pupil gets smaller and as the focal length of the lens gets larger. The two eyes of the observer are able to access the light beam simultaneously only with the aid of prisms. Since the stereo angle α is small, the stereoscopic effect is not very large (Fig. 4.13).

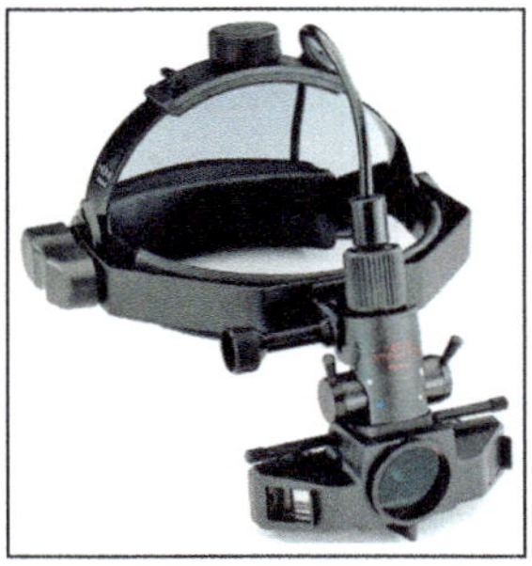

Fig. 4.13 Indirect ophthalmoscopy according to Schepens. The illumination is built-in

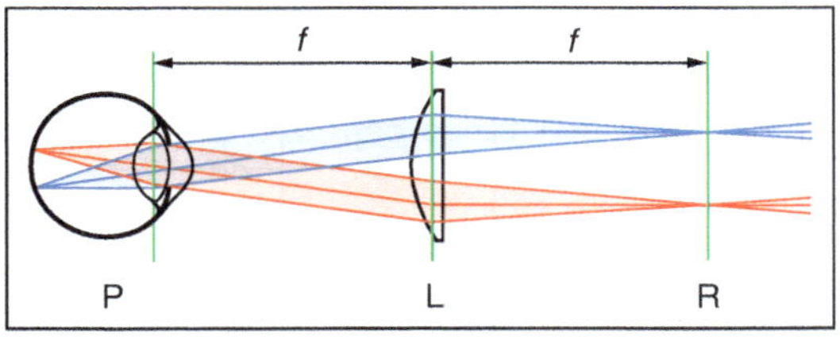

Fig. 4.11 Indirect ophthalmoscopy. The field of view is optimal when the distance of the convex lens (L) from the nodal point of the eye is the same as the focal length of L. The nodal point of the eye lies roughly in the center of the crystalline lens. In this arrangement, the field of view is determined only by the edge of L. A stronger lens yields a smaller image and a wider field of view

4.1.3 The Slit Lamp

The slit lamp was developed by Gullstrand[5] in 1911. The improvements made by Goldmann[6] contributed decisively to its popularization as a standard instrument. In all models, the microscope (observation unit) and the illumination unit are combined. The microscope can be positioned in all three dimensions, independent of the direction (and tilting angle) of the illumination (Fig. 4.14). Two modes of observation are possible: focal, i.e., direct observation of an illuminated structure, or confocal, whereby the scattered light that arises in

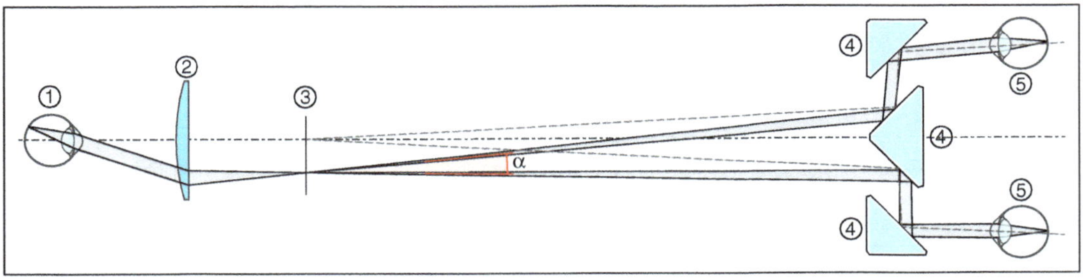

Fig. 4.12 Binocular (stereoscopic) indirect ophthalmoscopy. Just as in monocular indirect ophthalmoscopy, the lens (*2*) produces a real image (*3*) that is observed by the physician. The narrowing of the beam pathways with the aid of prisms or mirrors (*4*) is necessary so that the beam of light that emerges from the real intermediate image (*3*) reaches both eyes of the physician. Thus, the stereoscopic impression, although diminished, is still present and helpful for the surgeon. In this figure, the distance of the examiner from the intermediate image is shown too short; in reality, it amounts to about 50 cm. The magnification from retina to retina is given by the ratio of the focal length of the lens (*2*) to the distance of the ophthalmologist from the intermediate image. In this figure, it is roughly 0.15, approximately 6 times smaller than in direct ophthalmoscopy (1:1) – this is the price for the larger field of view. α stereo angle

[4]Charles L. Schepens (1912–2006). Belgian-American ophthalmologist.

[5]Allvar Gullstrand, referred to in the previous section.

[6]Hans Goldmann, referred to in Sect. 2.9.

the illuminated slit is observed obliquely from the side (see the figures in Sect. 2.7).

The slit lamp biomicroscope enables stereoscopic observation with variable magnification (typically 6× to 40×). The optics fulfill the requirements of an upright image while leaving enough space for manipulations in front of the examined eye. A large depth of focus is desirable. The optics correspond to the principle of the telescopic magnifier. The optics can be completely separated for the two eyes, but instruments with common object-side magnifiers are also possible (Fig. 4.15). In both cases, the two eyes of the observer view the object under a stereo angle. This produces the stereoscopic effect. In Sect. 4.1.5, we shall explore the application of the biomicroscope for observing the fundus.

The illumination unit produces a very bright slit light with adjustable elevation, width, position, and angular orientation. A very homogeneous illumination is desired; this is a task that is accomplished very well by the beam path, according to Köhler, known from the microscope (Fig. 4.16). The color of the slit light can be selected with filters. A large blue portion is able to produce sufficient stray light in the transparent media and also bring about fluorescence effects. The use of red-free light is further addressed in Sect. 4.1.

At this point, we are not going to discuss any of the highly developed accessories for slit lamps, such as tonometers, pachymeters, cameras, or the optics for introducing laser radiation. Gonioscopy and funduscopy are mentioned in

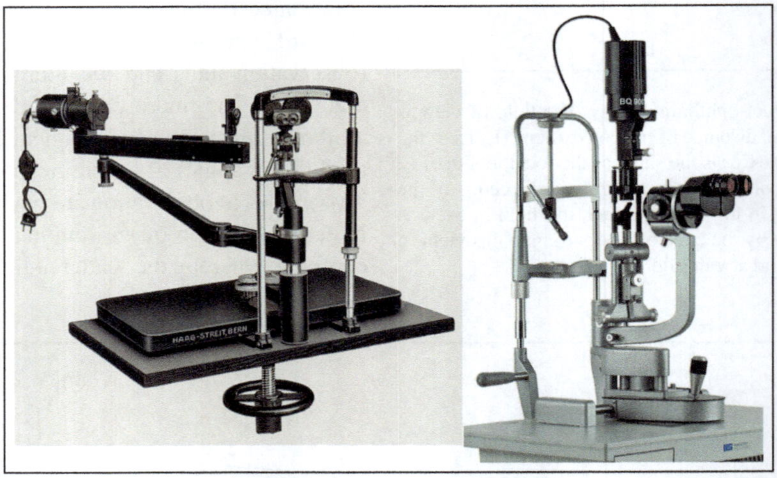

Fig. 4.14 *Left*: Goldmann slit lamp (1933). One recognizes the stereomicroscope and the projector for the slit light. *Right*: modern model. (Courtesy of Haag-Streit AG, Switzerland)

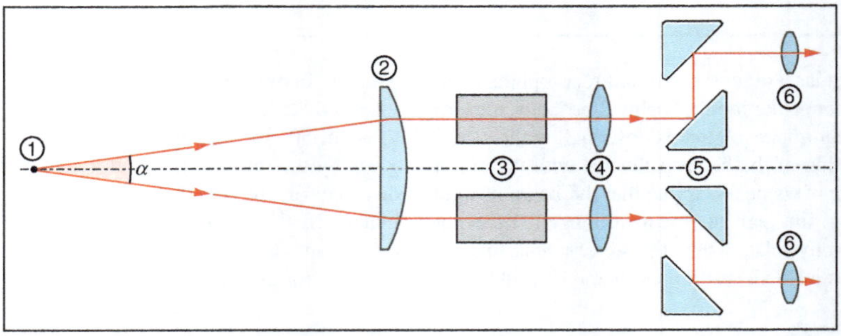

Fig. 4.15 Ray pathways of a slit lamp microscope according to the principle of the telescopic magnifier. In the construction sketched here, the beam paths for the two eyes of the observer separate only behind the magnifier (*2*). The magnifier (*2*) images the object (*1*) at infinity. Observation is performed with prisms (*5*) and with the telescope, consisting of an objective (*4*) and an ocular lens (*6*). The telescopic system (*3*) is used to alter the magnification. The stereo angle α equals 10° to 14° and corresponds to a stereoscopic view such as from a distance of 25 cm

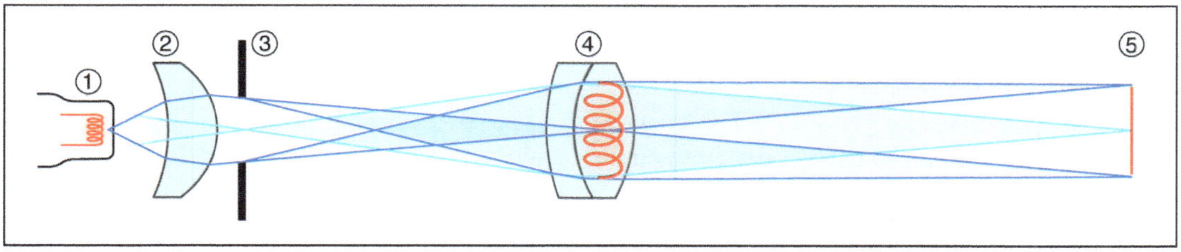

Fig. 4.16 Slit illumination according to Köhler. The collimator (*2*) images the filament (*1*) of an incandescent lamp through a slit aperture (*3*) into the objective lens (*4*). The objective images the slit, the dimensions and angular orientation of which are variable, into the slit light (*5*). The image of the filament does not have to lie precisely in the objective lens – what is important is that it is not imaged sharply by the objective (*4*) into the slit image (*5*)

separate chapters, and Goldmann's tonometry was discussed in detail in Sect. 2.9.

4.1.4 Contact Lenses

Contact lenses of different constructions permit the viewing of various regions and structures of the eye (Fig. 4.17). A contact lens essentially neutralizes the refractive power of the cornea. In Sect. 2.5, we showed how viewing into the anterior chamber angle is possible with a suitable contact lens (gonioscopy[7]). The Goldmann 3-mirror lens permits viewing angles on the whole inside of the eye (Fig. 4.18). The mirrors are tilted at angles of 59°, 66°, and 73°. The outermost periphery can be viewed only when the sclera is indented with the

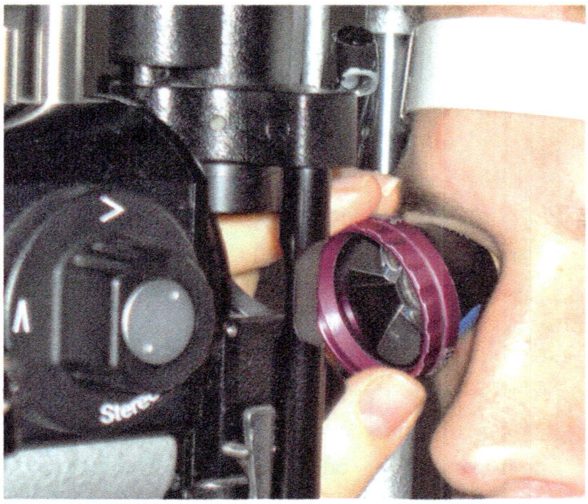

Fig. 4.17 Patient's eye is examined using the slit lamp microscope through a contact lens

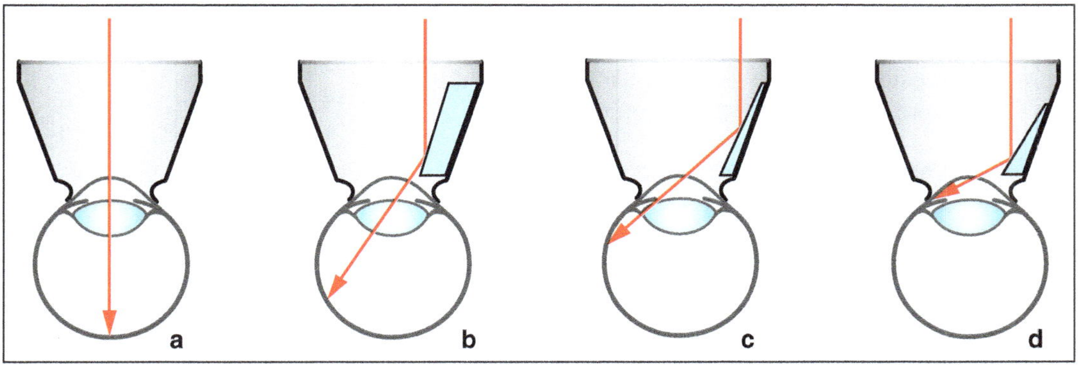

Fig. 4.18 Direct view and views through variously tilted mirror contact lenses. (**a**) Direct fundus examination. (**b**) With the 73° tilted mirror. (**c**) Mirror with a tilt of 66°; observation of the peripheral fundus. (**d**) Mirror with a tilt of 59°; observation of the very peripheral fundus and of the chamber angle (gonioscopy)

[7]Gonioscopy originates from the Greek (gonios = knee, skopein = seeing).

so-called indentation contact lens, which contains a scleral depressor. Contact lenses made of acrylic material have a somewhat higher index of refraction than the eye media ($n=1.49$). The contact lens neutralizes approximately the refractive power of both the cornea and the crystalline lens.

4.1.5 Funduscopy with the Slit Lamp

It is often said that the slit lamp was created mainly for examination of the anterior segment. This is true since strictly optical observations of the anterior segment are easy to perform and do not create a problem. Due to the strong refractive power of the cornea and lens, direct observation of the retina is not possible. Various methods have been developed to view the retina directly with the slit lamp. The direct view through a contact lens (e.g., Goldmann's 3-mirror lens) allows inspection of the fundus with the slit lamp microscope.

The retina can also be examined by the slit lamp when using stronger lenses with a refractive power of, for instance, 78 or 90 D that are not in direct contact with the patient's cornea (Fig. 4.19). Just as with indirect ophthalmoscopy, the lens then produces a real, mirrored, and inverted image (see Figs. 4.10 and 4.11). In contrast to indirect ophthalmoscopy, one views this image with a microscope. It is true that the 78 D lens minifies the retinal image compared to the original by a factor of 0.75 (60/78). However, the biomicroscope's 10-fold magnification results in a net 7.5-fold magnification.

Another possibility is the insertion of a minus lens (lens according to Hruby[8]). With a refractive power of about −60 D, the eye's refractive power is neutralized. The physician then sees an upright image of the retina.

The stereo angle of slit lamps amounts to about 12°–15°. There are models that permit a reduction of the angle to roughly 5°. This can make observations of the fundus easier – for example in a patient with a small pupil – and it enlarges the binocular field of view. However, the stereo effect is weakened.

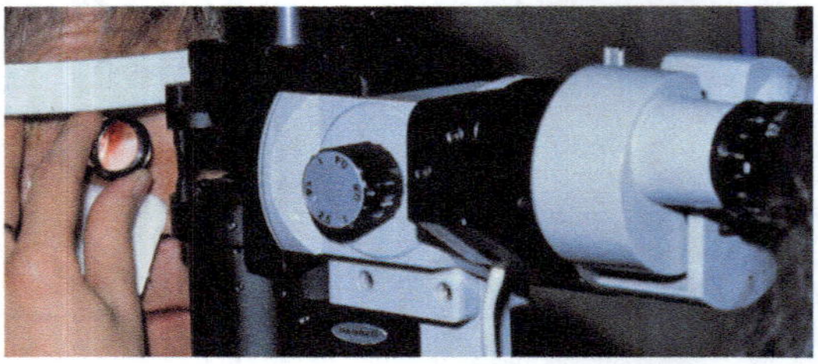

Fig. 4.19 Funduscopy with a 90 D magnifier

4.1.6 The Operating Microscope

In principle, an operating microscope (Fig. 4.20) has an optical ray path similar to that of a slit lamp microscope (Fig. 4.15). Several requirements are comparable, such as those of a free working area, depth of focus, variable enlargement, and a stereo effect. Unlike the slit lamp, though, the illumination does not have to be freely movable and, for this reason, the light source is firmly built into the microscope. A change between focal and confocal illumination is possible.

For the anterior segment, an operating microscope is ideally suited. A very interesting addition, the BIOM,[9] together with a SDI works according to the principle of indirect ophthalmoscopy

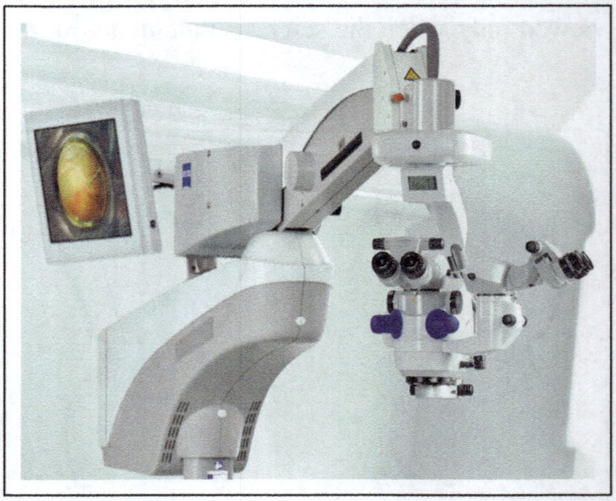

Fig. 4.20 Example of an operating microscope (Courtesy of Carl Zeiss AG, Switzerland)

[8]Karl Hruby (1912–1995), Austrian ophthalmologist.

[9]The BIOM (binocular indirect ophthalmoscope) and the SDI (stereoscopic diagonal inverter) were developed by M. Spitznas and J. Reiner in 1987.

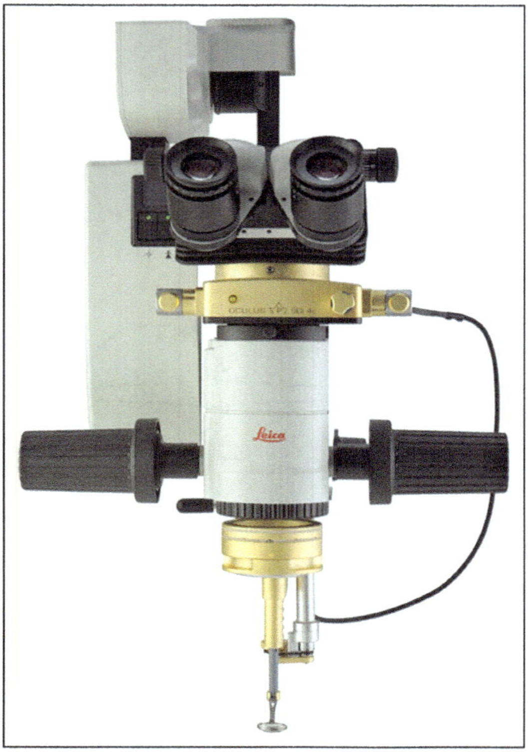

Fig. 4.21 BIOM combined with SDI (Courtesy of Oculus Optikgeräte GmbH, Germany)

(Fig. 4.21). It enables non-contact 70°–90° wide-angle observation of the fundus during vitreous retinal surgery. A 60–90 D lens in a BIOM produces a bright, inverted image of the fundus. This is then observed through the operating microscope whereby, though, an SDI is put in between that uprights the image by means of a passive optical system of mirrors and, in addition, exchanges the optical ray paths for the surgeon's eyes to ensure true stereoscopy.

4.1.7 Retinoscopy (Skiascopy, Shadow Test)

Retinoscopy is a method for objectively determining the refractive error of an eye (Fig. 4.22). Only minimal equipment is required to obtain the optimal information concerning the refractive status of an eye. The retinoscope (or skiascope[10]) shines a narrow streak of light onto the retina and the position of the light can then be shifted by slightly rotating the instrument (Fig. 4.23). The physician observes the light phenomena in the patient's pupil during this motion. With retinoscopy, one attempts to determine the correction for distance. To avoid errors due to accommodation of the patient, accommodation is usually temporarily blocked by the use of cycloplegic eye drops.

When the retinoscope sends parallel light into the patient's eye, an illuminated streak appears on the retina and, in cases of ametropia, it is a somewhat wider streak. This broadening is not essential for understanding retinoscopy. Here, the method

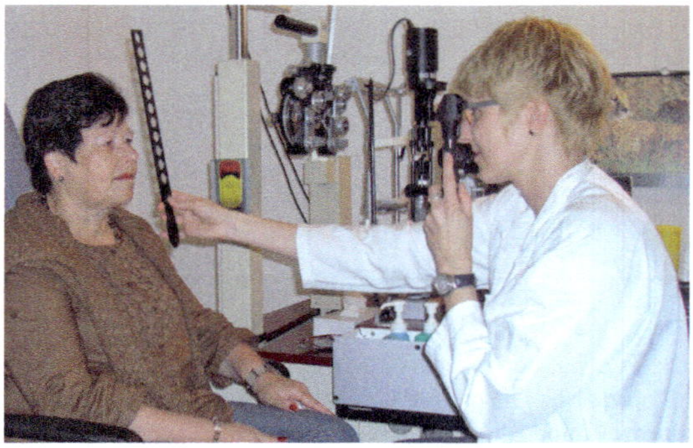

Fig. 4.22 Objective determination of the refractive error with a retinoscope

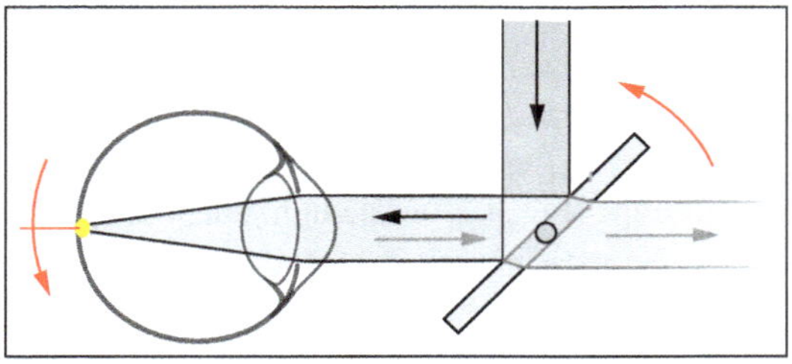

Fig. 4.23 The retinoscope creates a streak of light on the patient's retina that serves as a secondary light source. With small rotational motions of the retinoscope, the streak moves sideways on the retina. In this figure, the streak stands perpendicular to the diagram plane. The mirror is semi-transparent, making it possible for the physician to observe the appearance of the light inside the patient's pupil

[10]Skiascope originates from the Greek (skia = shadow and skopie = view).

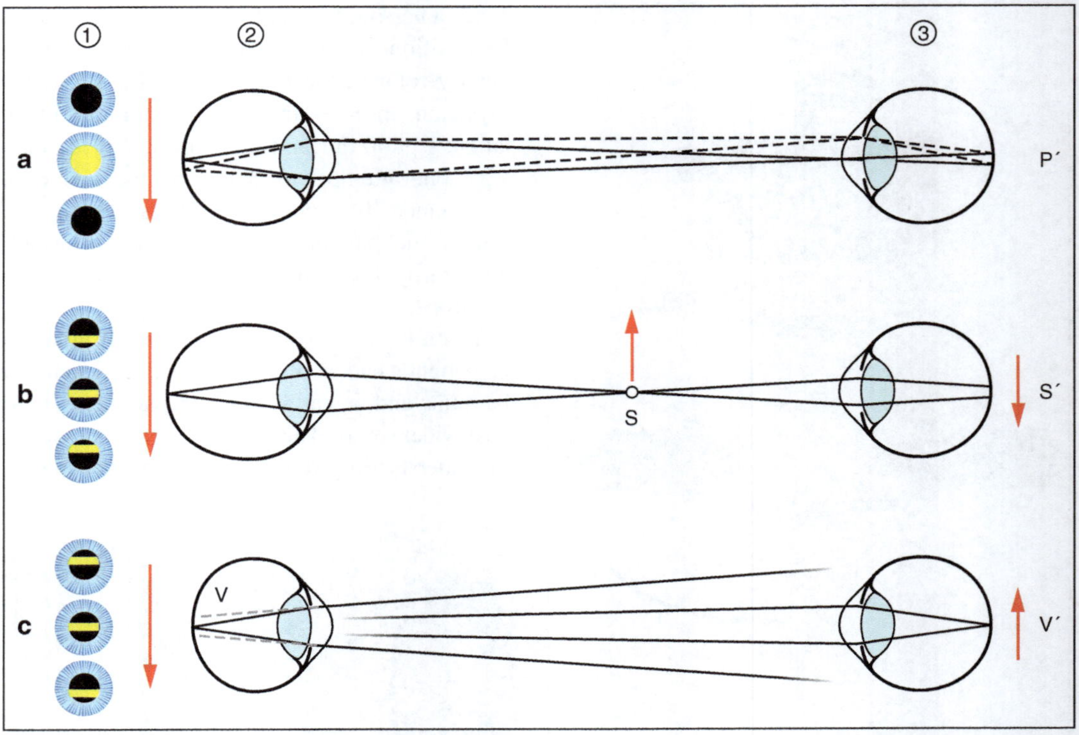

Fig. 4.24 Retinoscopy. Patient (*2*); observer (*3*). *Situation a*: The patient's eye is focused on the physician's eye. *Situation b*: The patient's eye is focused at a point between himself and the physician. *Situation c*: the regredient beam is divergent or the patient fixates at a point behind the physician. *Column 1* indicates which light phenomenon the physician sees in the patient's pupil when the retinoscope rotates in the same way as in Fig. 4.23. In *situation b*, the movement is in the opposite direction, while in *situation c*, it is in the same direction. *P′* Image of patient's pupil; *S′* image of S; *V′* image of V. *V* Virtual image of the light streak on the patient's retina (V is behind the retina)

will be discussed with the simplifying assumption that a streak-shaped secondary light source exists and moves on the patient's retina when rotating the retinoscope (Fig. 4.23). The light phenomena seen inside the patient's pupil depend on the refractive error (see Fig. 4.24):

A. If the patient's eye is focused to the distance of the examiner, a quick change between a bright and dark pupil can be observed in the patient. This is similar to a "flashing" movement, which is also known as the neutralization endpoint (Fig. 4.24a). If the patient is in a cycloplegic state, where accommodation is impossible, the distance refraction can be determined. If the examiner performs retinoscopy from a distance of 50 cm, a 2 D difference exists between the lenses used for neutralization of the reflex and the patient's true distance refraction.
B. If the patient's far point is located between patient and physician (as in cases of higher myopia), a glowing streak is seen inside the patient's pupil and it moves in a direction opposite to that of the retinoscope ("against motion"; Fig. 4.24b). The streak is narrow when the refractive error is large. A broader streak near the ideal correction is no longer perceived as a streak but as a dark-bright-dark alteration. The point of optical neutralization has then been achieved.
C. If the patient is hyperopic, the same images appear in the pupil as in (B) but with the inverse direction of motion; that is, the streak in the pupil moves in the same direction as the illumination ("with motion"; Fig. 4.24c).

The phenomena observed by the physician inside the patient's pupil can be understood in the following way (Fig. 4.24). It is assumed that the physician focuses onto the patient's pupillary plane:

A. The easiest case is when the patient focuses on the physician's pupil. Independent of the position of the light source, the physician always sees the same image: the glowing pupil of the patient. The optical ray path indicates that the image P' of the patient's pupil does not move on the retina of the observer while the light streak moves on the patient's retina. However, the patient's pupil switches to dark when the focused light is next to the physician's pupil. Therefore, an abrupt change is noted between the dark and fully illuminated pupil of the patient ("blinking").

B. The glowing streak on the patient's retina is seen to be at a location S, between the physician and the patient, and it functions there as a tertiary source of light. The physician sees its image S′ as blurred or the streak as widened. The movement of S is opposite to that of the retinoscope.

C. In hyperopia, the light beam leaves the pupil divergently. The observer then sees the backward extension of this beam as a virtual light source V. It moves in the same direction as the light streak on the patient's retina. (This is also true when the patient fixates on a point behind the head of the observer or at infinity. This situation is not depicted in the figure.)

So far, we have discussed the so-called streak retinoscopy. In contrast to spot retinoscopy, it has the advantage that the determination of astigmatic corrections is easier. The figures do not indicate a situation where the streak projected onto the retina by the retinoscope may be blurred. This is of subordinate importance because we have been concerned here with discerning movement directions.

4.1.8 Refractometry

Like retinoscopes, refractometers objectively determine an eye's refractive status. A special target is projected onto the retina and the refractive error of the eye is then judged based on the retinal image. The examiner observes this image through optics with a small aperture to circumvent aberrations of the examined eye. The lenses required to see the target in focus correspond to the patient's refractive error. The Raubitschek chart is often used as the test target (Fig. 4.25).

Today, refractometers are usually automated. Different methods are applied. Information can be obtained, e.g., by directing a secondary light source onto the retina. This light appears as pinpoint, thanks to a very small aperture, even in cases of refractive error. In emmetropic eyes, light that exits the eye forms a parallel beam. Deviations can be observed with the help of an aperture with multiple pinholes. In the ideal case, the passing rays can be imaged as a single point. A camera measures the actual positions of the point pattern and, thus, determines the amounts of the spherical and cylindrical corrections (Fig. 4.26).

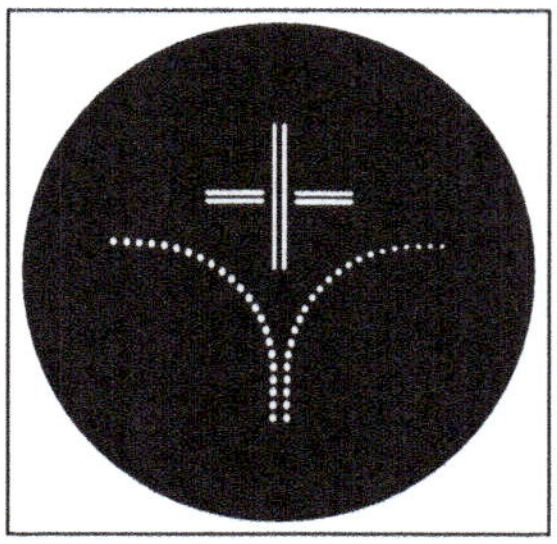

Fig. 4.25 Test target according to Raubitschek

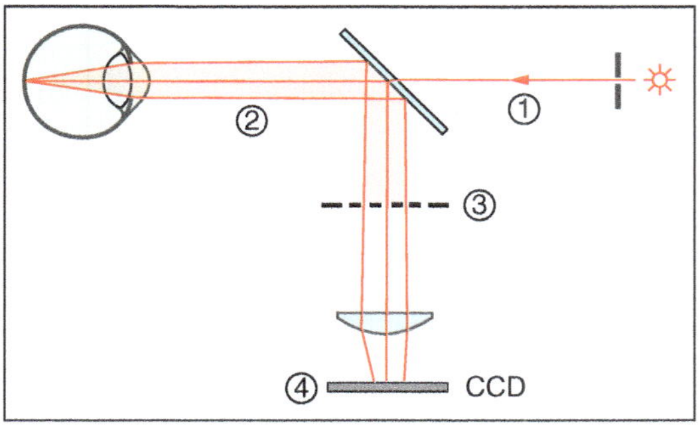

Fig. 4.26 Principle of an automatic refractometer. (*1*) Projection of a point onto the retina. (*2*) In the ideal case, the light reflected back is a parallel beam. (*3*) Multiple-pinhole aperture. (*4*) In the ideal case, the rays from the aperture converge to a single point at the camera's sensor

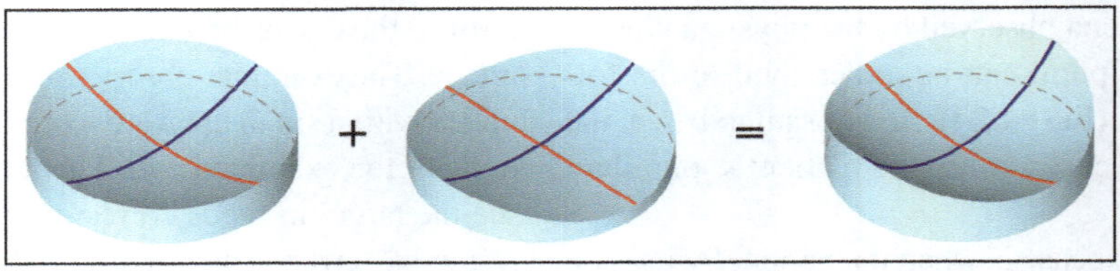

Fig. 4.27 A surface can be described approximately in terms of spherical (*left*) and cylindrical contributions (*middle*). In the superposition (*right*), the principal meridians with the smallest and greatest curvatures are perpendicular to each other. They correspond to the axes of the cylindrical contribution. The sphere has rotational symmetry; i.e., the curvature is the same in all directions

Various effects can influence the precision of automatic refractors. Due to the use of infrared light, the instruments need to correct the results for dispersion (the difference between the refractive powers in the infrared and visual ranges). For several reasons, the results of objective refractometry are not completely identical with those obtained subjectively.

4.1.9 Keratometry and Corneal Topography

Measuring the outer surface of the cornea serves various purposes.[11] The greatest part of the light refraction occurs at this surface – almost 50 D; therefore, it contributes far more than the corneal inner surface or the lens. Refractive errors, such as astigmatism, have their main origin at the corneal curvature. In addition, the geometry of the corneal outer surface is an important parameter in biometry for calculating the relationship between the refractive power of an artificial intraocular lens and the resulting refraction. A further application consists of the individual fitting of the back surface shape of a contact lens.

Two steps are to be distinguished in the geometric description of the corneal outer surface. The sphero-cylindrical description of a central region of 3–4 mm diameter often gives sufficient information for many purposes. This description is based on the assumption that the smallest and largest radii of curvature are found in two meridians that are perpendicular to each other (Fig. 4.27). The measurement of these three parameters – two radii and the angular orientation of the axes – is performed with a keratometer, either in the classical, manual implementation or in the automated version. In this description, the special case of a spherical cornea is included (the radius of curvature is the same for all angular orientations, with no axes). To go further, a more complete description of an individual cornea also takes into consideration deviations from this approach, such as the flattening away from the center (asphericity), irregular astigmatism, or, generally speaking, any behavior as a function of location. Today's modern instruments make such local measurements possible (corneal topographers, Fig. 4.32).

First, we discuss classical keratometry, which gives information on the sphero-cylindrical description of the corneal outer surface. In most cases, this description is adequate for the central portion.[12] The cylindrical contribution defines the two axes that are perpendicular to each other (Fig. 4.27). In these, we have the smallest and largest radii of curvature.

The keratometer is able to measure the radius of curvature for any given angular orientation of a meridian. Keratometers master this task with the observation of reflections from the tear film (reflections from the corneal inner surface are weaker due to the lower change in the indices of refraction). Figure 4.28 shows how a point source at a large distance produces a reflection in the

[11] More than 200 years ago, pioneers were interested in the question as to where the accommodation occurred. An alteration of the corneal curvature – besides lens position and lens form – was one of the theories discussed.

[12] Mathematically, this is the second-order description of the local behavior of any surface.

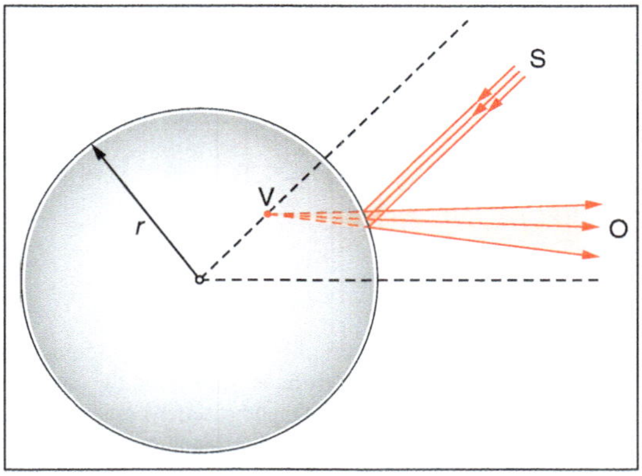

Fig. 4.28 Reflection on a convex spherical surface of radius r for a point source (S) located a large distance away. The reflecting rays appear to an observer (O) as if they come from a point (V), the virtual image behind the reflecting sphere. V is located on the line connecting the center with the source and is approximately in the middle between the center and the spherical surface.

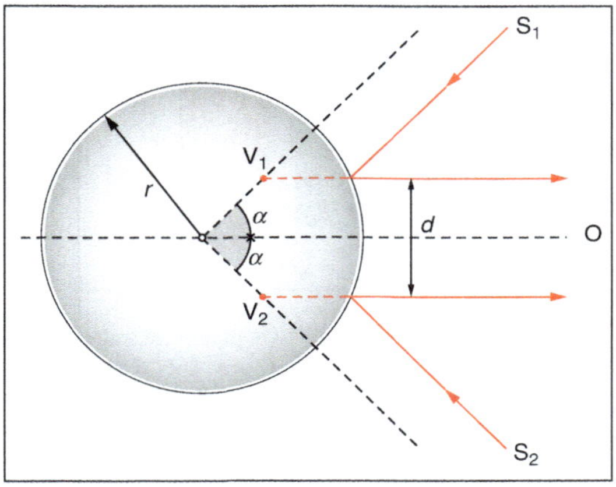

Fig. 4.29 Measurement of the radius of curvature by means of two point sources (S_1, S_2) at a large distance. The observer (O) sees two reflections (V_1, V_2). For a given angle of incidence α, the distance d between the two reflections is proportional to the radius of curvature r (Since $d=\sin(\alpha)\cdot r$, r can be calculated from measuring d)

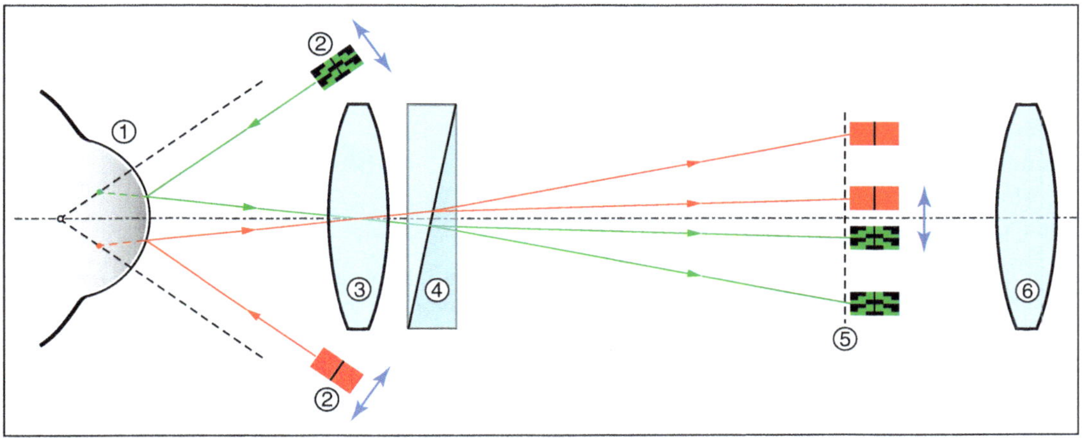

Fig. 4.30 Principle of the keratometer according to Javal-Schiøtz. The reflections on the cornea (*1*) of the mires (patterns) (*2*) are imaged by the objective lens (*3*) onto the plane (*5*). The optical element (*4*) doubles the images (it can be an arrangement of semi-transparent mirrors or a Wollaston prism). The plane (*5*) is viewed by the observer through the ocular (*6*). Through a symmetric change of the locations of the sources (*2*), the two innermost mires in the plane (*5*) must be brought to where they touch each other. From the positions of the sources (*2*), the instrument derives the radius of curvature of the cornea. The instrument can be rotated about the longitudinal axis and thereby measure the radius of curvature in the associated meridian. The mires are drawn in the same plane as the diagram; in reality, they are perpendicular to it

form of a point behind a spherical surface (a so-called virtual image, like our image that appears to be behind a plane mirror). Two symmetrically arranged point sources have two reflections. For a given direction of incidence, the distance between them is proportional to the radius of curvature of the reflecting surface (Fig. 4.29). Note that the radius of curvature refers to the meridian chosen by the angular orientation of the instrument.

A classical keratometer determines the distance between the two reflections. The principle, according to Javal-Schiøtz, is sketched in Fig. 4.30. Red and green mires (patterns) serve

Fig. 4.31 Keratometer. The two innermost mires in the plane (*5*) of Fig. 4.30 as seen by the observer. *Left*: the offset means that the current rotation does not coincide with one of the principal meridians. *Middle*: no offset, meaning that the angular orientation of the instrument coincides with one of the principal meridians. *Right*: the two sources have been shifted in such a way that the two mires touch. The radius of curvature for this axis can be read from a scale

as sources for the reflections. Their two reflections are imaged in a plane that can be observed through an ocular lens; however, the shakiness of the image from eye movements makes a direct reading of the distance between the two reflections difficult. This problem is overcome by an optical doubling of the images in the field of view of the observer. Figure 4.31 shows the two doubled reflections as they would present themselves for some angular orientation of the section.

Modern keratometers automate the measurement of the corneal outer surface by processing a video recording of the reflected image of a pattern of concentric rings (a Placido disk, Fig. 4.32). The observation of the reflections takes place through the center of the disk. In a spherical cornea, equidistant circles are seen in the reflected image. When astigmatism is present, an elliptical distortion occurs (Fig. 4.33). Further deviations due to local irregularities can also be discerned (corneal topographer).

The importance of keratometry for estimating the refractive power of the cornea has already been mentioned. In addition, the radius of curvature of the posterior corneal surface must also be known. In any case, the posterior surface contributes roughly −5 to −7 D to the refractive power of the cornea. Often, the relationship between the radii of the anterior and posterior surface was assumed to be fixed. To take this into account, one relies on the anterior curvature while replacing the true refractive index of the stroma ($n = 1.376$) with a hypothetical one (n'). However, not all instruments use the same value for n' (n' between 1.332 and 1.3375).

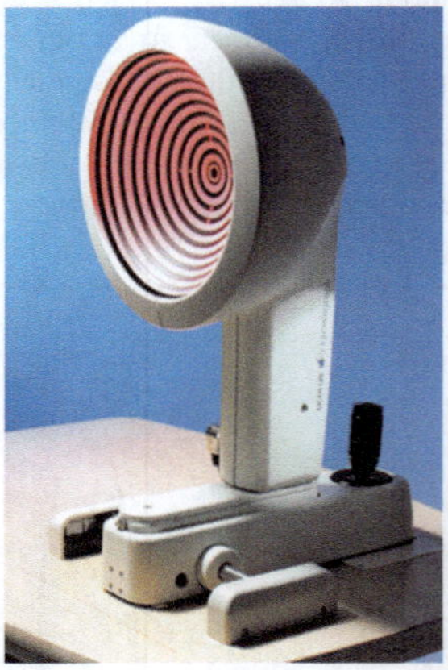

Fig. 4.32 Keratograph: the Placido disk is reflected by the cornea and recorded and automatically analyzed (Courtesy of Oculus Optikgeräte GmbH, Germany)

Thus far, we have discussed only the methods that utilize reflections from the cornea. Instruments can also be based on white light interferometry.[13] The principle of white light interferometry is treated in Sect. 4.2.

One particular instrument[14] goes beyond classical keratometry. It employs the Placido principle for a precise measurement of the anterior surface and, in addition, determines the profile of the corneal posterior surface relative to the

[13] OM-900 (Haag-Streit AG, Switzerland).

[14] GALILEI (Ziemer Ophthalmic Systems AG, Switzerland).

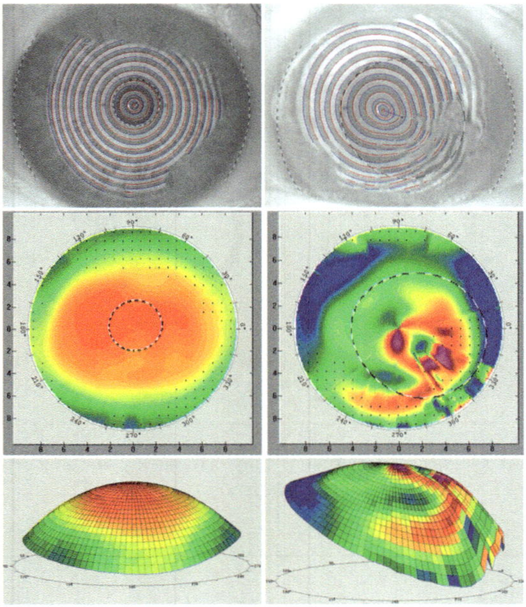

Fig. 4.33 *Top*: reflection of a Placido disk. *Middle*: topographic representation of corneal curvatures in false colors. *Bottom*: three-dimensional topography of the cornea

anterior one for all angular orientations from slit images (Scheimpflug imaging). This results in a true determination of the refractive power of the cornea.

Keratometry should not be confused with aberrometry, where local deviations from ideal optical imaging across the whole pupil can be obtained. This includes the influence of the outer surface of the cornea as well as all the imaging elements of the eye (including the lens and the posterior surface of the cornea). Aberrometry is addressed in Sect. 19.3.

4.1.10 Pachymetry

Corneal thickness can be measured with traditional optical methods, with ultrasound, or with white light interferometry. Numerous variants of the optical methods have been developed. Roughly the same precision of ±5 μm can be achieved with ultrasound as with optical methods. This is rather amazing because the wavelengths even of high frequency ultrasound (50 MHz) are still relatively large (ca. 30 μm, in water). The precision is achieved by the automatic averaging over many measurements. The ultrasound measurement basics are discussed in Chap. 5 (A-scan), while white light interferometry is addressed in Sect. 4.2.

4.1.11 Fundus Photography

The first photographs of the human retina toward the end of the 19th century required exposure times of several minutes and were disturbed by reflections. It was the reflection-free ophthalmoscopy, according to Gullstrand, that provided the prerequisites for the Nordenson-Zeiss camera (Fig. 4.34) that can be considered the basis of all the instruments in use today. The Littmann-Zeiss fundus camera of 1955 employed flash light and imaged the central 30° of the retina. Fluorescence angiography incorporated flash photos taken at intervals of less than a second. Today, electronic image recordings have replaced photographic film images. The earlier capturing of a series of images (e.g., in angiography) is being replaced more and more by video recordings.

4.1.12 Confocal Scanning Laser Ophthalmoscope

A scanning laser ophthalmoscope scans the retina with a laser beam. In a scan with a rapidly moving laser beam, the intensity of the scattered light is registered point for point. Thus, in less than 30 ms, a monochromatic image of about 15° in diameter is produced with a transversal resolution of 10 μm.[15]

With the confocal recording technology,[16] light stemming from a specific depth (Fig. 4.35) is preferentially registered. The light that comes

[15]Typical information, provided by the manufacturer.

[16]Marvin Minsky, American researcher in the field of artificial intelligence, invented the confocal microscope in 1957.

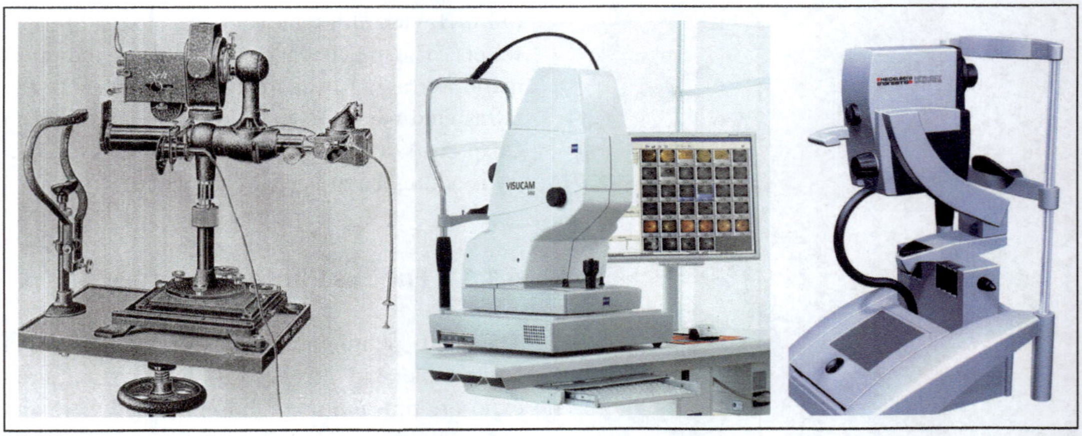

Fig. 4.34 *Left*: camera constructed in 1925 by J.W. Nordenson. *Middle*: modern fundus camera (courtesy of Carl Zeiss Archive). *Right*: fundus video camera (Courtesy of Heidelberg Engineering GmbH, Germany)

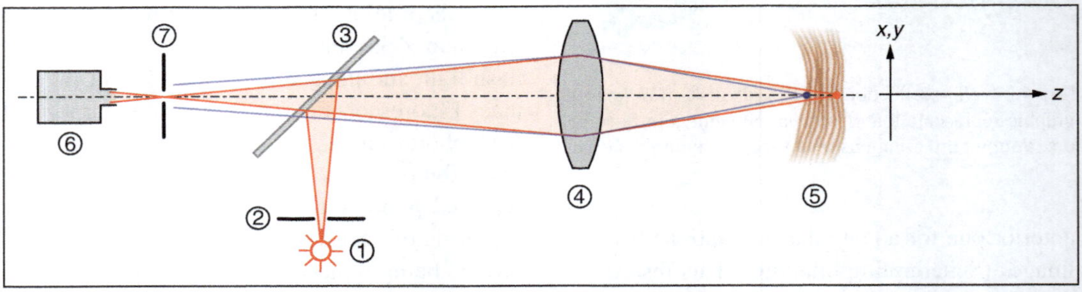

Fig. 4.35 Principle of confocal imaging. The point-shaped light source (*1*, *2*) is imaged by the semitransparent mirror (*3*) and the objective lens (*4*) into a point that is most intense at a certain depth (*red point*). The detector (*6*) registers the light reflected and scattered back as it passes preferentially through the aperture (*7*). The light (*displayed in blue*) originating from the wrong depth (*blue point*) is blocked by the aperture (*7*). The object is scanned point for point in three dimensions (*x*,*y*,*z*) using this approach. An image can be assembled from the pointwise registered intensities. Coordinates *x* and *y* stand for the transverse coordinates in the fundus and *z* for the coordinate into the retina

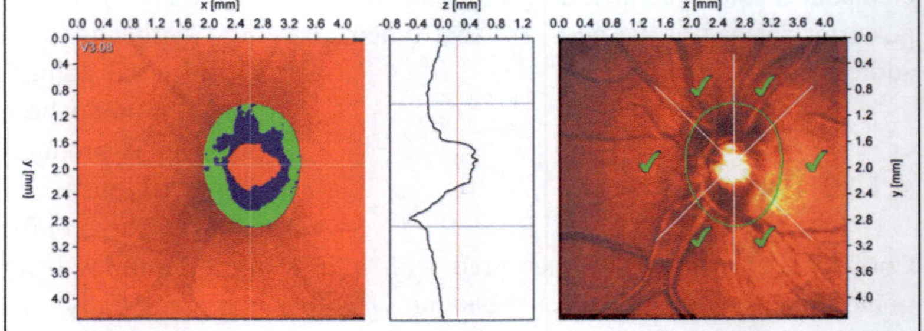

Fig. 4.36 Display of the three-dimensional structure of the retina and optic nerve head, based on an examination with the confocal scanning laser ophthalmoscope (HRT). In addition to the relief in false colors, both a horizontal and a vertical sectional slice are displayed

from other depths is imaged as blurred on the aperture and only a small part of it passes through it. Several recordings from various depths of the retina yield three-dimensional image data[17] in the form of the intensity of the scattered light. Typically, a depth range of 1 mm is scanned by 64 recordings. The axial resolution is, then, 60 μm. It is, however, limited by the diffraction at the confocal aperture.

A topographic image can be derived from the 3-D image data. It is based on the depths from which – for each retinal point – the maximum intensity of the reflected light is registered. Converted into false colors, a topographic image of the retina is generated (Fig. 4.36). Sectional slices representing the heights (z-coordinates) with respect to a local reference can also be constructed from the three dimensional data (Fig. 4.36).

4.1.13 Perimetry

More than two millennia ago, Galen[18] described various visual field disorders, among them the central defect, peripheral constrictions, and large scotomas. This is not the oldest evidence from the literature: Hippocrates[19] mentioned hemianopic deficits.

Von Graefe[20] recognized the importance of examining the visual field. Since then, instruments for the subjective determination of visual fields have been produced in numerous implementations. The term "perimeter" stems from the time when practitioners were still restricted to determining the outer boundaries of the visual field. Today, one can measure the sensitivity at test locations within the entire visual field or in the central 30°. Over the decades, differential light sensitivity has been generally accepted as one of various functions (e.g., visual acuity, flicker fusion frequency, movement, and color) that can be tested. Differential light sensitivity is a primary visual function. Its inspection is robust against various disorders and the task is clearly understood by the patients.

In testing differential light sensitivity, a uniformly illuminated background is superimposed by an additional light stimulus and the patient signals whether he or she has perceived it. In photopic vision, differential light sensitivity decreases from the center (fovea) to the periphery; i.e., the weakest stimuli are perceived in the center. There are two approaches for measuring broad areas of the visual field: kinetic and static perimetry (Fig. 4.37). Static perimetry determines the luminance of the stimulus that the patient is just able to recognize at a given test location. In kinetic perimetry, the examiner introduces a stimulus of constant

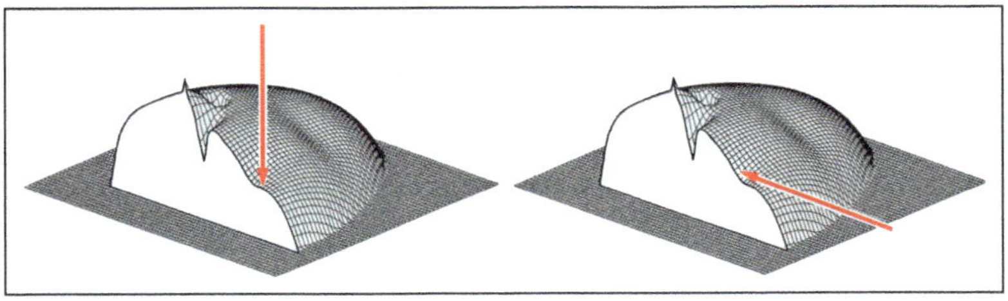

Fig. 4.37 Hill of vision: the differential light sensitivity across the visual field. Static perimetry seeks to determine the sensitivity thresholds at fixed, predetermined visual field locations (*left*). Kinetic perimetry operates with moving stimuli of constant luminance (*right*)

[17]The term "tomography" is derived from the Greek word "tomos = layer."

[18]Galen (131–210 AD).

[19]Hippocrates (460–370 BCE).

[20]Mentioned in Sect. 4.1.1.

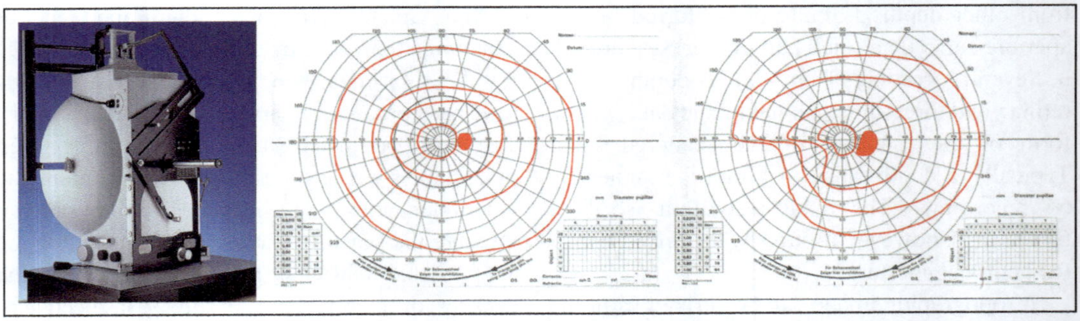

Fig. 4.38 *Left*: Goldmann perimeter (Courtesy of Haag-Streit AG, Switzerland). *Middle*: normal kinetic visual field. The isopters connect the same sensitivities. *Right*: glaucomatous visual field with a nasal step

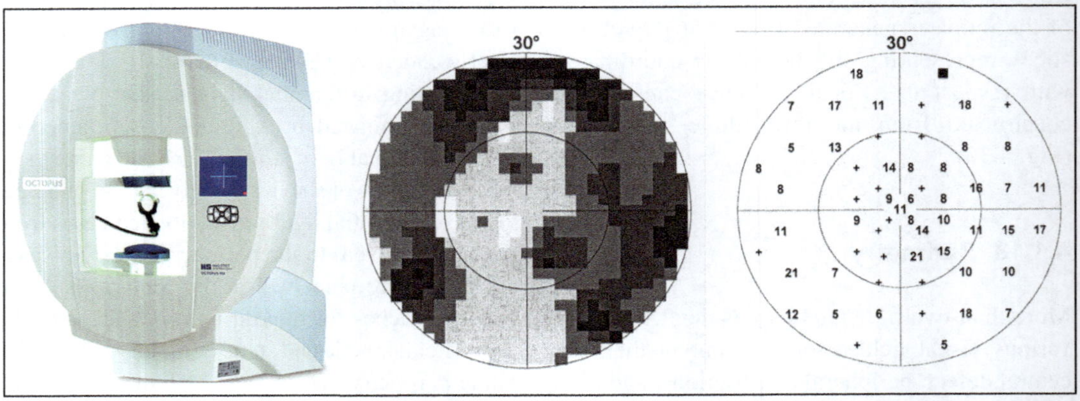

Fig. 4.39 *Left*: automated cupola perimeter (Courtesy of Haag-Streit AG, Switzerland). *Middle*: gray-scale representation (Octopus G1 program). *Right*: the same visual field; numerical display of the local defects (values in dB; see Table 4.1 for the scale). The value "11" in the center means that a sensitivity loss of 11 dB relative to the normal sensitivity at this location and for this age class has been measured

luminance and moves it from "not seen" to "seen," i.e., until the patient signals that it has been perceived (normally from the periphery inward).

Kinetic perimetry experienced a worldwide spread with the 1945 introduction of the Goldmann perimeter. The radial guidance of the same stimulus from the periphery to the center from various directions reveals locations of the same sensitivity. The examiner connects these with the so-called isopters. With variation in the size or luminance of the stimulus, multiple isopters result. For diffuse damage, the isopters are pushed inward (concentric constriction). The recognition and measurement of local defects are more difficult (Fig. 4.38).

Static perimetry experienced an upswing with the introduction of automated perimetry. A substantial impetus for its development came from Fankhauser[21] in the 1970s. Using a bracketing strategy with stimuli of differing luminance, the local thresholds of contrast sensitivity are determined at specified test locations. Theoretically, a stimulus that has a 50 % chance of being seen is considered the sensitivity threshold at that test location. Practically, though, the threshold is found to be in the middle between the seen (stronger) and not seen (weaker) stimulus.

Setups for an examination are (a) a background luminance of, e.g., 10 cd/m^2, (b) the angular diameter of the test stimulus of, e.g., 0.43°, and (c) a stimulus duration of, e.g., 100 ms. In the center of a normal visual field of a young patient, the luminance at the threshold amounts to about 0.5 cd/m^2. There are implementations of automatic perime-

[21]Mentioned in Sect. 7.2.

ters with a cupola (Fig. 4.39) or with direct projection of the background and stimulus into the patient's eye. The description of the defect at each test location is referenced to the normal visual field of the patient's age category. Defects are described as the factor by which the luminance of the local normal threshold must be increased to the level that the patient perceives it. The instrument indicates this factor in a logarithmic scale (Table 4.1).

With a static examination of the visual field, the local defects are displayed as numerical values at typically about 60 test locations. In graphical displays, the quality of a defect pattern can be ascertained in its entirety, but for judging the deviations from normal values or for detecting trends, statistical approaches are helpful. A simple, robust, and, in its importance, memorable data reduction consists of the display of the mean and standard deviation of the local defects (visual field indices[22]). Among

Table 4.1 Decibel scale for expressing local defect depths. *X* is the factor by which the luminance of the stimulus must be elevated in relation to the normal threshold so that the patient perceives it. The factor 1 corresponds to the normal threshold. Perimeters display the decibel values, defined as 10 times the logarithm of *X* with respect to base 10. In Fig. 4.39, for example, the central value of 11 (dB) means that the patient requires a 12.8-times brighter luminance of the stimulus to perceive it compared with the normal central threshold stimulus for this age class

Factor *X*	1000	100	10	3	1.28	1
$\log_{10}(X)$	3	2	1	0.5	0.1	0
Decibel value (dB)	30	20	10	5	1	0

numerous graphical displays, the cumulative defect distribution that displays the defects in order of their size is also valuable, particularly for the separation of local from diffuse damage (Fig. 4.40).

Despite the apparent precision of the results, the subjective character of the method should not be forgotten. Fluctuations express themselves in the scatter of the results in repeated measurements of the same threshold. Fluctuations already exist in normal visual fields and they increase with growing defect depths. Deviations from the normal values or temporal progressions are, thus, to be interpreted cautiously.

The cause of the fluctuations lies in the statistical nature of the patient's responses. At a given test location and under given examination conditions, for a certain patient, the probability of a perception increases as a function of the stimulus luminance (Fig. 4.41, upper left). The threshold is, thus, not a sharp separation between "perceivable" and "not perceivable" but, rather, a stimulus that has a 50 % chance of being perceived. In normal visual fields, the increase in the probability of perception from 10 to 90 % requires an increase in the luminance by roughly a factor of 4 (6 dB). Due to the statistical nature of the patient response, the results of immediately repeated measurements at the same visual field test location are subject to scatter (Fig. 4.41, lower left). In normal visual fields, the typical scatter amounts to ca. 1.5 dB.[23] Comparison of visual fields from the same eye, taken at different intervals, must take into consideration the fact

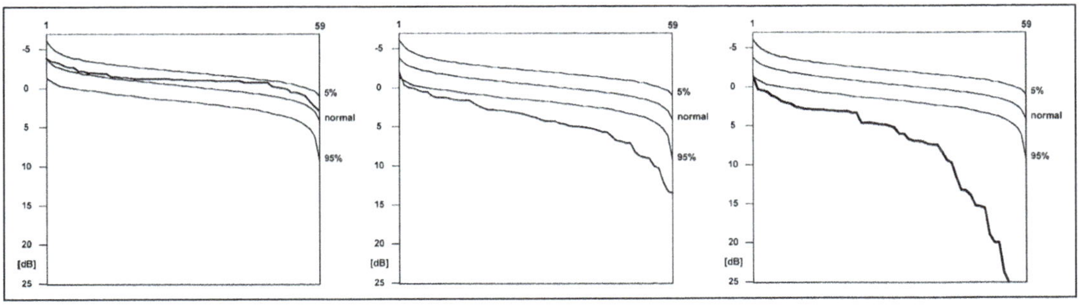

Fig. 4.40 Cumulative defect distribution (Bebie curve). Normal visual field (*left*); diffusely impaired visual field (*middle*); diffuse and local impairment (*right*)

[22]Flammer J (1986) Graefes Arch Clin Exp Ophthalmol, 224.

[23]Averaged (rms) over the visual field.

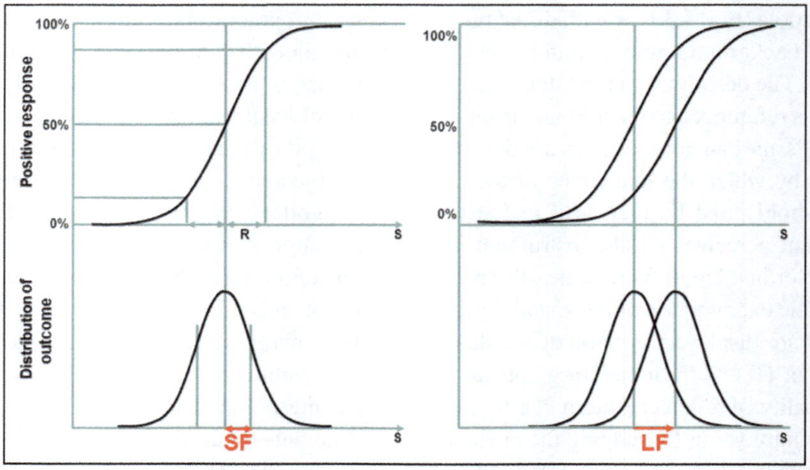

Fig. 4.41 The frequency-of-seeing curve (*upper left*) causes distribution of individual outcomes (*lower left*, so-called short-term fluctuation, SF). The frequency-of-seeing curve fluctuates over time (*upper right*), leading to so-called long-term fluctuation (LF) and, therefore, contributing to the total fluctuation (*lower right*) (modified after Flammer J. In: Drance SR, Anderson DR (Eds.) (1985) *Automated Perimetry in Glaucoma*. Grune and Stratton, New York)

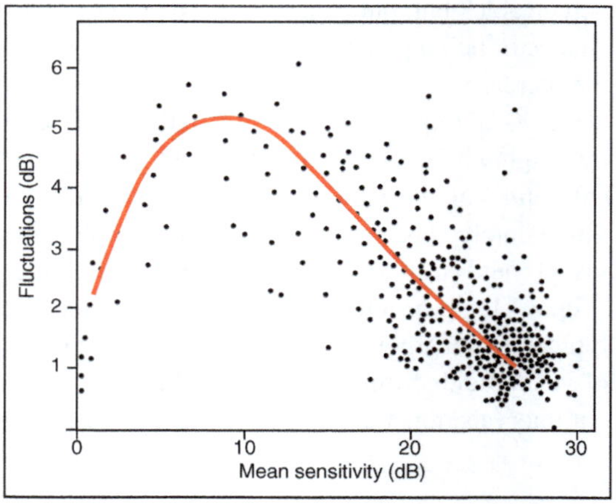

Fig. 4.42 The fluctuations increase with the depth of the defects. Consequently, the thresholds in impaired visual fields are less reproducible and the interpretation of alterations must be undertaken more cautiously than in normal visual field regions. This figure shows the relationship between the mean sensitivity of a visual field and the fluctuations. Here, fluctuations are defined as the short-term fluctuations, averaged over several test locations (modified after Flammer J, et al. (1984) *Arch Ophthalmol*, 102)

that the entire level of the differential light sensitivity fluctuates from day to day (Fig. 4.41, upper right). Interpretations are even more difficult due to the fact that fluctuations increase with the depth of the defects (Fig. 4.42).

4.2 Interferometric Methods

Light waves can interfere with and mutually strengthen or weaken each other, depending on whether wave peaks meet with wave peaks or with wave valleys. Laser speckles represent an impressive demonstration of interference: although an expanded laser beam uniformly illuminates a rough wall's surface, we see a pattern of bright and dark spots. The pattern does not really exist on the wall but, instead, arises on the retina through the phenomenon of interference. Dark spots appear in areas of the visual field, where all the contributions, arriving from all points of the wall, mutually extinguish the light's electrical field on the retina. Colored stripes on soap bubbles can also be understood in terms of light's capability for interference.

The phenomenon that we have already encountered in Sect. 1.2.1 in the double slit experiment can easily be understood: light extinguishes itself where the contributions from the two openings arrive with opposite direction of the electric field. This occurs when the two paths differ by one, three, five, etc., half wavelengths. Figure 1.8 could just as easily represent the interference behavior of water waves, sound waves, or light. Interference is based on the fact that it is not the

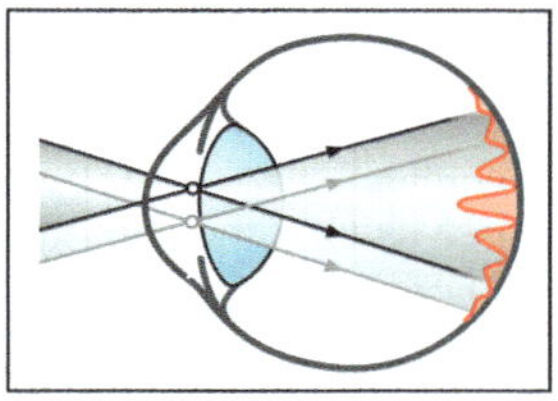

Fig. 4.43 Principle of retinometry. The light cones from two secondary point sources in the anterior segment are superimposed on the retina and create an interference pattern. Two sources with a lateral separation of 1 mm result in an interference pattern with a periodicity of about 2 min of arc (if the separation is 0.5 mm, then there will be 4 min of arc)

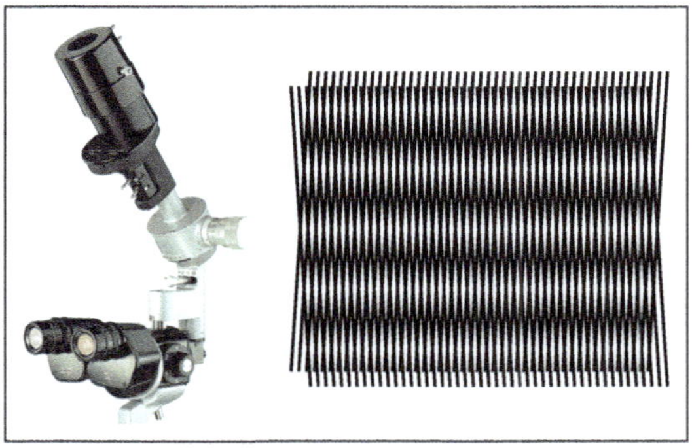

Fig. 4.44 Visometer according to Lotmar (Courtesy of Haag-Streit AG, Switzerland), based on the Moiré effect (*right*)

energy of the partial light waves that are additive (since this would always lead to purely positive contributions) but, rather, their amplitudes – regarding light, the electric field vector.

One of the first applications of interference in ophthalmology was so-called retinometry; i.e., the measurement of grating visual acuity, introduced in 1935 by Yves Le Grand. It functions according to the same principle that was elucidated in Fig. 1.9. Light comes into the eye through two separated openings, creating an interference pattern through the superimposition of the two light cones (Fig. 4.43). By altering the distances of the bundles of light, the distance between the interference stripes is changed, making a visual acuity examination possible ("interference test"). The result depends substantially on retinal function but much less on the refractive media because the light pattern on the retina is not formed by imaging. The contrast of the stripe pattern depends only relatively weakly on the transparency relationships of the lens and cornea at the two entry locations. A special implementation of a retinometer, the Visometer developed by W. Lotmar, utilizes the Moiré effect and has the advantage that it can be used with white light (Fig. 4.44).

4.2.1 Interferometry: The Principle

Interferometry, developed by Michelson,[24] is a highly precise optical measurement method that utilizes the interference of light and, among other applications, makes it possible to measure the distances and movements of objects with accuracies of fractions of wavelengths as well as imaging layer structures with a resolution of 1 μm. With his famous interferometric experiment in the 1880 s, Michelson wanted to verify the movement of the earth through the ether. The completely negative result of the experiment fully destroyed faith in the existence of an absolute space and an ether, which had been believed to be the carrier of electromagnetic waves, including light.[25] Interferometric methods have been further developed into a wide variety of applications. The measurement of the tiniest angular diameters of stars down to hundredths of seconds of arcs by means of interferometry was a sensational success. There are also modern applications in ophthalmology, such as optical coherence tomography (OCT) for the three-dimensional recording of retinal layers or optical low coherence reflectometry (OLCR) for the precise measurement of the axial separations of all the ocular interfaces, from the cornea down to the retina. In both applications, an axial resolution on the order of 1 μm is achieved.

Figure 4.45 shows the physical basic principle of interferometry. A beam of light (1) is separated into two beams when it passes through a

[24]Albert Michelson (1852–1931). American physicist. Nobel Prize 1907.

[25]We do not go into the details of this historic experiment.

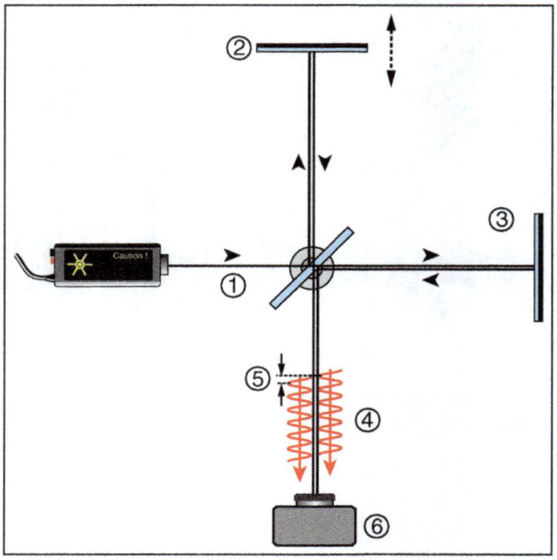

Fig. 4.45 The basic principle of an interferometer. *1* A beam of light is separated into two beams using a half-silvered mirror beamsplitter. *2* Reference mirror. *3* Object with unknown position. *4*, *5* The two light components that return from *2* and *3*. From the intensity of the superposition of 4 and 5, observed in the detector (*6*) in dependence on the movement of the reference mirror, distances can be measured and information about the structure of the object can be derived

semi-transmissible mirror (known as a beam splitter). One half of the beam travels to a reference mirror (2) and the other to the object to be measured (3). The light from the reference mirror, as well as that from the measurement object, travel back through the beam splitter, whereupon these two components interfere (4). The observation of this superposition allows conclusions regarding the distance to or the structure of the measured object. In various setups, the details of the arrangement can vary considerably.

4.2.2 For a Start: Interferometry with Monochromatic Light

Interferometry is easy to understand if we look at the ideal case of a monochromatic beam of light. If one "freezes" an electromagnetic wave, as in a snapshot, the electric field is purely sinusoidal over large spatial distances. When the object to be measured is a mirror located the same distance away from the beamsplitter as the reference mirror, the two waves coming back to the detector have the same phase (Fig. 4.45); i.e., their electric fields reinforce each other so their strength is combined (this is known as "constructive" interference). In the case where the object is now shifted a quarter of a wavelength along the beam axis, a difference of half a wavelength is created (in traveling there and back) and the electric fields of the two waves tend to cancel one another out in the detector ("destructive" interference). A further movement of a quarter wavelength causes them to reinforce each other again and so on. In this way, it is possible to detect very small shifts of the object on the order of fractions of a light wavelength. For an ideal monochromatic beam of light, this periodic alteration in the intensity can be observed for any length of displacement (Figs. 4.46 and 4.47): movements along the beam axis can be followed and distances can be measured with high precision. However, with monochromatic light, no information regarding the structure of an object can be obtained (see Sect. 4.2.3).

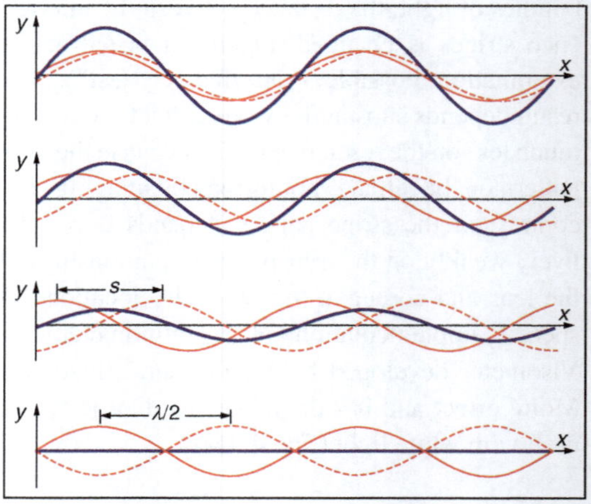

Fig. 4.46 Superposition (*blue*) of two waves (*red*) with differing displacements with respect to one another. *Top*: both components have almost the same phase, yielding maximum amplitude and large mean intensity (their electric fields add). *Bottom*: with a displacement of a half wavelength with respect to each other, the electric fields of the two waves tend to cancel, causing the intensity to diminish (going to zero if they have the same electric field amplitude)

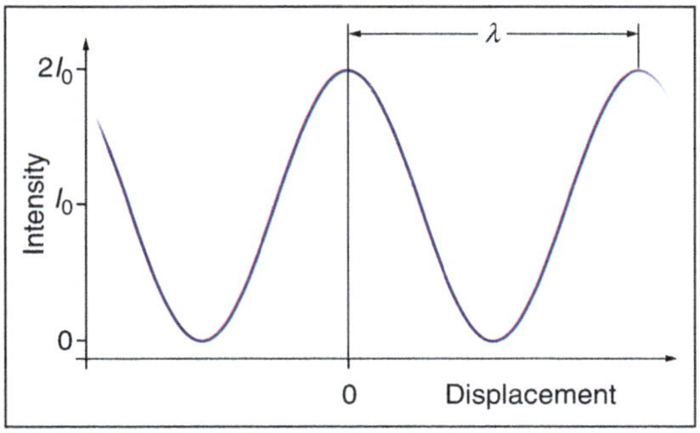

Fig. 4.47 The intensity of the superposition of two monochromatic waves of the same wavelength and same amplitude as a function of their mutual displacement. This interference exists for any displacement size. The ability of two components to interfere becomes apparent when there are intensity deviations with small displacements of fractions of a wavelength. *Ordinate*: intensity. *Abscissa*: displacement with respect to one another

4.2.3 White Light Interferometry

In reality, there is no ideally monochromatic light – even laser light has a certain bandwidth; i.e., it consists of an overlap of a narrow spread of wavelengths. Consequently, one cannot extend the longitudinal range of the interference effect indefinitely because the constructive and destructive interference of the various wavelengths tends to wash out. The interference ability is lost beyond a certain displacement and the intensity of the superimposed light beams no longer reacts to further displacements of the reference mirror or the object and remains constant. Figure 4.48 illustrates the interference ability of light with various spectra with itself; namely, the intensity to be expected as a function of the displacement when the interferometer is operated with light of one or the other spectrum.

The displacement over which the interference ability is possible is called the (temporal) coherence length *LC* of the light beam. For an ideal monochromatic light beam, the coherence length would be infinitely long, a situation we discussed in Sect. 4.2.2. Figure 4.48 illustrates the differing coherence lengths, beginning with quasi-monochromatic light (column a). For a continuous laser, the coherence length can amount to many wavelengths – for a simple He–Ne model, it can be a few tens of centimeters; with a frequency stabilized laser, it can be up to a kilometer. *LC* is, thus, shorter in proportion to the bandwidth $\Delta\lambda$ of

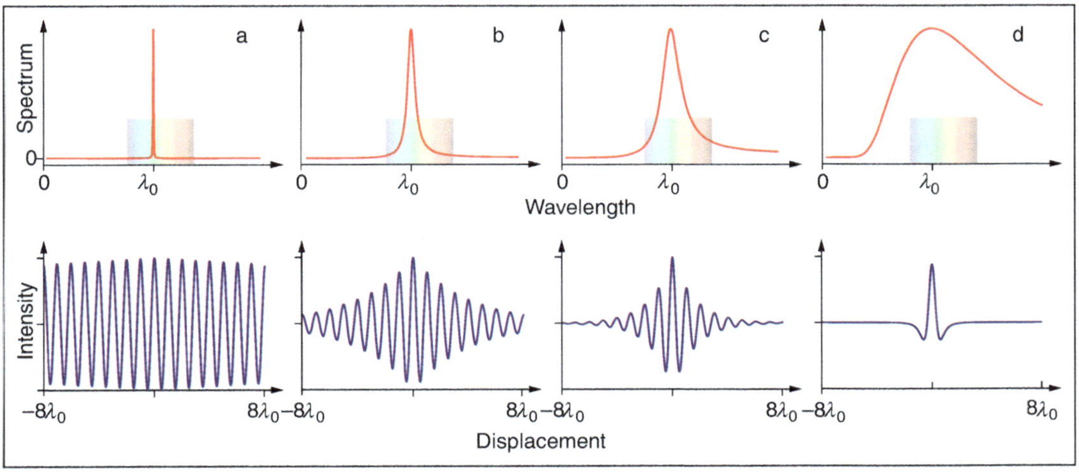

Fig. 4.48 Interference ability of light of a given spectrum with itself, for four spectra with different widths. *Upper row*: spectrum. *Abscissa*: wavelength λ. Center wavelength λ_0. *Lower row*: intensity as a function of the displacement. *Abscissa*: displacement, ranging from $-8\lambda_0$ to $+8\lambda_0$. *Column a*: quasi-monochromatic light. Interference ability is preserved over long distances. *Column b*: spectral width 4 % of the central wavelength λ_0. *Column c*: spectral width 10 % of the central wavelength λ_0. *Column d*: thermal light (e.g., unfiltered sunlight; the two minima of the intensity lie at displacements of ± 0.3 μm). With a broader spectrum, the coherence length L_C is reduced (see text)

the light beam.[26] For example, for $\lambda = 0.8$ µm and $\Delta\lambda = \lambda/10$, $LC = 10\ \lambda = 8$ µm (column c). One can even have interference with thermal light (column d), but it is very limited: the intensity increase requires a highly precise congruence of the two beams.

The limited coherence lengths of relatively broadband light in so-called white light interferometry – on which OCT and OLCR are based – are utilized in that the absolute distances of the reference mirror and the measurement object must exactly agree to within an order of L_C so that interference occurs. This can be observed as changing intensities in the detector with slight movement of the reference mirror. Technically, the measurement process takes place as the reference mirror is moved until interference happens (Fig. 4.50). If the measurement object consists of several partially reflecting surfaces, their positions can be scanned with high precision as the reference mirror is moved. This is the basis of OLCR (Sect. 4.2.4). In the same way, a layered structure can be scanned if the light from the various depths is scattered back (OCT of the retina, Sect. 4.2.5). Note that we are describing, so far, the procedure along one axis. We will return to the scanning of the retina along a section in Sects. 4.2.5 and 4.2.6.

4.2.4 Optical Low Coherence Reflectometry (OLCR)

Instruments for measuring the distance between reflecting surfaces along an eye's optical axis are based on the interference principle described above in Sect. 4.2.1 and, in more technical detail, in Sects. 4.2.2 and 4.2.3. The position of

[26]For light in the wavelength range $\lambda \pm \Delta\lambda$, LC is on the order of $\lambda^2/\Delta\lambda$. The reason is that the phase difference between two sine waves with the wavelengths λ *and* $\lambda + \Delta\lambda$ changes along the distance $\lambda^2/(2\Delta\lambda)$ by 180°. Along a distance of this size, the various components of the light in the wavelength range $\lambda \pm \Delta\lambda$ get out of phase.

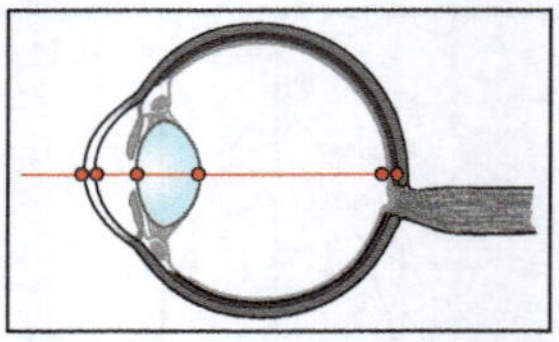

Fig. 4.49 Laser interference biometry. *Red points* indicate structures that reflect light

reflecting structures, such as the anterior or posterior surface of the cornea, can be obtained with high precision (Fig. 4.49). The corneal thickness is quantified (so-called optical coherence pachymetry) by the difference of the anterior corneal surface and Descemet's membrane. This is the basis of laser interference biometry for determining, e.g., the refractive power of the intraocular lens to be implanted. If positions are measured several times a second, one can detect pulsations and quantify them. Retinal pulsations (caused by the differences in blood perfusion of the underlying choroid during the cardiac cycle) enable an indirect quantification of the pulsatile component of ocular blood flow.

4.2.5 Time Domain Optical Coherence Tomography (TD-OCT)

Optical coherence tomography (OCT) creates an optical "cross-sectional" image through a tissue such as the retina (Fig. 4.51). This allows for more a detailed analysis, such as the quantification of the retinal nerve fiber layer (Fig. 4.52). The first commercial instrument became available in 1996. A beam of light with a short coherence length scans the retina and determines the intensity of light that is scattered back as a function of the depth in a large number of points. In this way, layers with a variety of scattering properties can be identified in the depths. If the scanning procedure explained in Sect. 4.2.3 is employed, it is referred to as *time domain optical coherence tomography*

(TD-OCT, Figs. 4.50, 4.51, and 4.52). For every analyzed point of the retina, the reference mirror must travel through the whole depth range (a so-called "A-scan"). The observed light intensity from a given depth can be displayed as a gray scale or in color code in the cross sectional image, whereby the differing light intensities display the structures. The typical characteristics of the TD-OCT are: axial (longitudinal) resolution of 10 μm, transverse resolution of 25 μm, and 400 A-scans per second. In contrast to the intensity of reflections, the intensity of the scatter depends strongly on the wavelength. OCT with various wavelength ranges provides additional information.

Multiple scattering represents a disturbing effect in OCT measurements, i.e., when light

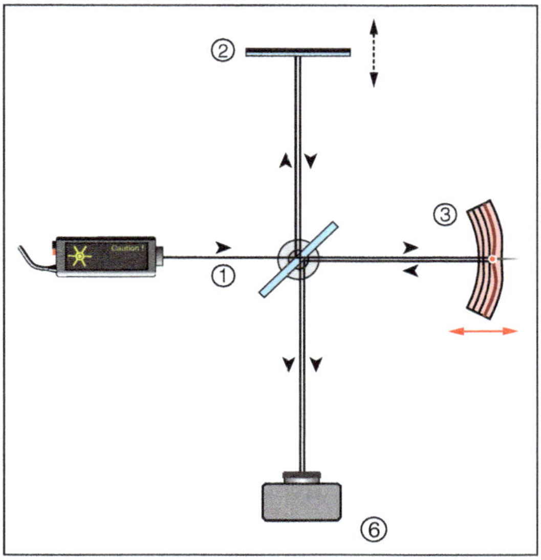

Fig. 4.50 Scheme of time domain optical coherence tomography (TD-OCT). For a given location (x,y) on the retina, the movement of the reference mirror (*2*) corresponds to the scanning of the retina in the depth z (see Fig. 4.35 for the coordinates x,y,z)

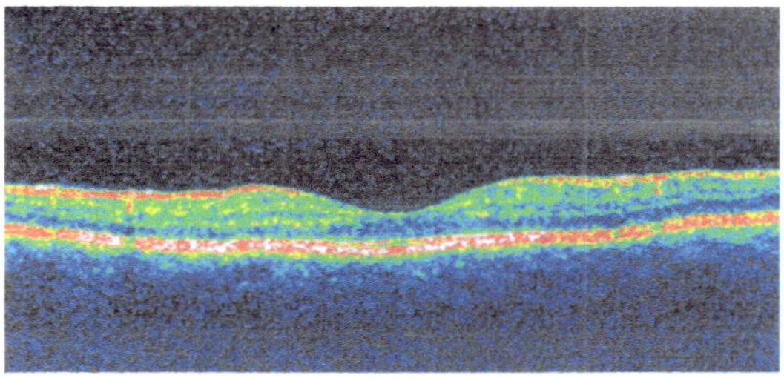

Fig. 4.51 Example of an OCT outcome

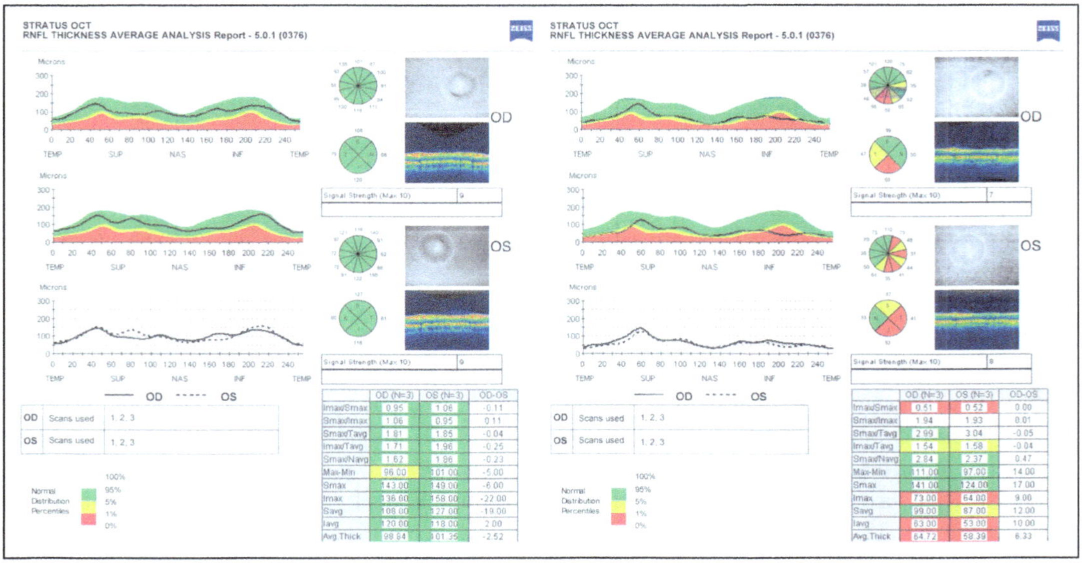

Fig. 4.52 RNFL thickness analysis. *Left*: normal eye. *Right*: glaucomatous eye (Stratus OCT)

from the illuminated retina indirectly returns to the detector via a second scatter. This affects resolution and contrast. For this reason, we would ideally like to measure only returning light that has been scattered once. Repeatedly scattered light returns temporally later and thereby contributes to a diffused signal.

4.2.6 Spectral Domain Optical Coherence Tomography (SD-OCT)

Shortly after the turn of the century, OCT experienced a powerful increase to higher scanning speeds with the development of *spectral domain* OCT. The same information as with TD-OCT is obtained but without a moving reference mirror and, therefore, in a considerably shorter time. Computation makes it possible to obtain the in-depth profile of the scattering sources from the spectrum of the light that is scattered back (Fig. 4.53). Thus, for a point (x,y) on the retina, the measurement process consists of superimposing the light scattered back from the penetrating beam with the reference beam from a fixed reference mirror and analyzing it spectrally, i.e., with a diffraction grating. The spectrum can then be registered for all wavelengths at the same time with an array of photodetectors. The temporal and serial scanning of the TD-OCT in the depths is, thus, replaced by a one-step registration. This has enabled a considerably more rapid 3-D scanning of the retina. An example is given in Fig. 4.54. Commercially available instruments perform 30,000 A-scans per second at an axial resolution of 5 μm.

It may seem surprising that the spectrum of the light is at all influenced by the depth profile. One can understand the origin of the effect with a simple example of two reflecting transparencies at differing depths: certain components of the broadband light with certain wavelengths return from the two transparencies with the same phase and reinforce each other; other wavelengths cancel one another. For example, a separation of 6 μm yields constructive interference for the wavelength $\lambda = 0.95$ μm but destructive interference for $\lambda = 0.76$ μm. The spectrum is, thus, modulated by the depth profile.

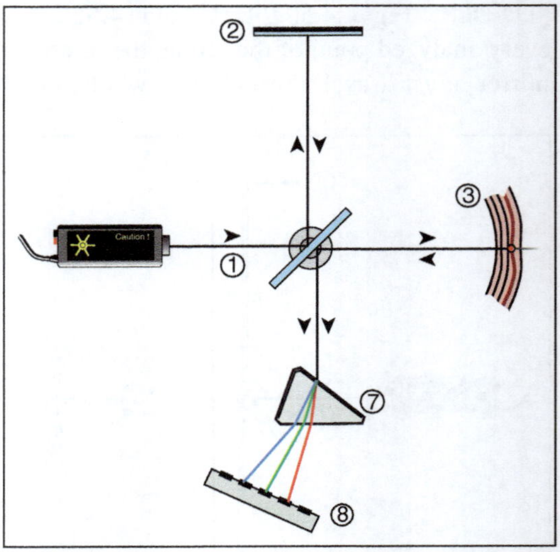

Fig. 4.53 Scheme of the spectral domain optical coherence tomography (SD-OCT). The reference mirror (*2*) is now fixed. The depth profile for one location of the retina is obtained via a spectral analysis of the light. SD-OCT is considerably more rapid than TD-OCT

4.2.7 Laser Speckles

Laser speckles (Fig. 4.55) have always been visually attractive in its fascinating liveliness. At the beginning of Sect. 4.2, we pointed out that a pattern of speckles – occurring, for example, when a plaster wall is illuminated with a broadened laser beam – does not exist on the wall itself but arises as an interference pattern on the retina. The phenomenon goes back to the spatial coherence of the light at the various locations of the wall (see Sect. 1.8). In an acoustic analogy, the points on the wall could be thought of as tiny loudspeakers that are driven by the same signal but with differing delays. The subjectively perceived pattern has an interesting property: it remains stable when an eye that is precisely focused on the wall moves laterally. If, on the other hand, the eye has

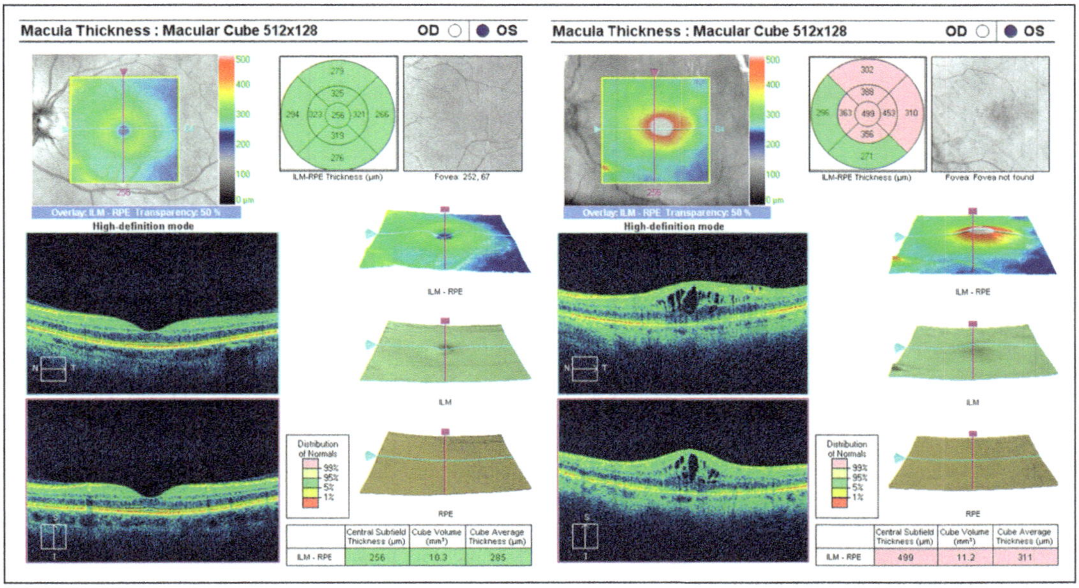

Fig. 4.54 Retinal cross-section image. *Left*: normal eye. *Right*: eye with macula edema (Cirrus HD-OCT)

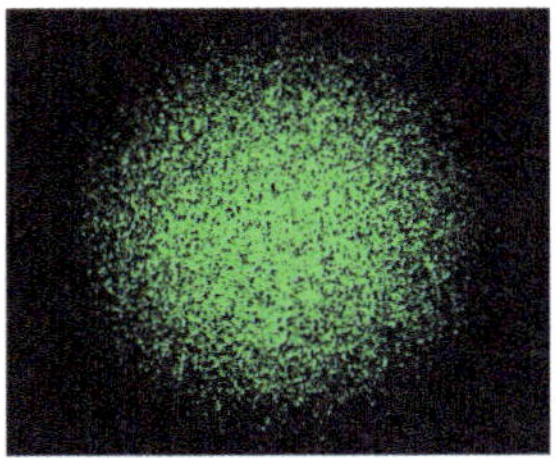

Fig. 4.55 Laser speckles

a focal length that is too short (myopic), the speckle pattern moves in a direction opposite to that of the head. For farsighted eyes (hyperopic), the reverse is true. The phenomenon is easily understood in terms of wave optics and is often shown in optometrist's shop windows. One can demonstrate it with a laser pointer by pointing its beam at a rough white surface from a relatively large distance.

If the illuminated area is moving (such as the blood in the vessels of a tissue), the light field fluctuates. In a photograph with longer exposure times, the variations from the moving regions are smoothed out, resulting in reduced contrast. In this way, one can indirectly obtain information about the blood flow; i.e., in the optic nerve head. The findings obtained in this way correlate well with those of laser Doppler flowmetry.

4.3 The Laser Doppler Principle

Everyone has encountered the Doppler effect when it comes to sound: the pitch of a fire truck's siren is higher or lower depending on whether it is approaching or receding from us.[27] The origin of this effect is not difficult to understand (Fig. 4.56). The explanation is the same for every kind of wave (sound, ultrasound, light). For a source that moves away from us, the constantly increasing distance is the cause for the stretching of the time between the arrival of the waves, one after the other, and this means a lower frequency. In this way, it has been discovered that all the galaxies move away from us and, indeed, with a

[27]Christian Doppler (1803–1853), Austrian mathematician. He predicted that the effect would occur for light. With his attempt to explain the colors of stars with it, though, he was mistaken.

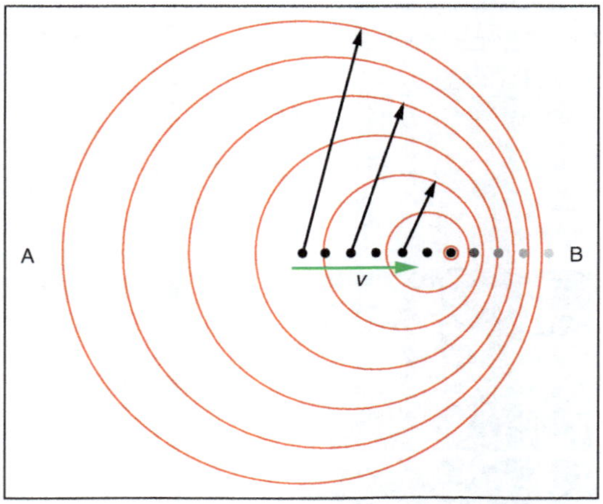

Fig. 4.56 Origin of the Doppler effect. The sender is moving to the right. The *circles* are the fronts of the waves that are sent out at various times. In *B*, the perceived frequency is higher than in *A*. The points indicate the locations of the sender at the times in which the displayed wave fronts were sent out

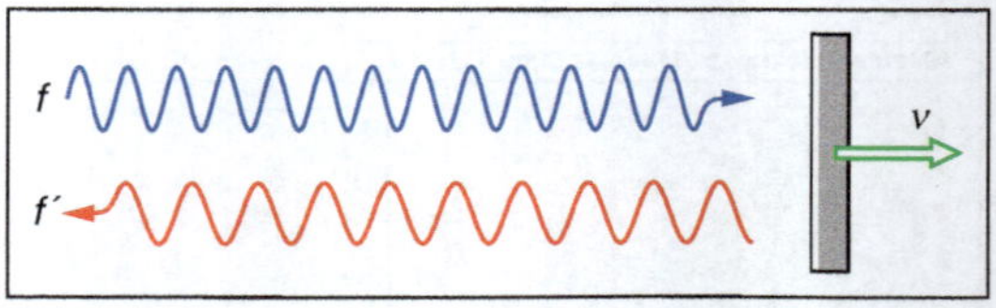

Fig. 4.57 Light reflected from an object that moves away with a velocity *v*. The frequency f′ of the light scattered back toward the source is lower than that which was sent out

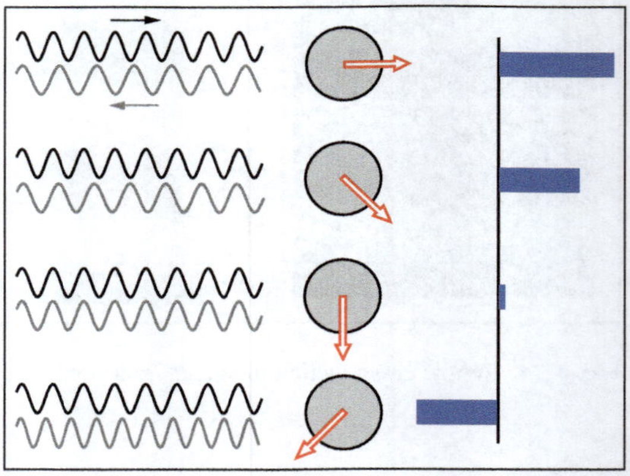

Fig. 4.58 The observed frequency shift depends on the movement direction of the reflecting object (*red arrows*). With the frequency shift, however, only the component of the velocity that is in the direction of the observer is determined. The *bars* indicate the size of the frequency shift under the assumption of the same values of the velocity

speed that is proportional to their distance away (the farther a galaxy is away from us, the farther its light spectrum is shifted toward red). This was the basis for the Big Bang hypothesis that forms our idea of how the universe began.

In medicine, one utilizes this effect both with light and with ultrasound to observe movements, primarily blood flow. In both cases, the instrument registers the frequency shift between the transmitted and the reflected waves. In this chapter, we take a look at the optical Doppler effect. The observation of blood flow in ocular vessels with the aid of ultrasound is discussed in Sect. 6.3.

In 1972, Riva et al. were able to measure the blood flow velocity in retinal vessels based on the optical Doppler effect. When light with a certain frequency f strikes an object that is moving away from the beam of light, the light scattered back from it has a lower frequency f' (Fig. 4.57) and $f'>f$ when the movement is toward the light beam. The frequency shift is given by $\Delta f = f' - f = \pm 2v/\lambda$, where v stands for the velocity of the object and λ for the wavelength. To give a numerical example, a velocity of 10 mm/s away from the observer and a wavelength of 0.5 μm correspond to a frequency shift of 40 kHz. The frequency shift can be precisely observed and used to calculate the velocity.[28]

However, it needs to be noted that the observed frequency shift does not encompass the value of the velocity but only the component that is in the direction of the observer (Fig. 4.58). For an angle of 45°, the frequency shift is reduced to 70 %. For movements perpendicular to the direction of observation, no shift in frequency occurs. Because the value of the velocity is the factor of interest, the angle between the direction of the vessel and the observation direction should be known. Normally, this is not the

[28]The intensity of the returning light oscillates with the frequency Δf. In this simple situation, the returning light consists of the sum of two components: frequency shifted light from the moving object and light with the unaltered frequency, stemming from structures at rest such as vessel walls.

case. Nevertheless, determination of the velocity value without knowing this angle is possible using the bidirectional laser Doppler technique, where frequency shifts from two differing directions of the reflected light are measured.

Usually, the reflected light shows more than a single frequency shift because the beam captures objects moving with a variety of velocities. A spectrum of velocities is obtained, meaning the distribution of the encompassed moving objects with their various velocities. For an individual vessel, for example, the velocities of the blood corpuscles range from 0 at the vessel wall up to a maximum value along the vessel's axis. The scattered light can even originate from several vessels with varying velocities and flow directions. Therefore, the returning light received by the instrument is a superposition of many different wavelengths, originating partly from moving objects (mainly blood) and partly from structures at rest, in and around the blood vessels. From the interference between the different components, a variation of the light intensity results and from this fluctuation, the computer calculates the velocity spectrum with the aid of a Fourier analysis (explained in Sect. 19.5). Above, we have used the simple situation of a single velocity[28] to explain that a mixture of different frequency shifts causes fluctuations of the intensity of the returning light.

5 Ultrasound Diagnostics

Ophthalmologists mainly use light for examinations and therapies, but not all tissues are accessible to light. A typical example is the contents of the orbit, behind and beside the eye. This is why ultrasound examinations also have an important place in our area of expertise. Ultrasound imaging relies on the detection of ultrasound reflected back from inhomogeneities. High frequencies are used to attain good resolution that can today approach fractions of a millimeter. Limits are set by the fact that higher frequencies are more strongly attenuated in water and tissue than lower frequencies.

5.1 Sound and Ultrasound

Physically speaking, audible sound and ultrasound represent the same phenomenon, with the only difference being the frequency. As an example, we know about reflected sound in the audible range (Greek mythology gave it the name of the mountain nymph Echo); bats employ ultrasound echoes for their orientation. Sound and ultrasound spread out in gases and fluids by alternating compression and decompression phases. Here, the individual molecules vibrate longitudinally about their rest locations (Fig. 5.1). At boundary surfaces, the vibrations go over from one medium into another one.

Beyond the audible range (16 Hz to 10–20 kHz, depending on age) is the ultrasound range, starting at frequencies above 20 kHz. The propagation velocity depends only very slightly on the frequency and is, thus, the same for audible sound and for ultrasound. It is highly dependent on the medium: 340 m/s in air, 1,480 m/s in water, and approximately 3,600 m/s in bones. In analogy to optics, with sound, one also encounters typical wave phenomena such as reflection, refraction, diffraction, scattering, absorption, and the Doppler effect. In ultrasound diagnostics, these all play a role in locating structures and movements. Thus, reflections are of particular importance.

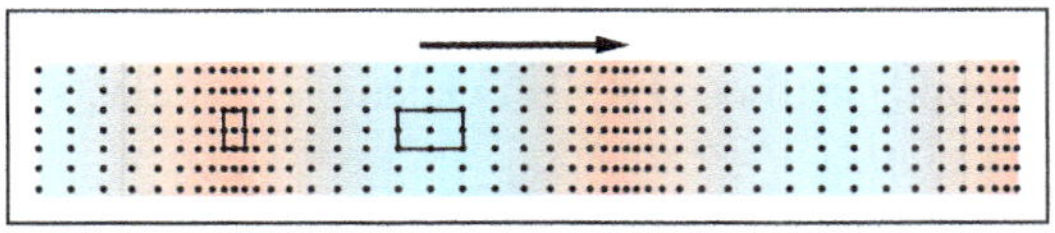

Fig. 5.1 Snapshot of sound vibrations. The compressed (*red*) and decompressed (*blue*) regions arise through the vibrations of the molecules about their resting positions. The dots indicate the positions of molecules

Ultrasound is interesting for applications in ophthalmology because locations in the eye can be observed that are not accessible optically (Fig. 5.2). Locating a sound-reflecting object is possible only with a certain limited precision, on the order of the wavelength of the sound signal that is used. The wavelength of sound or ultrasound is given by the sound velocity divided by the frequency. Thus, the higher the frequency, the better the spatial resolution will be. For 10 MHz ultrasound, the wavelength in a watery medium is about 0.15 mm. The use of much higher frequencies – very attractive from the point of view of resolution – is limited to applications in the anterior chamber due to strongly increasing absorption (see Sect. 5.1.1).

J. Flammer et al., *Basic Sciences in Ophthalmology*,
DOI 10.1007/978-3-642-32261-7_5,

Bats and dolphins are among the best known animals that utilize ultrasound for perception in their environment (Fig. 5.3). Their analysis of the echo is so perfect that not only is an object located but it is also possible for them to recognize the shape of the object reflecting the ultrasound waves. Bats typically send out ultrasound pulses several times per second in a frequency domain of 40–140 kHz. Their ability to hunt insects even in total darkness was long a mystery. In 1795, the Italian researcher Spallanzani observed how bats, in spite of being blinded, could orient themselves perfectly, and in 1798, Ludwig Jurin, a physician in Geneva, saw that bats lost their orientation when their ears were plugged. It was only in 1938 that Donald Griffin and Robert Galambos – students at Harvard University at the time – were able to uncover the secret as they were able to register the bats' sound signals electrically and, based on the frequencies, identify them as ultrasound.

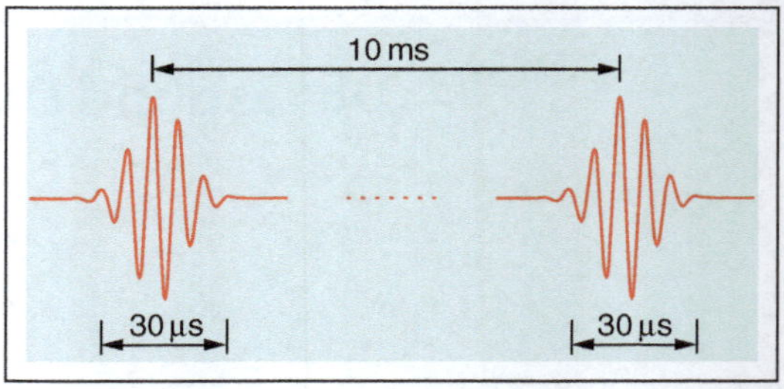

Fig. 5.3 Among other signals, dolphins send out sequences of ultrasound pulses to locate and scan objects based on their echoes. The figure shows two typical pulses from a series with regular intervals. The intervals between the pulses are long enough that an echo has time to return. Repetition frequency $\approx 10^2$ Hz; ultrasound frequency $\approx 10^5$ Hz. Number and pulse forms are just examples; depending on the species, they can vary a lot. In addition, dolphins in groups use lower frequencies to whistle to each other. The pulse-echo technique is also used in medicine

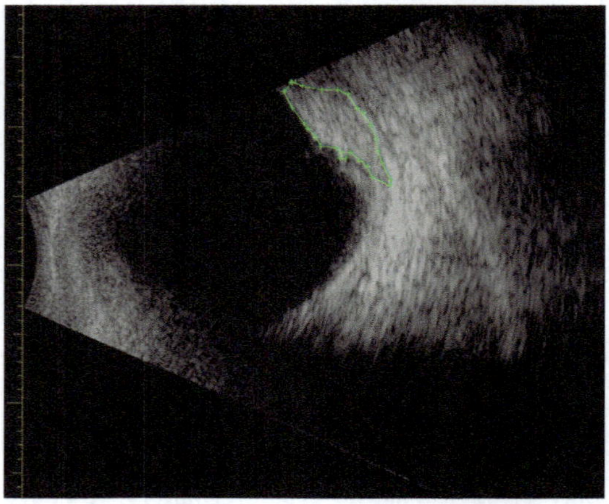

Fig. 5.2 Ultrasound image of a tumor in the eye (melanoma). Frequency 7.5 MHz (corresponding to a wavelength of 0.2 mm)

The first technical applications after 1920 were employed for locating icebergs and submarines as well as material testing. Austrian neurologist Karl Dussik was the first physician to use ultrasound for diagnostic purposes. He sent waves of 1.5 MHz through the head from one side and registered the local attenuations on the other side, producing a two-dimensional display (Fig. 5.4). His publications of 1942 and 1947 mark the beginning of ultrasound applications in medical diagnostics. Following a discussion of physical aspects, we shall treat applications in the eye.

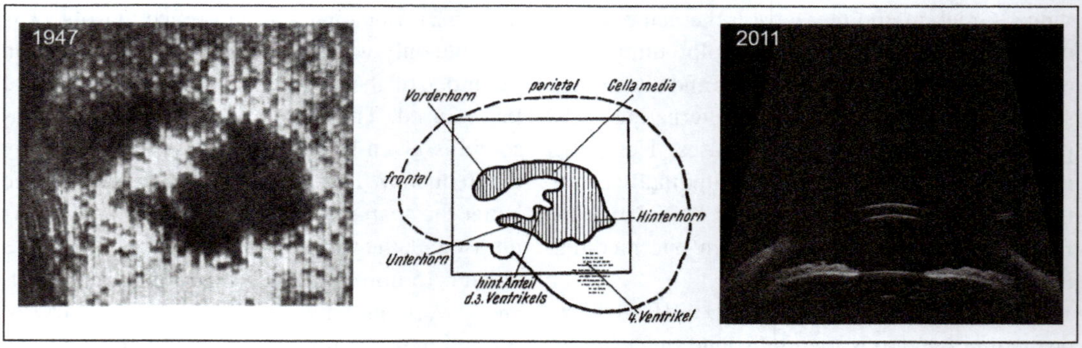

Fig. 5.4 *Left*: one of the first ultrasound images was taken of the human skull by Karl Dussik (1947). When passing through the ventricle, the beam is less attenuated. *Middle*: the explanation from the original publication. *Right*: modern ultrasound image with high resolution (biomicroscopic image of the anterior segment of the eye, 50 MHz)

5.1.1 Frequency, Wavelength, Resolution, Attenuation

Normally, ultrasound for diagnostic purposes is created by a longitudinally vibrating crystal (see Sect. 5.1.4). A sound emitter (as shown in Fig. 5.5) creates a beam that initially has a cylindrical shape and then begins gradually to widen. Imaging in the eye uses focused beams. Focal diameters down to an order of magnitude of one wavelength are possible.

Today, technology permits a wide range of wavelengths (and thus frequencies) to be selected.[1] The spatial resolution is limited to the order of one wavelength. High frequency means good resolution. The price paid for using a shorter wavelength is an increase in attenuation (Fig. 5.6). The decrease in sound intensity[2] with increasing penetration depth is exponential. Note that the reflected wave on its way back to the receiver is subject to the same attenuation once again. The attenuation in water is much less than in air, while in biological tissues, it is intermediate.

A conventional ultrasound in ocular diagnostics of 10 MHz can be used up to a depth of 30–40 mm and produces a resolution on the order of 0.5 mm. In so-called ultrasound biomicroscopy, frequencies of 50 MHz or more are applied. Thus, a resolution down to approximately 50 μm is possible but at the cost of reduced working depth, which then amounts to only a few millimeters.

In a homogenous medium, the attenuation is mainly due to the absorption of sound energy and its conversion into heat. This process is energetically far more important than the relatively weak scatter due to small inhomogeneities such as blood, for example. While the heat can be therapeutically useful, only the backscattered ultrasound is diagnostically important.

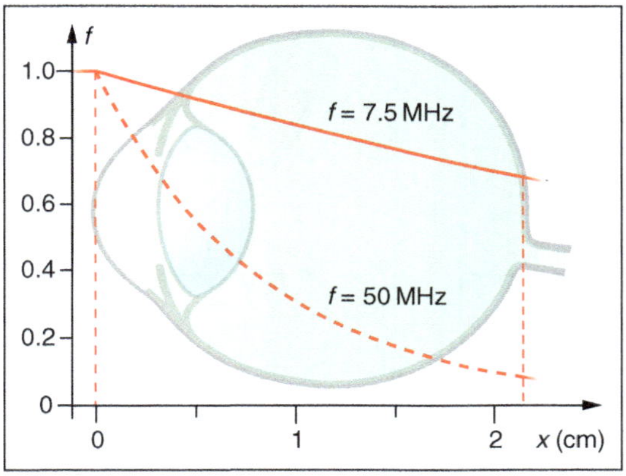

Fig. 5.6 Decrease in the intensity with increasing penetration depth. Higher-frequency ultrasound is more attenuated. Attenuation is exponential: if, at a certain frequency, the intensity decreases to half after a distance of 6 mm, it will again be halved along the next 6 mm, and so forth

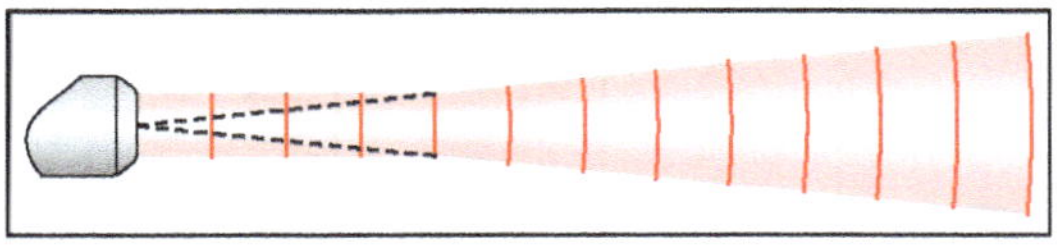

Fig. 5.5 Transducer with a plane exit surface. In the near field, the beam has a cylindrical shape and then begins gradually to widen

5.1.2 Reflection, Refraction, Scattering, and Diffraction of Ultrasound

As with the interaction of light with matter, with ultrasound, the phenomena of reflection, refraction, scattering, and diffraction also occur and waves arise with new propagation directions. Of primary interest are the processes that send sound back into the sensor. These include reflection at a smooth interface between two media, reflection at an uneven or rough interface, or scattering at small particles and inhomogeneities (Fig. 5.7). Only the small portion of energy that returns to the sensor can be used diagnostically.

Motionless inhomogeneities leave the frequency unaltered, while moving objects (such as erythrocytes in flowing blood) cause a Doppler shift in the frequency of the scattered sound (see Sect. 5.3).

For a reflection at a flat, smooth interface, the angle of incidence and reflection are the same

[1]Wavelength = sound velocity/frequency. In water, sound velocity is about 1,500 m/s. At 7.5 MHz, the wavelength is 0.2 mm; at 50 MHz it is 0.03 mm.

[2]Intensity is measured as power per unit area (W/m^2).

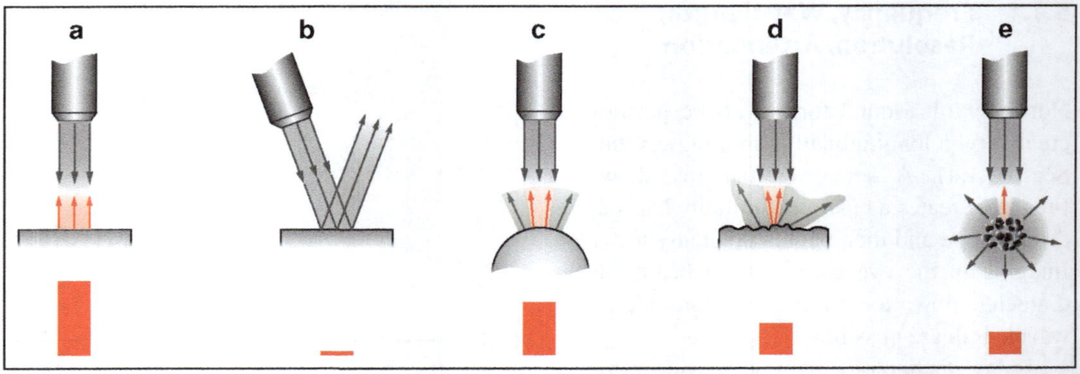

Fig. 5.7 Reflection of ultrasound. (**a**) Smooth interface between two media, perpendicular incidence. (**b**) Slanting incidence. (**c**) A curved interface. (**d**) Rough surface or interface. (**e**) Scattering at particles or inhomogeneities. The differing portions of energy that return to the sensor (built into the probe) are indicated by the *red bars*

(Fig. 5.7a and b). A slanting incident wave of sound may not return to the signal producer and is lost as an echo signal (if the sensor is built into the probe). When the reflecting surface is smooth but curved, the reflected wave is widened (Fig. 5.7c) and the amount of power that can return to the sensor is less than with a perpendicular incidence on a flat interface. In diffuse reflection at an uneven, rough interface, the sound is sent off in all directions (Fig. 5.7d) and the portion of the reflected power that arrives at the sensor is relatively small. Finally, particles or inhomogeneities present in a three-dimensional medium also create a diffuse reflection in all directions (Figs. 5.7e and 5.8). Even an individual cell represents a number of tiny reflectors. Small impediments are set to vibrate by the incident sound and, for their part, act as active sources of sound. Because the sound from these sources is sent out in all directions, the amount that returns to the sensor is small; the echo can, thus, be relatively weak. Note that not only the interfaces but also three-dimensional structures can act as sources of echoes.

Why does one have to put the sound probe either directly on the lid or introduce the sound into the eye via a fluid-filled funnel (Fig. 5.9)? In the transition of the ultrasound from the air into the eye, almost all the energy would be reflected by the cornea for a purely physical reason. In a transition from air to water or vice versa, the differing densities are mainly responsible for the high degree of reflection (a digression with a more precise explanation can be found in Sect. 5.1.3).

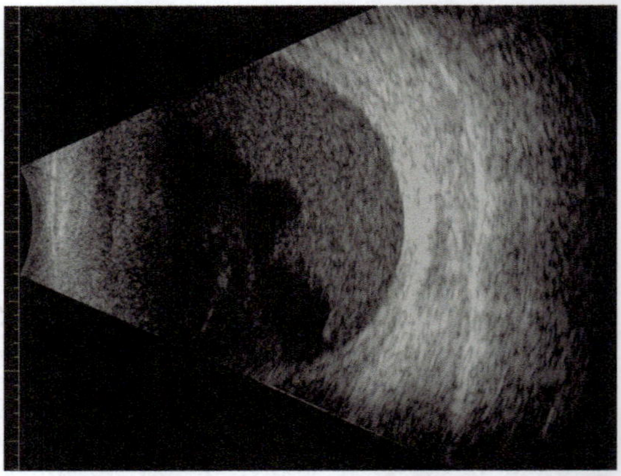

Fig. 5.8 B mode in an eye with vitreous hemorrhage. The local strength of the echo is coded as a gray value

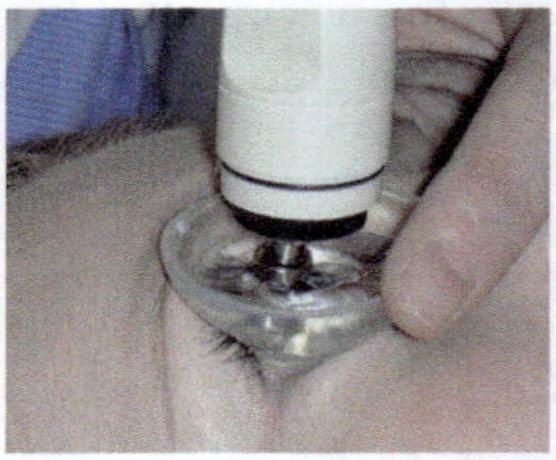

Fig. 5.9 Funnel in place

At the interface between two media of differing sound velocities (Fig. 5.10), the sound wave is refracted; that is, it continues in an altered direction. The refracted beam is closer to the surface normal when the new medium has a lower sound velocity; otherwise, it is positioned away from the vertical. This means that a certain distortion cannot be avoided. For applications in the eye, this is of secondary importance. In daily experience, for an attentive observer, a refraction of audible sound occurs in situations such as when a train passes behind a flat hill. The train will not be visible but can be heard with greatly differing volume, depending on whether the temperature gradient in the air above the hill points in an upward or downward direction – which depends on the weather conditions (Fig. 5.11). Acoustic lenses that can focus sound represent an application of refraction, analogous to optical lenses.

Further phenomena are very naturally connected with the wave nature of sound and ultrasound. We have all experienced diffraction with audible sound: behind a house, we hear a call from the front side – the sound travels around corners of the house (Fig. 5.12).

5.1.3 Digression: Impedance

At this point, we shall explain why sound is almost always reflected from interfaces. This can be well understood based on an example: the transition of sound from water into the air (Fig. 5.13b). The longitudinal vibrations of the water molecules also make themselves evident in the back-and-forth movements of the water's surface and it occurs with the same amplitude. With this same amplitude, the air molecules are now moved in longitudinal vibrations. Due to its far lower density, the air can transport only a fraction of the energy at this

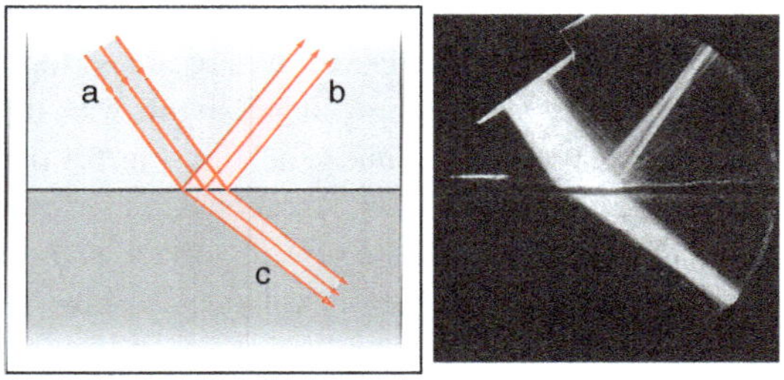

Fig. 5.10 Reflection and refraction of ultrasound at the interface between two media. *Left*: (*a*) Incoming sound beam. (*b*) Reflected beam. (*c*) Refracted beam. The velocity of sound is greater in the lower medium. Therefore, refraction is away from the normal to the surface. *Right*: Historical image from A. Giacomini (1949). *Upper medium*: water. *Lower medium*: petroleum ether. (From Giacomini A (1949) Il Nuovo Cimento, 6. With permission)

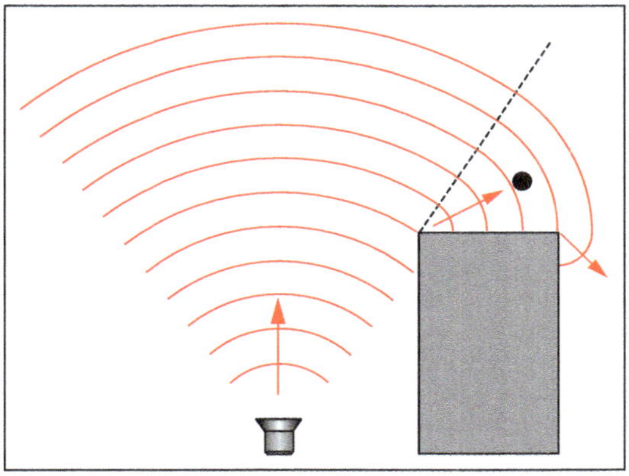

Fig. 5.12 Sound diffraction around a house corner (view from above). In addition, echoes from neighboring buildings often contribute to the audibility

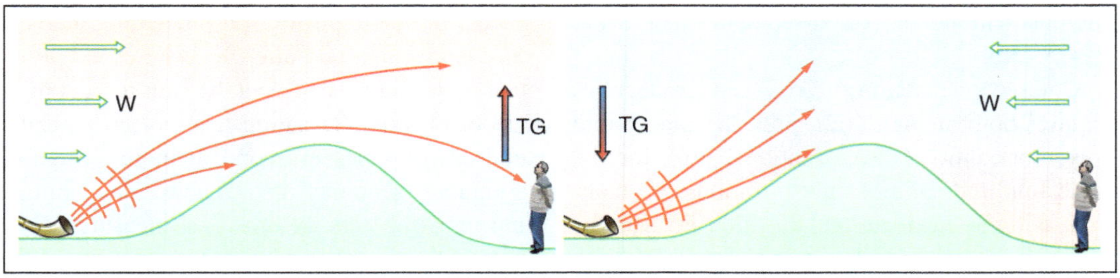

Fig. 5.11 Sound refraction due to air temperature gradients. *Left*: air temperature increasing upward (temperature gradient TG); sound source audible behind the hill. On flat ground, the wind also contributes to the phenomenon when it blows toward the observer and the wind velocity increases upward. *Right*: the contrary effect of a temperature gradient and wind direction

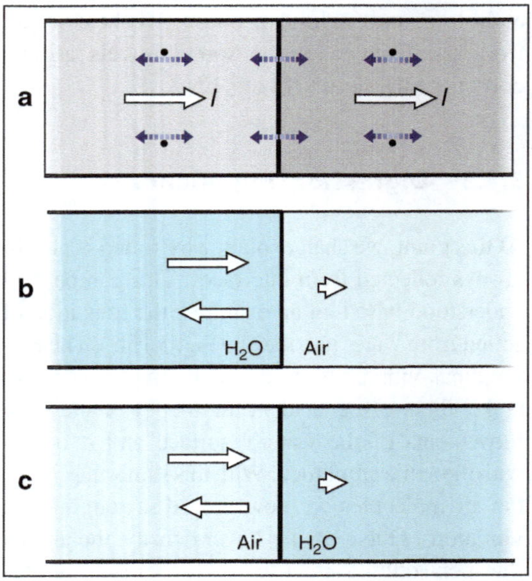

Fig. 5.13 Transition of sound from one medium to another. In any case, in both media, the vibratory amplitudes of the molecules must be the same. (**a**) Ideal matching, i.e., reflection-free transition, presupposes that both media transport the same power per area for the same vibratory amplitudes of the molecules, a condition that is only rarely met. (**b**) Transition of sound from water into air. At the same amplitudes, the air can take only over a fraction of the sound energy due to its lower density. The greatest portion is reflected. (**c**) Transition of sound from air into water. Transmission of the wave across the air–water interface without any reflection is impossible since the energy carried by the wave in air is not sufficient to generate the same vibratory amplitudes of the water molecules. Perfect match at the interface can be achieved only by means of reflection. The superposition of incoming and reflected waves decreases the vibratory amplitudes

amplitude, meaning that the largest part must be reflected. The same fraction of the reflected energy results for sound traveling in the opposite direction, but this will not be discussed in further detail (Fig. 5.13c).

Complete matching (i.e., no reflection) requires both media to transport the same sound power per unit of cross-sectional area for the same amplitude of molecule vibrations. We mention, without justification, that this is then the case when the so-called impedances[3] of both media are the same (Fig. 5.13a).

[3]The impedance of a medium is the product of its density and the associated sound velocity.

5.1.4 Sound Probe and Receiver

In medical instruments, piezo-electric crystals function both as ultrasound senders and also as receivers. In 1880, the brothers Pierre and Jacques Curie discovered that mechanical pressure on certain materials (some crystals, e.g., tourmaline, today often lead-zirconate-titanate) generates an electrical voltage across opposite outer surfaces.[4] Thus, sound waves, which strike the crystal, produce electrical signals that exactly follow the pressure changes and can then be electronically amplified (Fig. 5.14a). This effect finds applications in the pick-ups of record players and in microphones. For ultrasound, piezo crystals are very appropriate because they are able to follow very high-frequency sound signals and can be constructed in suitable forms.

If, on the other hand, an electrical voltage is applied to a piezo-electrical material, its volume is changed. The crystal surface moves back and forth in step with the alternating voltage. In this way, the crystal surface acts as a sound generator (Fig. 5.14b). The amplitudes of the air vibrations follow precisely the vibrations of the crystal surface that, in turn, follow the changes in voltage. Besides its applications as an ultrasound generator in medical diagnostic instruments, numerous further ones can be identified, such as in high-frequency loudspeakers, kidney-stone shattering, teeth cleaning, cleaning surgical instruments, or nebulizers in inhalers.

As mentioned previously, in the diagnostic applications of ultrasound, piezo crystals function as both senders and receivers. For a short time, the probe produces a pulse with a specified frequency and then stops for an interval to detect echoes. In this impulse-echo procedure, the next pulse is sent out only after the echo of the previous one has arrived. Depending on the depth of the object, this alternation takes place typically hundreds of times a second or more, so quasi-continuous scanning is possible. The waiting time until the echo returns is a measure of the depth of the reflecting object. The intensity of the echo serves as additional information.

[4]Piezoelectric originates from the Greek (piezo = I press).

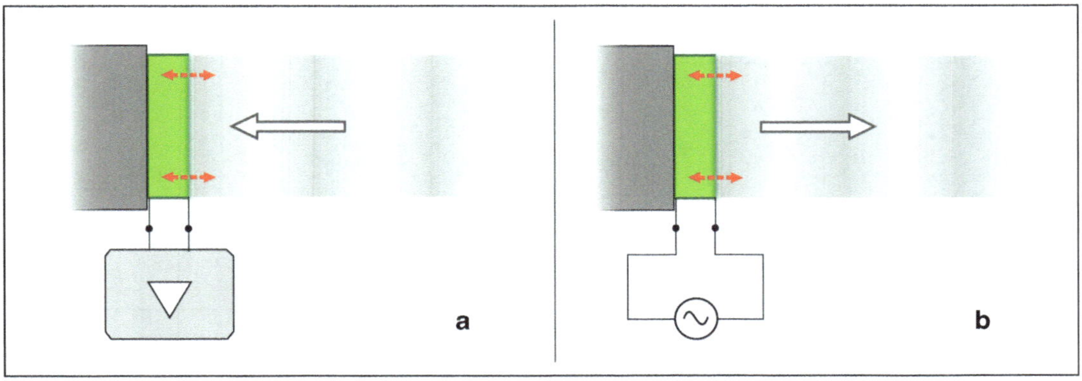

Fig. 5.14 Piezo crystal (*green*) as both a receiver and generator of sound. (**a**) Piezo crystal as a receiver. Pressure on the crystal exerted by the sound wave engenders an electrical voltage between the opposite outer surfaces. The voltage can be picked up and amplified. (**b**) Piezo crystal as a sound producer. The applied alternating voltage changes the crystal's volume. The movements of the surface transmit the sound into the adjacent medium. The support of the crystal is depicted in gray. *Red arrows*: oscillation of crystal surface. *White arrows*: direction of the sound propagation

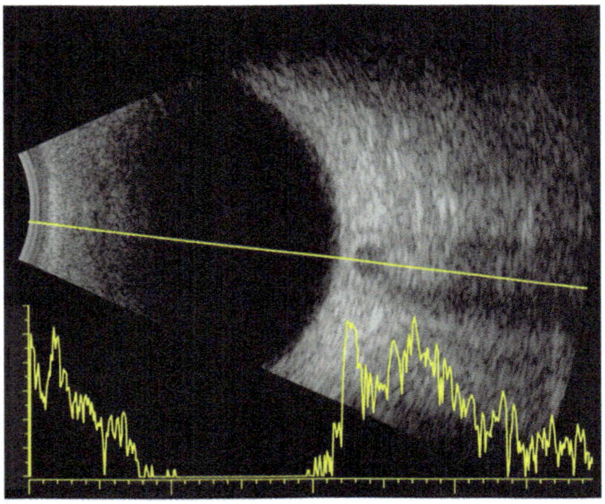

Fig. 5.15 B-scan (*top*) and A-scan (*bottom*) of a normal eye. The A-scan is limited to information along a single beam. The B-scan represents a two-dimensional cross-section

5.2 Sonography

Sonography (ultrasound examination) of the eye is based on the foundations shown in Sect. 5.1. It can be categorized into A-scans and B-scans. The historically older A-scan represents a one-dimensional inspection of the tissue: the intensity of the echo as a function of the depth along a single beam (Fig. 5.15, bottom). With the B-scan, the physician obtains cross-sectional images that convey a spatial impression of the size, form, and structure of the examined organs, soft tissue, and vessels (Fig. 5.15, top). In Sect. 5.3, we shall go into greater detail about a combination of these representations with information about blood flow in the examined cross-section.

5.2.1 A-Scan

The A-scan ("A" for "amplitude") displays the echo as a function of time elapsed since the pulse emission (Fig. 5.15, bottom, Fig. 5.16). The time to the organ and back to the receiver is a measure of its distance from the ultrasound probe. Strict proportionality is present only when the sound velocity in the body being examined is the same everywhere. The echo strength (y-axis) is registered as a function of time (x-axis). The higher the measurement curve is, the more echo-generating the tissue is at the given depth (Fig. 5.16). This represents a spatial one-dimensional finding along the penetrating ultrasound beam. This method is applied, e.g., in pachymetry and biometry. In ophthalmology, *biometry* involves the measurement of the eye's length and its corneal radius to calculate the power of the intraocular lens to be implanted.

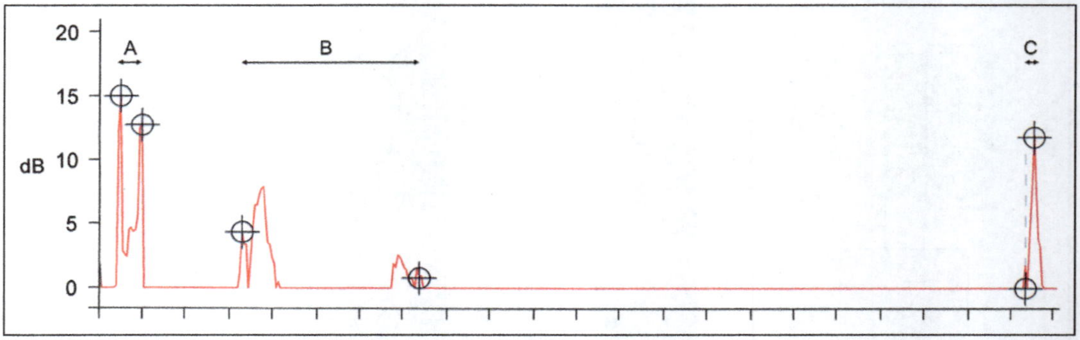

Fig. 5.16 A-scan: biometry of an eye along an axial beam. *Abscissa*: depth of the echo source (proportional to the temporal delay). *Ordinate*: intensity of the echo as a function of the depth of the reflecting structure. *A* thickness of the cornea. *B* thickness of the lens. *C* thickness of the retina

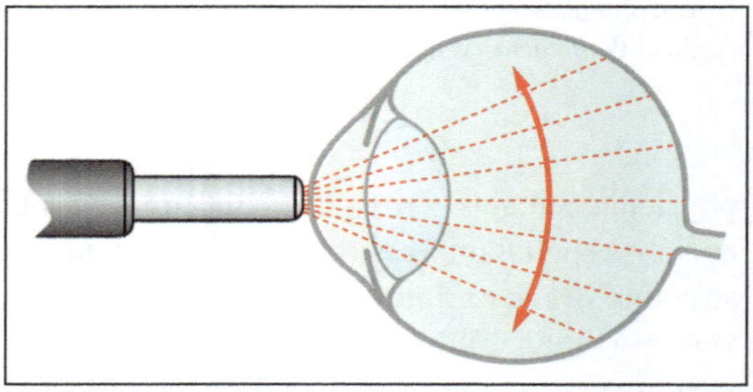

Fig. 5.17 In the B mode, the ultrasound examination generates a plane section through the tissue. The rapid deflection of the beam is controlled electronically

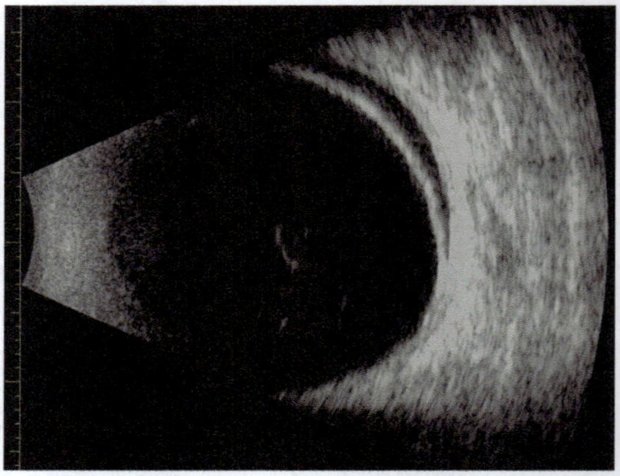

Fig. 5.18 B-scan in an eye with amotio retinae

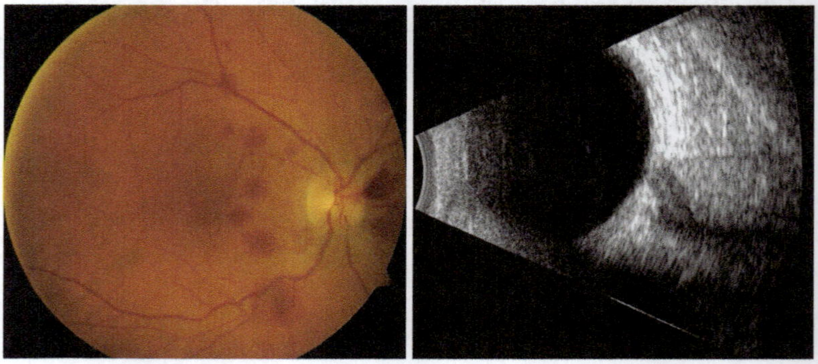

Fig. 5.19 Patient with Terson syndrome. *Left*: retinal hemorrhages. *Right*: optic nerve sheet hemorrhage observed on a B-scan

5.2.2 B-Scan

In a B-scan ("B" for "brightness"), a two-dimensional cross-section of the examined tissue is produced in real time (Fig. 5.17). The amplitude of an echo modulates the gray value of a picture element ("pixel") on the monitor screen. Thus, in this display, the localization and intensity of the echo sources are recognizable. This approach is applied for the eye when the fundus is not visible due to clouding (Fig. 5.18). The method is also applied in assessing the contents of the orbit (Fig. 5.19). Another application consists of the high-resolution display of the anterior eye segment (Fig. 5.20).

5.2.3 Ultrasound Biomicroscopy (UBM)

In the early 1990s, Pavlin et al. developed the so-called ultrasound biomicroscope, which is now available in several commercial versions. This instrument records B-scans with frequencies in the range of 40–100 MHz. While a frequency of 10 MHz yields a resolution of ca. 0.2 mm at a working depth of up to 40 mm, a typical instrument with 50 MHz can attain an axial resolution of about 50 μm and a lateral one of 80 μm, with a working depth limited to about 5 mm. The UBM is suited for detailed examination of the anterior segment of the eye (Fig. 5.20).

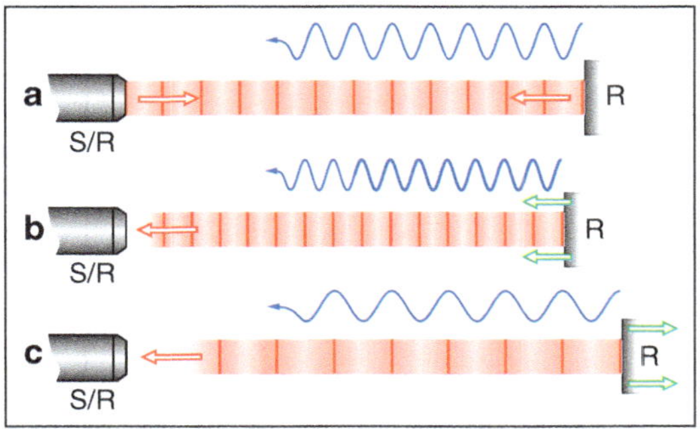

Fig. 5.21 Doppler effect (principle). *S/R* sender/receiver. (**a**) Reflector (*R*) at rest. Incident and reflected waves have the same frequency and wavelength. (**b**) Reflector moves toward the ultrasound source; the frequency of the reflected wave increases. (**c**) Reflector moves away from the ultrasound source; the frequency of the reflected wave decreases

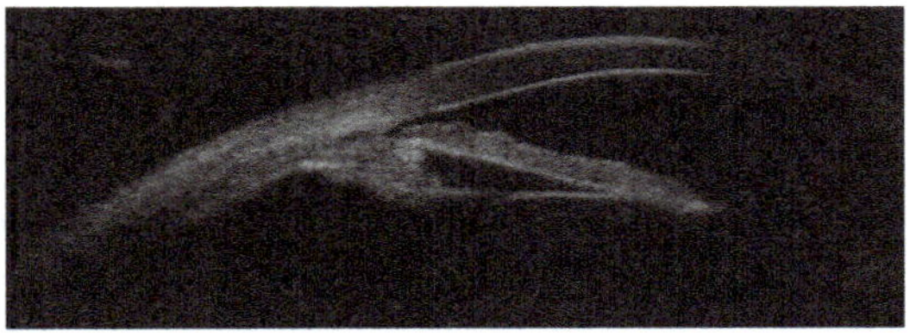

Fig. 5.20 Ultrasound biomicroscopic image of the anterior eye segment of a patient with a narrow angle

5.3 Doppler Sonography

So far, we have described static images. Various forms of Doppler sonography deliver information on movement in the examined regions. In the orbital region, this will mainly concern blood flow. The technique is based on the Doppler effect, which means that the movement of an object changes the frequency of a reflected wave. The basics of the Doppler effect are explained in Sect. 4.3 in relation to laser light. Depending on whether the reflecting object moves toward the probe or away from it, the frequency of the ultrasound echo increases or decreases (Fig. 5.21). It must be kept in mind that Doppler sonography captures only the motion component that approaches or withdraws from the observer (Fig. 5.22).

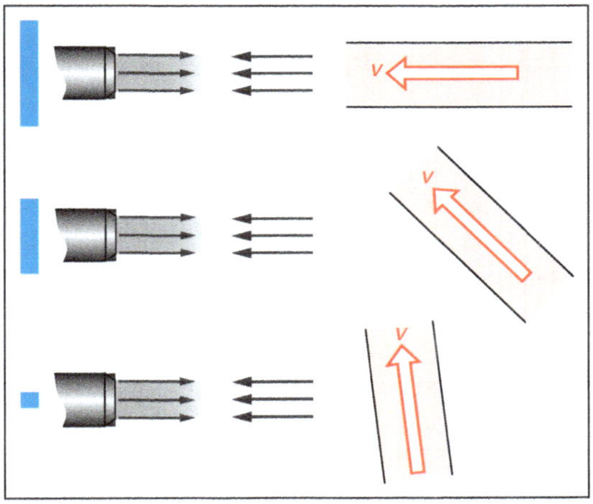

Fig. 5.22 The observed Doppler signal (Doppler-frequency shift) is determined by the component of the velocity that is in the direction of the sensor. *Top*: ideal case. Motion toward or away from the sensor; the observed velocity is equal to the true velocity. *Middle*: true velocity is larger than the observed velocity. *Bottom*: very small Doppler signal; motion is nearly undetectable. The *blue bars* indicate the size of the Doppler signal

If an overview of movement throughout a whole section (B-scan) is desired, one must be satisfied with the local averages of the different velocities that may be present at a given location of the tissue. On the other hand, if detailed information about the spectrum of velocities is desired (composite of movements at various velocities), one must sacrifice the spatial overview. Correspondingly, there are two different approaches for Doppler-based observations with differing priorities of information and differing displays: color duplex sonography and spectral Doppler sonography.

Color duplex sonography combines the two-dimensional anatomic information of a B-scan with local movement information. This limits it to the local averages of the velocities. These are color-coded and usually overlaid on the B-scan. To obtain depth resolution, the information is gained in a pulse-echo procedure. The details follow in Sect. 5.3.1.

Spectral Doppler sonography incorporates movement information along a single beam. The velocity information is refined compared to what is obtained with color duplex sonography. This approach delivers the so-called velocity spectrum in the vessel of interest as a function of time. Details follow in Sect. 5.3.2.

5.3.1 Color Duplex Sonography

Figure 5.23 shows the result of an examination with this technique. In its geometry, it resembles the display of a B-scan. In the regions shown in gray, no movement was detected. Movements are displayed with color codes, where red indicates a flow toward the probe and blue away from it. The velocity can be coded by means of the hue: saturated colors for small velocities and an increasing whitening with increasing velocities. The details depend on the instrument.

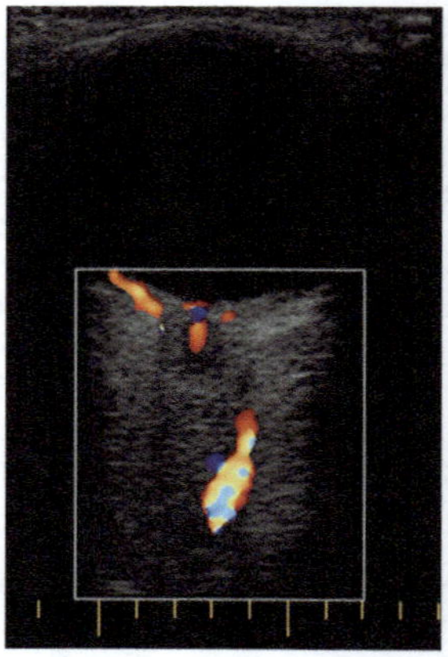

Fig. 5.23 Example of color duplex display of retroocular vessels. *Red* represents flow components toward the receiver and *blue* away from the receiver

The quality of the velocity information is subject to certain limitations: (a) at each point, only one (positive or negative) velocity can be displayed; the differing velocities within the cross-section of a small vessel are, thus, reduced to a local mean value and (b) in the pulse-echo approach upon which the B-scan is based, the pulses take up only a small part of the time – the module spends the majority of time waiting for an echo. Consequently, the sensitivity for detecting tiny blood flows is restricted in comparison with the spectral Doppler method.

5.3.2 Spectral Doppler Ultrasound

The spectral Doppler approach works with a continuous (not pulsed) ultrasound beam of constant frequency. During the measurement, the orientation of the beam in the tissue is fixed, often determined by the object being studied. In addition, with this method, the depth resolution is lost, but the velocity information is more differentiated. From the echo, a histogram – the so-called velocity spectrum – can be produced that can be recorded along the beam and in its cross-section. In that this spectrum can be observed in fractions of seconds, it can be recorded as a function of time and, thus, yields a very interesting display, e.g. for a pulsating flow of blood (Fig. 5.24). The horizontal axis represents the ongoing time, while the vertical axis represents the velocity spectrum. The intensity of a pixel along a vertical at a certain point in time indicates how high the associated velocity is represented in the echo. The momentary histogram of the velocities is, thus, read in gray codes (along a vertical). In particular, periodic alteration of the spectrum with the pulses can be spotted instantly. That almost all the velocities – from zero up to the momentary maximum – can be recognized in a typical spectrum is due to the fact that, in the vessel's cross-section, all velocities from zero (vessel wall) to the maximum (axis) are present. One concentrates primarily on the maximum velocity at each time (upper end of the spectrum) and on its maxima and minima over the course of time.

By virtue of its continuous send-receive operation, this approach has a further advantage besides the dynamic recording of the velocity spectrum. It reacts considerably more sensitively to tiny moving volumes than the pulse-echo approach of color duplex sonography. Due to their movements, tiny flows of blood can be recorded even when the associated vessel is not displayed in the B-scan. Although the signal collects all velocities encountered along the beam, a velocity spectrum typically stems from a single vessel or from a group of vessels in a certain anatomic structure. It is conceivable that blood flowing both toward and away in two different vessels along the beam produces a positive and a negative velocity spectrum at the same time. Another principle advantage consists of the impossibility of aliasing.

The mathematical principle of obtaining the velocity spectrum is touched on in Sect. 4.3 in connection with laser Doppler sonography. The principles of spectral Doppler ultrasound and of laser Doppler sonography have much in common.

5.3.3 Indices

The evaluation of the temporal behavior of a velocity spectrum, as shown in Fig. 5.24, requires a lot of experience. It is best to concentrate on the temporal progression of the upper boundary curve of the velocity spectrum, which corresponds to the momentary maximum of all the detected velocities. If the ultrasound beam hits a single vessel the maximum corresponds to the velocity along the vessel axis. The essential parameters at the upper edge of the curves are the maximum (systolic), minimum (end-diastolic), and mean flow velocity (v_s, v_d, v_m; Figs. 5.25 and 5.26), from which certain indices can be derived – among them the so-called resistance index *RI*.

The resistance index is defined by $RI = (v_s - v_d)/v_s$. It expresses the velocity difference between systole and diastole as a fraction of the systolic velocity. This index has the advantage that it is independent of the angle between the direction of the flow inside the vessel and the

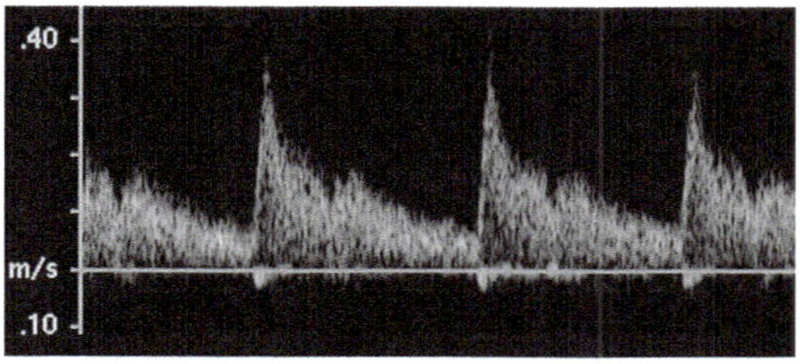

Fig. 5.24 Doppler spectrum (cerebral artery). *Abscissa*: time. *Ordinate*: velocities found along the observing beam; their weight (frequency) is coded as a gray value. Recognizable are the periodic systolic and diastolic distributions of the flow velocities in the vessel being observed. It is easy to see the maxima of the velocities along the upper edges of the shading

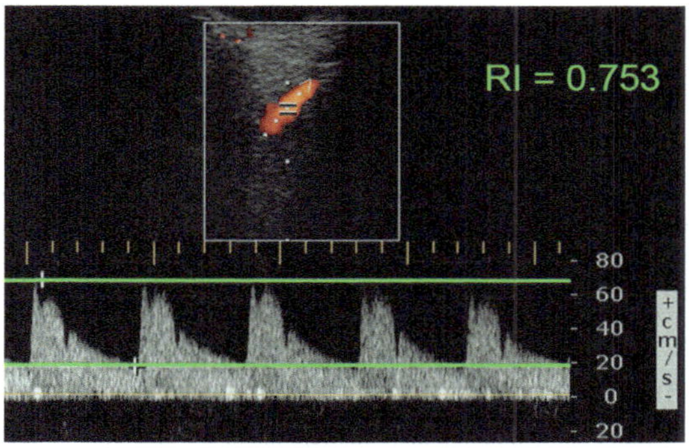

Fig. 5.25 A spectral Doppler diagram of the ophthalmic artery with a normal resistance index *RI*. The *upper green line* represents the peak systolic velocity and the *lower green line* the end diastolic velocity

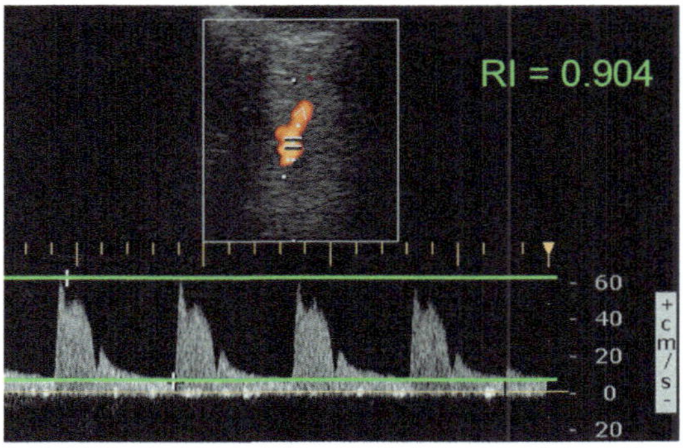

Fig. 5.26 A spectral Doppler diagram of the ophthalmic artery with an elevated resistance index *RI*. The *upper green line* represents the peak systolic velocity and the *lower green line* the end diastolic velocity

direction of observation – therefore, the complication is avoided where the Doppler observation only yields the velocity component along the observation direction. This independence is due to the fact that *RI* is defined in terms of the ratio of velocities and that the ratio of the velocity components in the observation direction is the same as the ratio of the true velocities. *RI* is expected to provide certain indications as to the resistance further downstream, i.e., in perfused blood vessels in which the blood flows away from the measurement window in the direction of the venous system. The interpretation of *RI* is still debated.

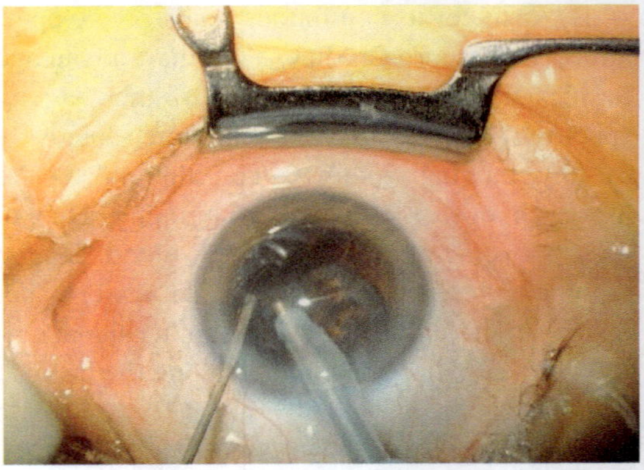

Fig. 5.27 Fragmentation of the lens core with ultrasound (phacoemulsification)

5.4 Ultrasound in Therapy

Ultrasound radiation of tissue can also be used therapeutically. Occasional applications occur in the treatment of neuralgia, chronic arthropathies, and phantom pains. Among other indications, ultrasound therapy is utilized in ophthalmology for cyclodestruction and, thus, for intraocular pressure reduction. Nowadays, the ciliary body is more frequently destroyed with cold (cryocoagulation) or with a laser (cyclophotocoagulation).

The most important therapeutic field of application for ultrasound in the eye is phacoemulsification.[5] In 1960, H. G. Schaye developed a method for cataract surgery (lens aspiration), primarily applied to soft congenital and traumatic cataracts. In 1967, Keelmann introduced the fragmentation of hard lens cores using ultrasound. A cannula (hollow needle) moves backward and forward with a frequency of 40 KHz and fragments the hardened tissue (lens). Combined with this needle, a channel sucks out the fragments (Fig. 5.27). The cannula has a diameter of about 1 mm and a wall thickness of 0.1 mm.

To conceptualize the process of fragmentation, a quantitative consideration is useful. The key concept is the acceleration of the cannula. Its axial movement is across a distance of approximately 0.1 mm. For the frequency stated above, the acceleration of the needle amounts to more as 10^5 times that of the gravity.[6] With its forward movements, the needle penetrates with this acceleration into the tissue. The material outside and inside the needle is unable to follow the needle's motion due to inertia: an incision results. With its backward movements, tiny cavitations (hollow spaces) arise that generate pressure waves during collapse and, thus, act upon the adjacent material. However, the relative importance of the various processes in vivo has not yet been completely clarified.

[5]Phacoemulsification: phakos (Greek) = lens, emulgere (Lat.) = emulsify.

[6]Maximum acceleration = $(2\pi)^2$ · amplitude · frequency^2.

6 Further Imaging Procedures

An ophthalmologist has the advantage that the most frequent pathological findings can be seen directly or imaged by means of light. Nevertheless, ophthalmologists sometimes also need to use additional imaging procedures, though less frequently than a neurologist, for example. Here, we will discuss conventional radiography, digital radiography, computed tomography (CT), and magnetic resonance imaging (MRI).

Radiography and CT are based on X-rays. These are "hard" electromagnetic radiation with photon energies far beyond those of visible light. They pass through the body and, while doing so, are more strongly attenuated by bony structures than by other tissues. MRI is based on the interactions between "soft" electromagnetic pulses and atomic nuclei, primarily hydrogen nuclei.

6.1 Analog Radiography

The discovery of X-rays in 1895 by Röntgen[1] was one of the greatest advances in medical technology (Fig. 6.1). For the first time, it was possible to image bones in the bodies in vivo. In very short order, his discovery received a great deal of attention worldwide. The image was recorded by a photosensitive layer (emulsions on the glass plates used in black-and-white photography at the time). The function of the photographic plate – and later film – as the detector and for long-term documentation continues today.

Expressed in today's language, Röntgen experimented with electrons that were accelerated in a vacuum tube by a high electrical voltage to make them collide with a positively charged electrode (anode), where they engender electromagnetic radiation of very short wavelengths (photons with high energies). This principle has remained unchanged right up to the present time. Röntgen's electrons originated from the gas remaining in the tubes and the negative electrode (cathode) was cold. Considerably higher and adjustable power resulted later with hot cathodes from which electrons could easily emerge (Fig. 6.2).

Typical operating voltages are in the range from 30 to 120 kV. Ideally, an electron transfers its whole energy to a single photon. The maximum energy of the photons created with a voltage of 30 kV, thus, amounts to 30,000 eV, approximately 10,000 times higher than the energy of a blue light photon. Correspondingly, the wavelength is shorter than that of blue light by the same factor (Table 6.1). In the language of electrodynamics, the generation of photons is based on the acceleration of electrons. Here, it is the deceleration (sudden stopping) that is involved; i.e., a negative acceleration. Part of the kinetic energy of the electrons is transformed into electromagnetic radiation (photons) and part contributes to heating of the anode.

[1]Wilhelm C. Röntgen (1845–1923), German physicist. He received the first Nobel Prize in physics (1901). The term "X-ray" was coined by him and its use continues in many countries. In German-speaking countries (and in others), X-rays are referred to as Röntgen rays.

J. Flammer et al., *Basic Sciences in Ophthalmology*,
DOI 10.1007/978-3-642-32261-7_6, © Springer-Verlag Berlin Heidelberg 2013

Fig. 6.1 The title of Röntgen's first publication, *A New Kind of Radiation*, indicates that the nature of the X-rays was unexplained in the beginning. *Right*: One of the first X-ray photos

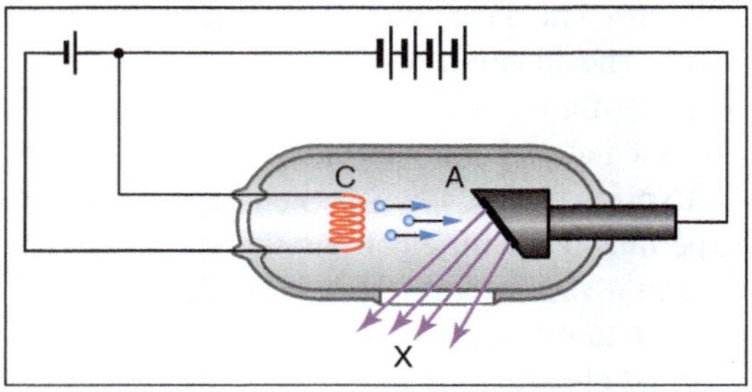

Fig. 6.2 Principle of the X-ray tube. The cathode (*C*) glows. A high voltage between the anode (*A*) and the cathode accelerates electrons. Their collisions with the anode give rise to high-energy photons (X-rays)

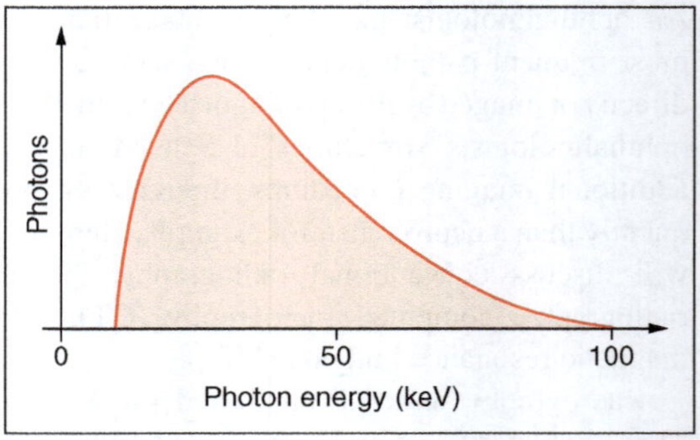

Fig. 6.3 Typical distribution of photon energies. The maximum is determined by the operating voltage of the X-ray tube (Table 6.1). Operating voltage: 100 kV

Table 6.1 X-rays. The relationship between the electrical operating voltage, wavelength, and maximum energy of the photons. 1 kV = 1000 V (for comparison with blue light photons: energy = 3 eV, wavelength = 400 nm)

Operating voltage (kV)	Wavelength (nm)	Maximum photon energy (eV)
30	0.04	30,000
120	0.01	120,000

Not all photons exhibit the maximum possible energy. Figure 6.3 shows a typical spectrum. The choice of voltage between anode and cathode influences the mean energy of the Röntgen quanta: the higher their energies are, the better they penetrate through sections of the body; i.e., the attenuation is less. X-rays damage tissue. Studies have shown that the additional risk for cancer due to X-ray examinations differs in industrial nations but may be on the order of roughly 1 %.

6.2 Digital Radiography

Digital radiography is slowly replacing the analog form. The difference does not lie in the way that X-rays are generated or in the interactions between the rays and the tissues but in how the X-ray images are recorded. Instead of photosensitive film, reusable flat-panel detectors are employed.

Among the various technologies, we will discuss the elements of a system that functions with an X-ray storage panel (Fig. 6.4). The X-rays are generated in the traditional way. They pass through the body that is being examined and the whole image cross-section simultaneously falls onto the X-ray flat-panel detector that has replaced photosensitive film. Rather than a photosensitive emulsion, the panel carries an approximately 0.1 mm thick photostimulatable phosphor layer that is

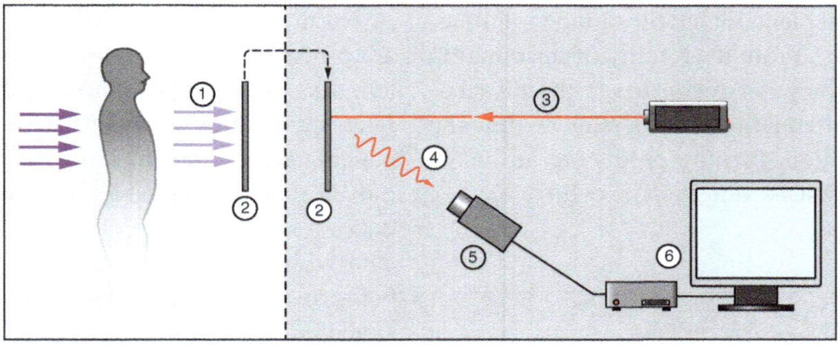

Fig. 6.4 Principle of digital radiography. The X-ray picture (*1*) is acquired and stored on a plate (*2*) containing crystals of a photostimulatable phosphor (e.g., BaFBr). Later, the image can be read with a scanning laser beam (*3*). Pulses of visible photons are produced, point by point, depending on the X-rays that were received (*4*). They are then detected by a photomultiplier or photodiode (*5*). A computer (*6*) records the image in the form of the point-by-point number of these photons. The resolution is on the order of 2,000 × 2,000 pixels with a dynamic range of 10 bits

able to receive an X-ray intensity image and store it for a period of time. This amazing process begins with the layer's electrons being pushed up to a higher energy level by the X-ray photons. If more X-rays arrive in parts of the image, a greater number of electrons undergo this transition. Instead of returning spontaneously to the previous level, these electrons are held captive in an excited state by the phosphor layer. This state can last for hours. It is only when the panel is illuminated with light that the electrons return to their original state, thereby immediately emitting fluorescent light. This reading process occurs in a separate reading unit where a laser beam scans the X-ray storage sheet line by line. A light-sensitive photodetector registers the temporal sequence of the fluorescent light. This signal is digitized and the information is stored in a computer.

The X-ray storage sheet is reusable. The storage of large X-ray films is no longer necessary. The digitally stored data permit post-processing of the image, e.g., contrast improvement and, thus, a certain independence from insufficient exposure. Compared to film radiography, the radiation dosage can be reduced to some extent while retaining similar image quality. The disadvantages include the investment required for the reading unit and the dependence on computer systems.

In Fig. 6.5, a dacryocystography is shown as an example of a classical radiograph, enhanced with a contrast dye.

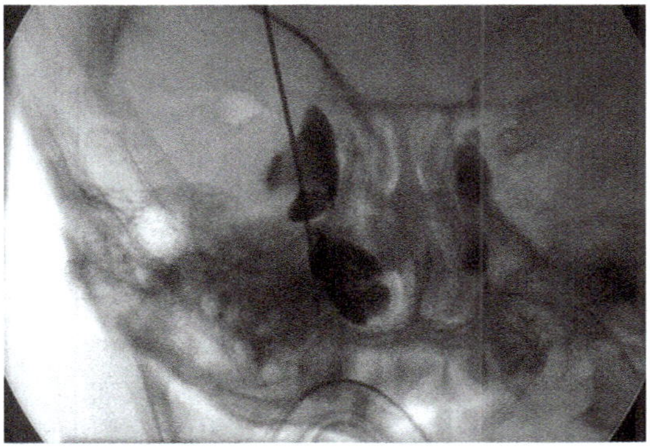

Fig. 6.5 Distal occlusion of the lacrimal duct visualized by dacryocystography. (Courtesy of L. Remonda, Kantonsspital Aarau, Switzerland)

6.3 Computed Tomography (CT)

The invention of computed tomography in the 1970s was another great breakthrough in diagnostics. Instead of silhouettes, sectional images through body parts are generated in which the forms and positions of the organs are displayed.

The basis of this method is also X-rays. The information for a section image is obtained by a thin sheet of X-rays that initially pass through the body in a certain direction. For this direction of irradiation, a series of detectors records a one-dimensional image of the attenuation of the radiation as a function of the transverse coordinate. These recordings are then repeated in the same sectional plane for a large number of directions (Fig. 6.6). From the totality of information acquired, a computer reconstructs the distribution of the attenuation coefficients within the irradiated section, thereby producing an image of the tissue distribution in that section.

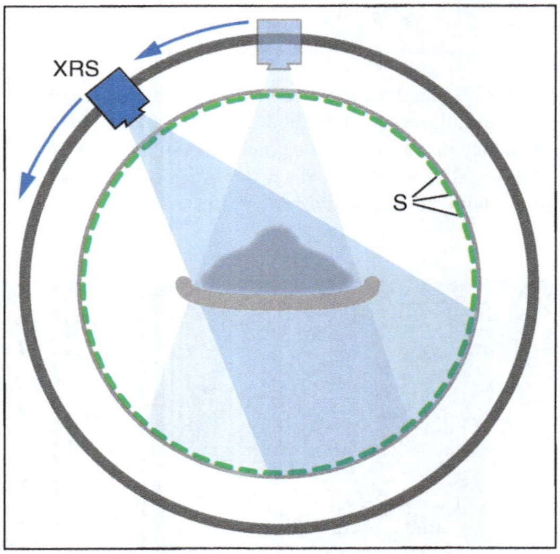

Fig. 6.6 Recording method in computed tomography. A sheet of X-rays passes through the body from a given direction. One-dimensional recording as a function of the transverse coordinate. Repetitions from many directions result in a section image. *XRS*: X-ray source. *S*: Sensors

Normally, several adjacent sectional images are procured. With spiral-CT, the recording system rotates with continuous radiation without interruption during the entire examination of the patient. At the same time, the patient table is also moved along continuously. This type of scan produces a spiral-shaped collection of transmission data with an unbroken recording of the region being examined. The data carry three-dimensional information that can be displayed as a calculated section image, where any position of the section plane can be chosen within the examined region. A clinical example of a CT of the orbita in a patient with head injury is shown in Fig. 6.7. The CT allows particularly clear visualization of the bones.

The development of computed tomography can be traced back to the independent work of Cormack and Hounsfield.[2] The problem of calculating the two-dimensional structure of the absorption values from a series of one-dimensional information acquired from the various directions had already been solved by Radon.[3] The construction of an image demands elaborate calculations. The improvement of algorithms over the years has led to a reduction in radiation exposure. Nevertheless, the radiation dose due to computed tomography is considerably higher than that from a single analog or digital radiographic examination (Table 6.2).

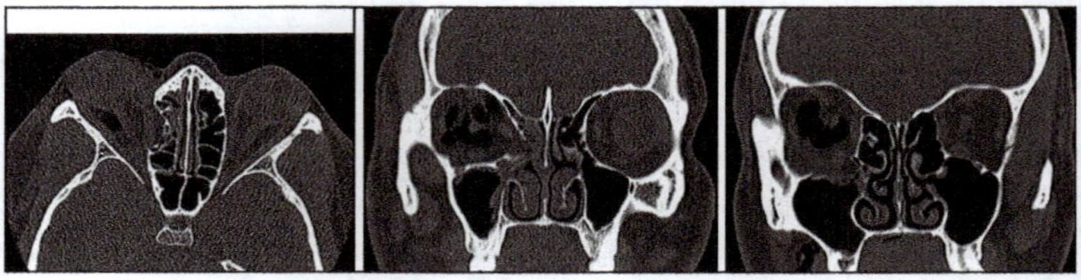

Fig. 6.7 Orbita CT revealing impression of the lamina papyracea, impression of the orbita floor, and an intraorbital hematoma. *Left*: axial section. *Middle and right*: coronal sections (Courtesy of L. Remonda, Kantonsspital Aarau, Switzerland)

[2] Allan M. Cormack (1924–1998), South African–U.S.-American physicist. Godfrey N. Hounsfield (1919–2004), English engineer. Joint Nobel Prize for physiology or medicine (1979).

[3] Johann Radon (1887–1956), Austrian mathematician.

Table 6.2 Radiation dosage due to various examinations with X-rays compared with the natural radiation dosage. The numbers are to be understood as rough indications

Examination	Effective dosage (mSv)
Natural radiation dosage per year	2.0
X-ray image of the thorax	0.02–0.1
Head CT	2
Screening mammography	3

6.4 Magnetic Resonance Tomography (MRT or MRI)

Magnetic resonance tomography[4] has an exceptional position in medical imaging. Like CT, it is able to record the form and position of organs in three dimensions. However, MRI differs from CT by providing a much better differentiation of soft tissues. With MRI, we can not only detect the distribution of hydrogen but also the embedding of the hydrogen atoms in their molecular environment.

Like CT, MRI is also based on electromagnetic interactions with tissues. While CT works with photons of energies roughly greater by a factor of 10^4 than the energies of visible photons, in MRI, the reverse is true: the photon energies are 10^7 times lower than those in the visible range. The frequencies lie in the range of radio waves (Table 6.3). There are also no ionizing effects on the tissues. Another difference is that, in CT, the interactions take place with the electrons, whereas, in MRI, the interactions occur with the atomic nuclei, especially with hydrogen nuclei (protons).

MRI is based on the principle of magnetic nuclear spin resonance. Around 1945, independently of one another, Bloch and Purcell[5] created the foundation for the later applications of nuclear spin resonance in chemistry, physics, biology, physiology, and medicine. At that time, spatial resolution (imaging) was not yet a topic. The advance – the step to MRI – was due to the successful work of Lauterbur,[6] who produced sectional images in the 1970s. Important contributions toward the development of MRI were also made by Mansfield.[6]

[4]MRT or MRI, acronyms for magnetic resonance tomography or magnetic resonance imaging.

[5]Felix Bloch (1905–1983), Swiss-American physicist. Edward M. Purcell (1912–1997), U.S. physicist. Joint Nobel Prize for physics (1952).

Table 6.3 Electromagnetic scanning of matter. Order of magnitudes of the characteristics of the waves that are used

	Photon energies (eV)	Frequency (s^{-1})	Wavelength (m)
X-rays	10^5	10^{19}	10^{-11}
Light (visual range)	10^0	10^{15}	10^{-6}
MRI	10^{-7}	10^8	10^1

The two key topics for understanding MRI are the following: on the one hand, eliciting a signal with low-power radio waves from the examined object (nuclear spin resonance) and, on the other, the localization of the origin of the signal (imaging). We shall first discuss the phenomenon of nuclear spin resonance and then localization.

6.4.1 Nuclear Spin Resonance: The Phenomenon

Experimentally, nuclear spin resonance manifests itself as follows (Fig. 6.8). An object containing hydrogen is situated in a homogeneous, temporally constant magnetic field B_0. In practice, strong magnetic fields are involved that are generated by either permanent magnets or super-conductive coils. Onto this static field, a second weaker oscillating electromagnetic field is superimposed – a wave with a specific frequency f in the range of 10^8 Hz, similar to waves produced by radio transmitters. This wave can be produced with an alternating current with frequency f in a coil. If one now continuously changes this frequency, something very surprising occurs at a very specific frequency f_0: the material being studied emits electromagnetic waves of precisely this same frequency in all directions. If the input frequency f is increased slightly beyond

[6]Paul C. Lauterbur (1929–2007), U.S. chemist. Peter Mansfield, British physicist. Joint Nobel Prize for physiology or medicine (2003).

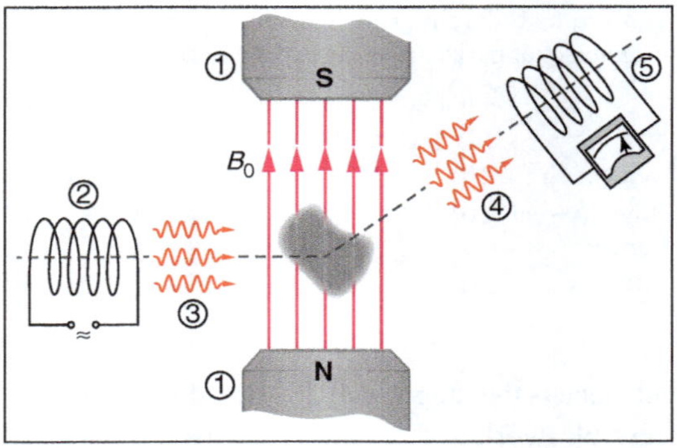

Fig. 6.8 Principle of nuclear spin resonance. Protons as hydrogen nuclei in some molecules find themselves in a homogeneous magnetic field (B_0), created here by permanent magnets (*1*). An alternating current with a frequency *f* in a coil (*2*) creates an electromagnetic wave (*3*). At a certain frequency f_0, the hydrogen nuclei in the sample give off electromagnetic radiation of this same frequency (*4*), which can be detected by a receiver such as a radio (*5*)

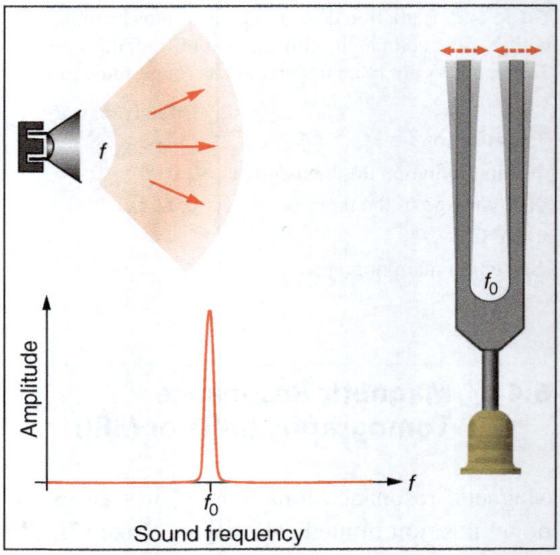

Fig. 6.9 A tuning fork with a resonant frequency $f_0=432$ Hz is exposed to sound waves arriving from another source. The fork vibrates along with them with large amplitude when the frequency *f* of the driving sound is the same at its own resonant frequency. In the resonant case ($f=f_0$), the forces of the sound waves drive the tuning fork with the right rhythm. *Lower left*: amplitude of the fork vibration as a function of the driving frequency *f*

f_0, the effect disappears again. For a magnetic field with a strength[7] $B_0=1$ T, $f_0=42.57$ MHz.

The dependence of the effect on the transmitter frequency *f* reminds one of the phenomena of resonance: a tuning fork resonates when it is exposed to sound with its own frequency f_0 (Fig. 6.9). For a frequency *f*, which differs from f_0, the tuning fork will also vibrate with that frequency (*f*), but the amplitude of this vibration is very low.

The intensity of the nuclear spin resonance signal depends primarily on the number of hydrogen atoms in the sample. Other influences, such as relaxation times, will be discussed later. The frequency f_0 is proportional to the magnetic field strength B_0. Nuclear spin resonance is not restricted to hydrogen nuclei. Instead, it is observed with all nuclei that exhibit an intrinsic magnetic moment, whereby each atom or isotope has its own characteristic resonant frequency f_0 in a given magnetic field.

[7]T is the abbreviation for Tesla. 1 T is the unit of magnetic flux density. MRI instruments work with magnetic fields of 0.3–3 T. The superconducting dipole magnets of the Large Hadron Collider at CERN produce field strengths of 8.6 T, while a small horseshoe magnet produces strengths of 0.1 T. The strength of the earth's magnetic field is about 0.00006 T.

6.4.2 Nuclear Spin Resonance: A Brief Explanation

We restrict ourselves to the hydrogen nucleus (proton), which is the most important case for medical applications. A proton carries an intrinsic magnetic moment, meaning that it reacts to magnetic fields like a tiny compass needle. Like mass and electric charge, the magnetic moment is an invariant property of the proton.[8] In thermal equilibrium and in the absence of a magnetic field, the directions of the protons' magnetic moment vectors are completely random (Fig. 6.10) and the energy of a proton does not depend on the direction of its magnetic moment.

The situation changes in the presence of an external magnetic field, which tries to align the magnetic moments of the protons in favor of its

[8]The magnetic moment of the proton is $\mu_p=1.41\cdot10^{-26}$ J/T. It is due to the spinning of the proton with its electrical charge around its own axis.

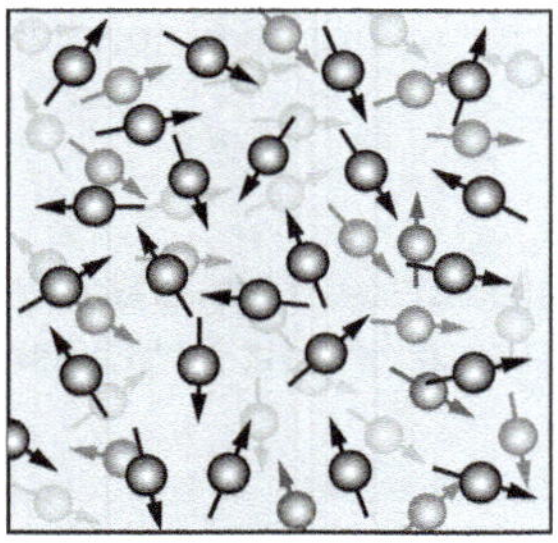

Fig. 6.10 Random alignments of the magnetic moments of the protons in the absence of an external magnetic field

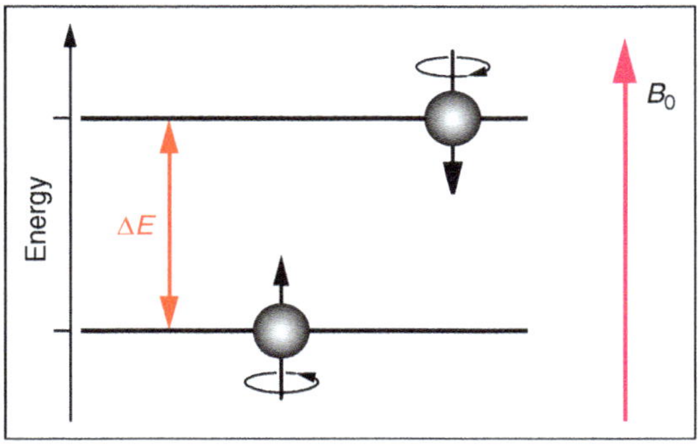

Fig. 6.11 In an external magnetic field B_0, the magnetic moments of the protons align themselves either parallel or anti-parallel to the field direction. The two states have different energies; that of the parallel alignment is lower

own alignment. At this point, the quantum mechanical description of the proton's behavior becomes quite simple and will turn out to be very helpful in view of the resonance to be explained. The key is the energy of the proton in the magnetic field. Quantum mechanics asserts that the proton has two possible energy levels in the given magnetic field (Fig. 6.11): the magnetic moment can be aligned either along the field's direction or in the opposite direction. For a proton to turn from a parallel alignment into the opposite direction, energy must be supplied (turning a magnet in a given magnetic field also requires energy). The energy difference ΔE between the two levels is proportional to the strength B_0 of the magnetic field.[9]

[9]It is given by $\Delta E = 2 \cdot \mu_p \cdot B_0$, where μ_p is the proton's magnetic moment.

For a proton in a field of strength $B_0 = 1$ T, this amounts to $\Delta E = 2.82 \cdot 10^{-26}\,J \approx 2 \cdot 10^{-7}$ eV, which is an extremely small energy (many orders of magnitude below the energy of visible photons).

The two energy levels of the proton remind us of an electron in an atomic shell. Now we are getting closer to an explanation of nuclear spin resonance. How can such a system be forced from one state into the other? We know the answer from the absorption of light (upward transition, Fig. 2.3) and stimulated emission (downward transition, Fig. 3.10), namely by exposure to photons with the proper energy or – in terms of waves – by exposure to an electromagnetic wave of the proper frequency. Resonance means that the driving wave has to have the matching frequency for the transitions to occur.

With this knowledge, the key to understanding nuclear spin resonance has been found. The photons of an electromagnetic wave of frequency f_0 can induce transitions between the two energy levels of the proton if their energy $h \cdot f_0$ equals the energy difference ΔE between the two levels. In a magnetic field of 1 T, the frequency matching this condition[10] is $f_0 = 42.57$ MHz. This value corresponds precisely to the actually observed resonant frequency. The explanation given here applies as well to any nuclei with any other magnetic moment – with the only difference being that f_0 will not be the same.

So far, we have explained the origin and frequency of nuclear spin resonance. The mechanism of the emission of electromagnetic radiation in all directions is more difficult to understand. A simple approach is to say that the protons absorb and re-emit electromagnetic radiation by their oscillation between the two levels. However, the deeper secret underlying the phenomenon is the coherence of the behavior of all protons. More precisely, their magnetic moments change direction in phase with the driving external wave. This creates a huge total magnetic moment that spins at the resonant frequency and

[10]According to the formulae given in the text and in the previous footnote, we have $f_0 = \Delta E/h = 2 \cdot \mu_p \cdot B_0/h$. Here, h is Planck's constant (its numerical value is given in the Appendix).

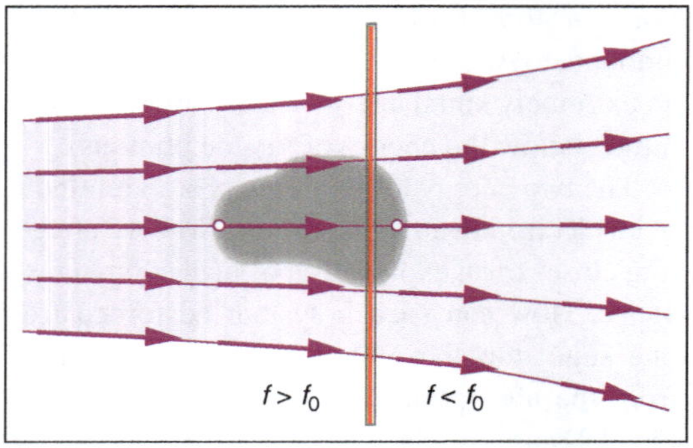

Fig. 6.12 Magnetic field with gradients. The resonant frequency decreases from left to right. Resonance with a given transmitted frequency f_0 arises only in a certain plane (in reality, in a thin layer)

goes on emitting radiation even after the driving pulse is switched off.

6.4.3 From Nuclear Spin Resonance to MRI: Location Coding

We understand that the strength of how variously composed materials react depends on the proportion of hydrogen present. We still have not explained how the MRI creates a spatial resolution to produce an image. Lauterbur discovered the tricks. For example, the probe is set into an inhomogeneous magnetic field of the kind that decreases from left to right (Fig. 6.12). We recollect that the energy difference between the two proton states (Fig. 6.11) is proportional to the strength of the external field. The resonant frequency is, thus, not the same everywhere. For a certain transmitted frequency, the resonance is generated only in a very specific plane. By changing the transmitted frequency, it is possible to scan the sample.

The signal from a plane (or, in reality, from a thin layer) stems from the sum of all activities in the layer. A further spatial dimension can be resolved by employing unequally directed field gradients during the transmitted pulses and during the subsequent detection of the signals. At this point, we omit further explanations of the complicated processes required for a complete spatial reconstruction. The problem reminds one of the reconstructions of spatial structures in computed tomography from projections.

6.4.4 Relaxation Times and Associated Measurement Processes

Nuclear spin resonance is triggered by a transmitted pulse. This changes the populations in the two levels or, more precisely, the quantum states of the protons. A so-called 90° pulse results in a maximum coherent signal with the resonance frequency. Following the end of the pulse, the emission of these signals by the protons continues. It fades gradually until the protons lose their special state due to the random thermal influences from their molecular environment. This takes place in two ways.

With the so-called spin–lattice relaxation, the population numbers of the two levels return to thermal equilibrium. These are the processes that occur with every return from a deviation from thermodynamic equilibrium, e.g., balancing the temperature when a small volume element has a temperature that differs from the immediate environment. These processes result in an exponential decrease of the signal. The corresponding time constant *T1* is known as the spin–lattice relaxation time (after an interval *T1,* the signal is smaller by the factor $e=2.72$). Typical values in humans lie between 0.1 s (body fat) and a few seconds for fluids (blood, aqueous). *T1* relaxation times depend on the strength of the magnetic field.

In so-called spin–spin relaxation, the same-phase behavior of the individual protons is lost. The reasons are the random magnetic influences of neighboring atoms. The individual protons no longer have the same resonant frequency and they get out of phase; i.e., the fields emitted by the individual protons no longer interfere constructively. The received signal decays exponentially with a time constant *T2*. The *T2* relaxation times depend only slightly on the magnetic field. They are smaller than *T1*.

The local relaxation times *T1* and *T2* and the proton density represent the information variables of nuclear spin resonance that finally influence the pictorial display of magnetic resonance tomography. With a specific stimulating series of pulses, images can be obtained that emphasize the differences in *T1*, while, with other pulse series, greater sensitivity to differences in *T2* can be attained (Figs. 6.13, 6.14, 6.15). For example, *T2* differentiates quite clearly between oxygen-rich and oxygen-depleted blood, while *T1* does not react to this difference at all.

Contrast media can be an aid in distinguishing between organs that are depicted similarly. Paramagnetic compounds containing gadolinium are the ones mainly used. They themselves are not depicted, but they influence the relaxation times *T1* and *T2*.

6.4.5 Examples of Clinical Applications of MRI

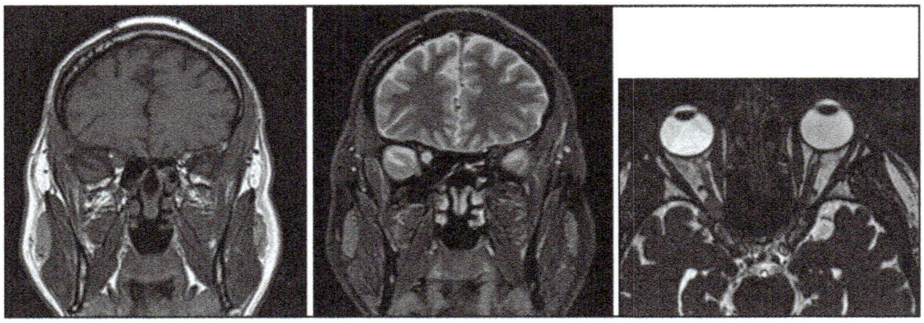

Fig. 6.13 MRI of a patient with an optic nerve sheet meningioma. *Left*: *T1*-weighted coronal section. *Middle*: *T2*-weighted coronal section with fat suppression. *Right*: *T2*-weighted axial section (Courtesy of L. Remonda, Kantonsspital Aarau, Switzerland)

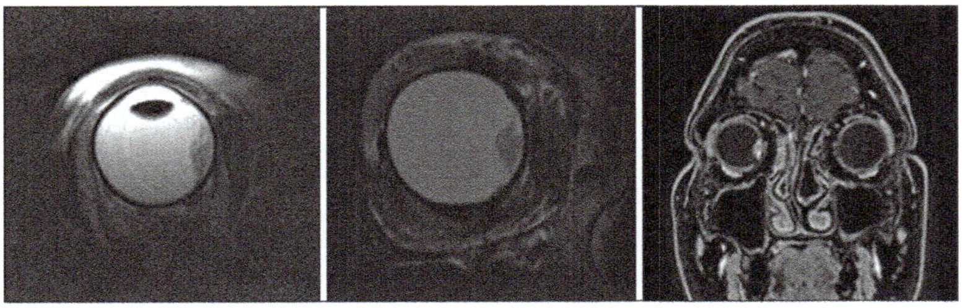

Fig. 6.14 MRI of a patient with a choroidal melanoma. *Left*: *T2*-weighted axial image taken with a surface coil. *Middle*: another *T2*-weighted coronal image taken with a surface coil. *Right*: *T1*-weighted coronal section with the paramagnetic contrast agent gadolinium (Courtesy of L. Remonda, Kantonsspital Aarau, Switzerland)

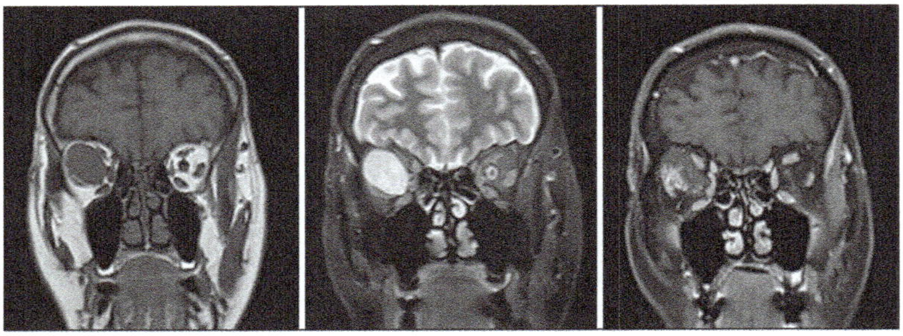

Fig. 6.15 MRI of a patient with an orbital hemangioma. *Left*: *T1*-weighted coronal section. *Middle*: *T2*-weighted coronal section with fat suppression. *Right*: *T1*-weighted coronal section with fat suppression and gadolinium (Courtesy of L. Remonda, Kantonsspital Aarau, Switzerland)

Interventions with Laser Light

7

In interventions in the eye, various qualities of laser radiation play a role. Laser light can easily be focused and managed temporally; therefore, it permits enormous leeway in the choice of power and intensity. It also has a defined wavelength. The various tissues have different absorption spectra. This also influences the depth of the penetration of light. Depending on the type of laser settings (wavelength, pulse duration, energy, fluence, and irradiance[1]), light interacts differently with tissue. Since we are not attempting to present a complete list but, instead, wish to explain the main mechanisms of light interactions with matter, we shall be limiting all explanations to the following categories, listed in Table 7.1:

1. Pure heating. This process is based on absorption by colored matter; its main application lies in coagulation of, e.g., the retina (Sect. 7.1).
2. Optical breakdown inside transparent tissue, based on concentrating the pulse energy in very short pulse durations (Sect. 7.2). The associated high power densities lead to a kind of microexplosion with mechanical effects (forming a cavitation). One of its applications is capsulotomy.
3. Removal of tissue at a surface. The process takes place via the splitting of molecules into small fragments that escape from the surface. One application is corneal refractive surgery (Sect. 7.3).
4. Optical breakdown with extremely short pulse durations, also producing small vesicles but without thermal effects in the environment. Corneal incisions represent one application (Sect. 7.4). With extremely short pulse durations, the process is effective even with very low pulse energies.

The choice of wavelength – in the range of UV to IR – is partially dictated by the absorption properties of the tissue and its pigment(s) (details are provided in Sect. 2.8). For example, ultraviolet light from an excimer laser (ArF, 193 nm) is already absorbed within the superficial layers of the cornea and is, thus, well suited for tissue removal but unfeasible for applications in deeper layers. For other applications, the different absorption properties of pigments determine the optimum wavelength. Both the argon laser (514 nm) and the ruby laser (684 nm), each employed in its own way for applications in the ocular background, are optimal for very specific interventions since blood absorbs green-wavelength light (514 nm). Apart from the absorption spectra, there are also technical aspects: the Nd:YAG laser (1,064 nm) is suited for photodisruption due to its short and intense pulses. Figure 7.1 provides an overview of the wavelengths in various applications.

While pulse durations range over an enormous extent from 1 s down to extremely short pulses of 10^{-13} s (i.e., over 13 orders of magnitude), there is one parameter that differs far less

[1]Power = pulse energy/pulse duration (W). Irradiance = power/irradiated area (W/cm^2). Fluence = pulse energy/irradiated area (J/cm^2). Irradiance is sometimes called intensity, but this usage leads to confusion.

J. Flammer et al., *Basic Sciences in Ophthalmology*,
DOI 10.1007/978-3-642-32261-7_7, © Springer-Verlag Berlin Heidelberg 2013

Table 7.1 Categories of laser light interactions with ocular media. λ wavelength

Category	Physical processes	Typical application	Typical laser (λ)	Section
Photocoagulation	Heating, coagulation, evaporating, charring	Photocoagulation in diabetic retinopathy	Argon (514 nm)	Sect. 7.1
Photodisruption	Optical breakdown, cavitation, shock waves	Capsulotomy	Nd:YAG (1,064 nm)	Sect. 7.2
Photoablation	Splitting of molecules, removal of tissue	Refractive correction of the cornea	Excimer (193 nm)	Sect. 7.3
Photocutting	Optical breakdown, cavitation	Lamination of the cornea	Femto (1,040 nm)	Sect. 7.4

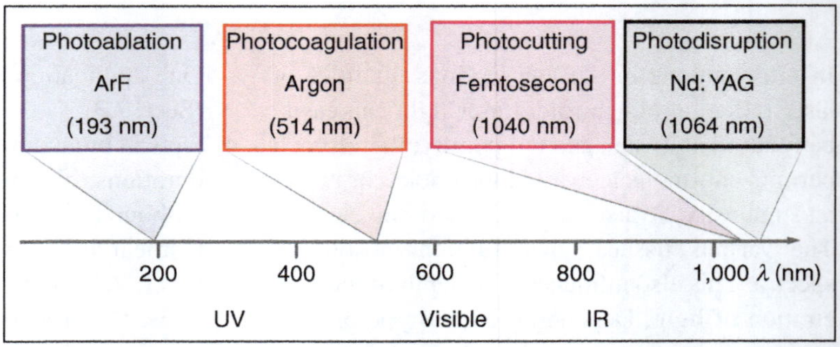

Fig. 7.1 Interventions at various wavelengths with typical lasers that are utilized (overview)

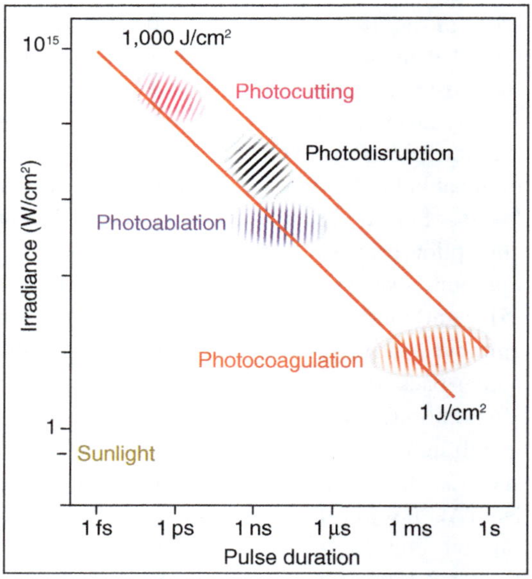

Fig. 7.2 Approximate pulse duration and irradiance of the various modalities. The precise values depend on the details of the application (e.g., the repetition frequency of the pulses). For comparison, the irradiance of intense sun radiation of the earth amounts to 0.1 W/cm^2. Note that both scales are logarithmic

Table 7.2 The pulse duration in laser treatments of the eye extends over 13 powers of ten Here, the various time units are defined

1 fs	1 ps	1 ns	1 μs	1 ms	1 s
10^{-15} s	10^{-12} s	10^{-9} s	10^{-6} s	10^{-3} s	10^{0} s

in the various applications: the pulse energy per area at the focus, stated in J/cm^2 (energy density, fluence). Typically, it lies in the range between 1 and 1,000 J/cm^2 and, thus, varies by only about three orders of magnitude. The absorbed power per area in W/cm^2 (irradiance) is, therefore, approximately indirectly proportional to the pulse duration, meaning that the choice of pulse duration more or less determines the power per area and, thus, essentially, the type of interaction between light and tissue. Figure 7.2 illustrates how the four principal modalities are clearly separated by pulse duration and irradiance. In comparison to the irradiance of sunlight (on the earth's surface), these are extremely large values. The time units are listed in Table 7.2.

The purely physical relationship between pulse power per area (irradiance) at the focus and

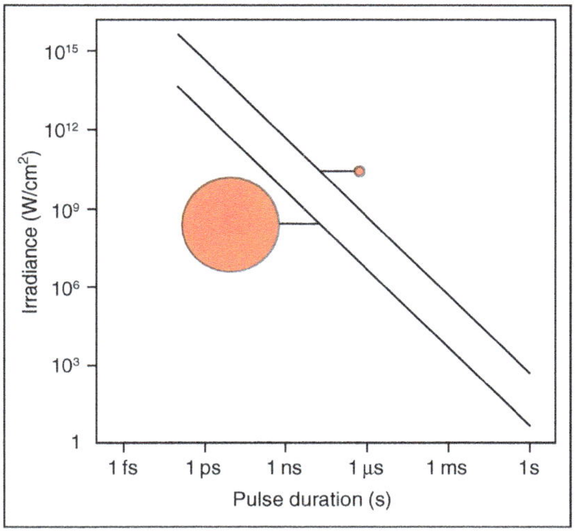

Fig. 7.3 The shorter the pulse duration is, the higher the irradiance will be. Irradiance is indirectly proportional to the pulse duration for a given pulse energy (3 mJ). The two curves refer to spot diameters of 30 and 300 µm. The smaller the spot diameter is, the larger the irradiance will be

the pulse duration for a fixed pulse energy of 3 mJ is shown in Fig. 7.3. The irradiance is the determining variable for the type of interaction of light with matter (Fig. 7.2). It is not only determined by the pulse energy but also depends on the diameter of the focus.

7.1 Photocoagulation

The coagulation of the retinal tissue for the prophylactic treatment of diabetic retinopathy to prevent neovascularisation represents one of the oldest, most well-established interventions with light in the eye. The treatment effect arises via the heating of the retinal pigment epithelium and choroid through the absorption of light energy. Historically, the interventions with light began before the invention of the laser with the experiments of Meyer-Schwickerath[2] carried out in Hamburg in 1949. He proceeded from the observation "that a progressive retinal detachment comes to stop at a retinal scar." He was stimulated by his observation of a patient who had viewed a solar eclipse too long and thereby developed a central retinal burn (Fig. 7.4). At that time, the sun was the light source with the highest irradiance. Meyer-Schwickerath developed an apparatus that directed sun rays into the eye being treated: the sunlight coagulator. A heliostat (two slowly turning mirrors) compensated for solar movements. He was then involved in the development of the Zeiss light coagulator, which worked with a xenon high-pressure lamp and was marketed in 1957 (Fig. 7.5). Fresh impacts of retinal coagulation with a xenon lamp are depicted in Fig. 7.6. A broad discussion about the indication of photocoagulation started. Peripheral retinal lattice degenerations (Fig. 7.7) were treated by some physicians but not by others.

One year after the development of the first laser, Campbell (U.S.A.) introduced it to ophthalmology (ruby laser, 684 nm; see Sect. 3.4.1). The wavelength of the ruby laser is only weakly absorbed by blood, which makes it suitable for irradiating retinal tissue or choroidal alterations near blood vessels without the risk of hemorrhages. However, it turned out to be unsuitable for occluding blood vessels. The main absorbing structure on the fundus is the pigment epithelium. The heated pigment epithelium leads to a secondary heating and, thus, to retinal coagulation (Figs. 7.8 and 7.9).

In 1968, L'Esperance introduced the argon laser to ophthalmology. This is a gas laser that emits blue-green light (488 and 514 nm), which is well absorbed by hemoglobin.

The first irradiation systems consisted of a coupling of the laser beam with a direct

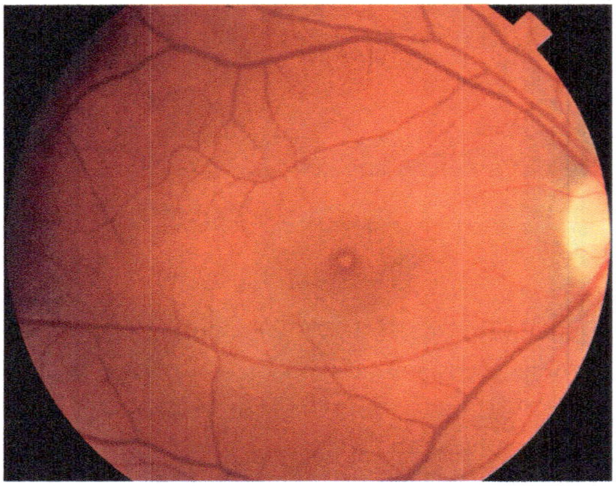

Fig. 7.4 Central retinal burn from viewing the sun

[2]Gerhard Meyer-Schwickerath, German ophthalmologist (1920–1992).

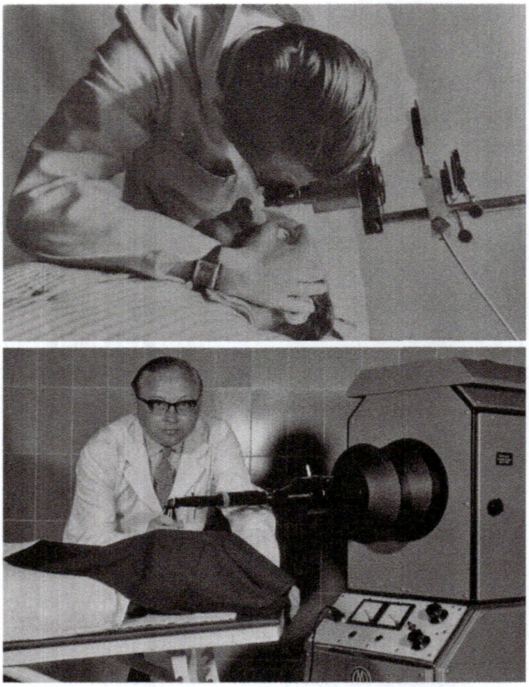

Fig. 7.5 Gerhard Meyer-Schwickerath coagulating the retina with sunlight (*top*) and a xenon coagulator (*bottom*) (Courtesy of R. Meyer-Schwickerath, Ahaus, Germany)

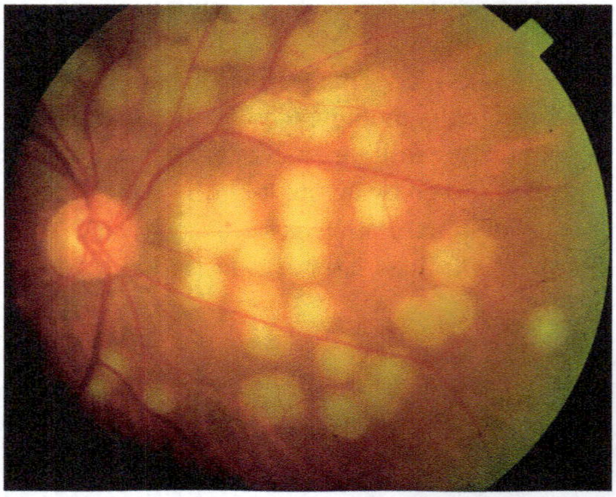

Fig. 7.6 Fresh retinal impacts following xenon light therapy

ophthalmoscope. In 1967, Fankhauser and Lotmar introduced irradiation with a laser coupled to a slit lamp (Fig. 7.10) and first used contact lenses for this procedure. Today, most laser instruments are slit lamp-based (Fig. 7.11). In subsequent years, the argon laser established itself as the "workhorse" in ophthalmology. Later, the

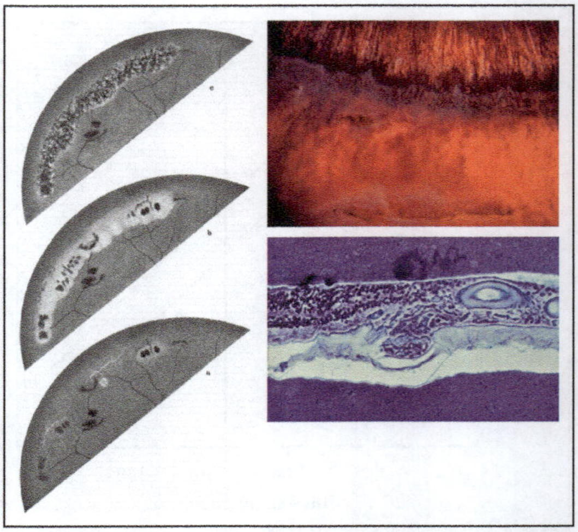

Fig. 7.7 Peripheral retinal lattice degeneration. *Left*: historical drawing (From Meyer-Schwickerath G (1954) Albrecht von Graefes Arch Ophthalmol, 156. With permission). *Right*: contemporary photo and histological picture (Courtesy of P. Meyer, University of Basel)

Fig. 7.8 Charles J. Campbell

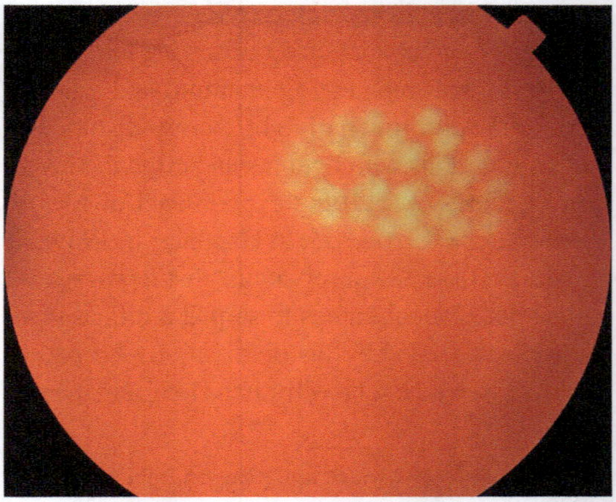

Fig. 7.9 Fresh laser impacts in the peripheral retina

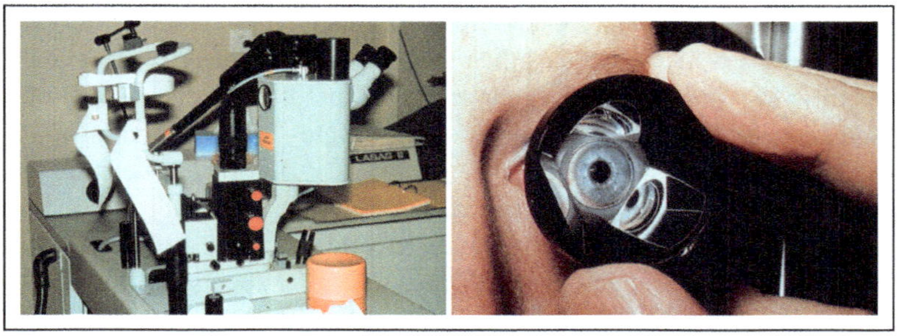

Fig. 7.10 Fankhauser and Lotmar first used a slit lamp-coupled laser and applied the laser beam through a Goldmann contact lens (Instrument: Lasag AG, Switzerland)

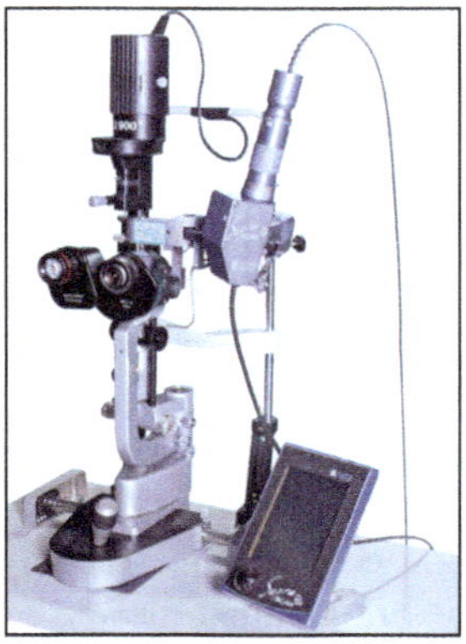

Fig. 7.11 Example of a contemporary slit lamp-based laser (Instrument: SupraScan, Quantel)

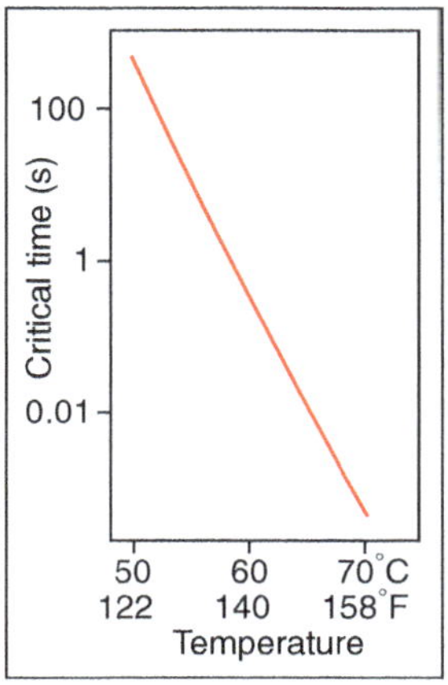

Fig. 7.12 The rapidity of the coagulation of the retina depends very strongly on the temperature. *Ordinate*: Time in which half the protein coagulates (logarithmic scale). Arrhenius curve, obtained with the parameters given by Welch AJ and Polhamus G (1984) BME, 31. Other parameters have also been published. The relation between critical time and temperature depends strongly on the parameters used

frequency-doubled[3] Nd:YAG laser became more and more popular and finally took its place (532 nm).

7.1.1 Biological Effects of Heating

The temperature increase depends on the radiation power, its duration, and the volume of the absorbing tissue. The reaction depends very strongly on the attained temperature and also on the duration of the heat. In the temperature range of 42–50 °C, hyperthermy occurs. At this temperature, the first relevant alterations of membranes and organic molecules begin. The irreversible denaturing of certain proteins in fever occur naturally, e.g., in fevers above 42 °C. At 50 °C, even vital enzymes are destroyed; irreversible damage occurs after a few minutes.[4] At 60 °C, the denaturing of proteins begins with the consequence of coagulation and cellular necrosis. Irreversible damage starts after a few seconds. Figure 7.12 illustrates the strong temperature dependence of these processes.

What is the coagulation of protein? Everyone knows how a hen's egg changes when heated. In their normal state, proteins – chains of amino acids folded into secondary and tertiary structures – are held together by weak bonds. During the heating process, the absorbed heat energy is sufficient to dissolve these bonds, i.e., to unfold

[3]The basic wavelength of the Nd:YAG laser is 1,064 nm. By non-linear processes, the frequency can be doubled; i.e., the wavelength can be halved to 532 nm.

[4]A measure of the effectiveness of a given temperature increase and its duration is given by the Arrhenius equation.

the molecule. Covalent bonds along the chain remain preserved. New bonds then arise between neighboring proteins and randomly structured water-insoluble shapes form with water inclusions. In this irreversible process, a gel forms.

However, coagulation is achieved not only by heat but also by cold. Cryocoagulation (treatment with cryoprobes) is mentioned in Chap. 16.

While it was assumed until now that the thermal destruction of cells was decisive for clinical results (e.g., in diabetic retinopathy), newer findings have indicated that biological effects also arise without cellular necrosis. The cells of the retinal pigment epithelium, when irradiated below threshold (i.e., without direct, visible effects), are assumed to release factors that inhibit neovascularization (so-called selective retina therapy, or SRT).

We shall not enter into a discussion of a further application, such as trabeculoplasty, that attempts to improve drainage of the aqueous humor by producing tiny thermal lesions in the anterior chamber angle.

7.1.2 Heating and Heat Diffusion

Local heating of the retina by irradiation with light is dictated by two processes: primarily the absorption of energy by the retinal pigment epithelium and the choroid and, secondarily, the diffusion of heat into the surroundings, especially into the retinal tissue. During the pulses, the temperature increases in the absorbing tissue and declines immediately after the irradiation is complete due to heat diffusion. For pulse durations below approximately 100 μs, the temperature increases practically without the relevant influence of diffusion. However, for pulse durations longer than 1 ms, the diffusion is already substantial during the irradiation process and thereby slows the temperature increase.

Heating up to approximately 60–80 °C denatures the irradiated spots – recognizable by the immediate blanching (loss of transparency) – and leads to cellular necrosis. The pigment is released and taken up by phagocytosing cells. This results in an inhomogeneous, speckled distribution of pigment in the scar. The typical values of parameters in a classical laser photocoagulation are: diameter of the irradiated area $d = 100$–400 μm, irradiation time $\tau = 100$–500 ms, and radiation power $N = 50$–500 mW. The preferred wavelengths are 500–600 nm.

The spatial and temporal course of the heating process depends on the absorbed energy E, the diameter d of the irradiated spot, and the irradiation duration τ. Roughly half of the irradiation energy is absorbed in the retinal pigment epithelium (RPE) and heats it up. The largest temperature increase results in the center of the irradiated spot in the pigment epithelium. We now discuss the basic dependencies in a simplified model that only takes the absorption in the RPE into consideration. It does not account for the minor absorption of light in the retina and absorption in the choroid or the cooling due to the blood flow in the choroid with irradiations of longer durations.

Figures 7.13 and 7.14 represent the temporal course of the temperature in the absorption center for various values of d and τ. The temperature in the absorption center increases only marginally with extended irradiation time and constant beam power due to the damping influence of the heat diffusion (Fig. 7.13). In contrast, the temperature increases considerably stronger when the same power is concentrated on a smaller spot (Fig. 7.14). After the end of the irradiation, the temperature in the absorbing region declines rapidly. For given values of d and τ, temperature increases are proportional to the power of the laser beam as long as no other processes are involved, such as evaporation or a change in the color of the absorbing tissue.

To what extent does the heat penetrate into the retina? Figure 7.15 shows temperature profiles

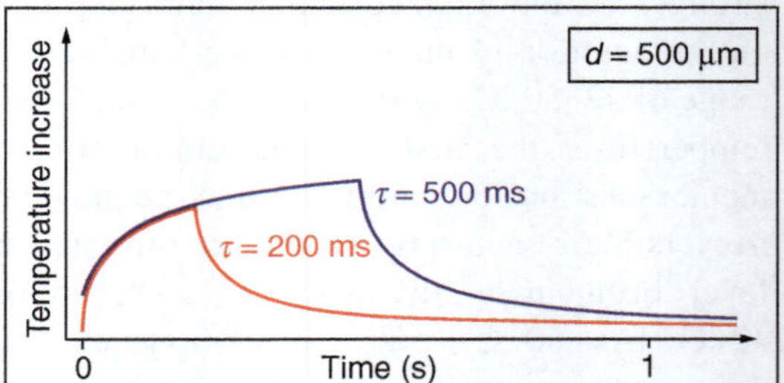

Fig. 7.13 Temperature of the RPE. A longer exposition time increases temperature only marginally. Beam diameter $d = 500$ μm. A constant irradiance (power per irradiated area) is applied

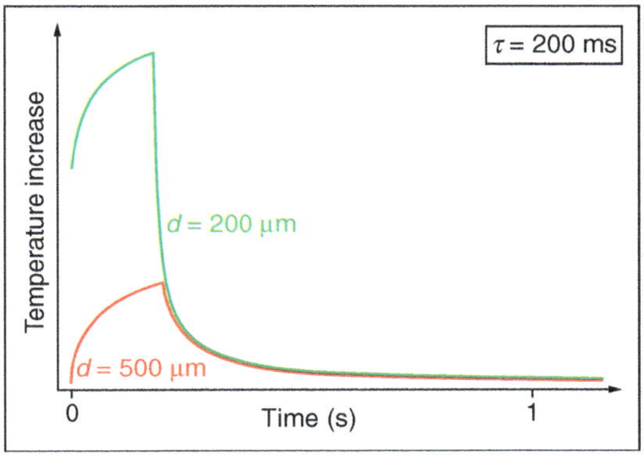

Fig. 7.14 Temperature of the RPE. The same beam power is applied to different spot sizes. The smaller the diameter of the irradiated spot is, the larger the peak temperature will be because the same energy is concentrated in a smaller absorbing volume. Exposition time $\tau = 200$ ms. Immediately following the irradiation, the temperature decreases very rapidly

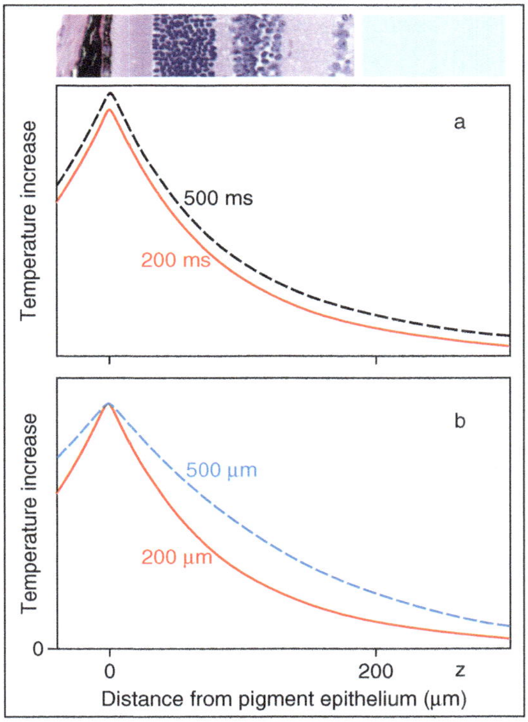

Fig. 7.15 Laser-induced increase of temperature across the retina. The maximum temperature is in the PE. *Abscissa*: Perpendicular distance z from the retinal pigment epithelium. *Ordinate*: Temperature increase at the end of the pulse. (**a**) Extending the pulse duration for the same irradiance changes the temperature profile only marginally. Focus diameter 200 µm. (**b**) Temperature profile for various focus diameters. Pulse duration 200 ms. Pulse power adjusted for the same peak temperature

along the z axis perpendicular to the retinal pigment epithelium at the end of irradiation. At this time, the temperature reaches its maximum.[5] A lengthening of the pulse duration with constant irradiance increases the temperatures only slightly[6] (Fig. 7.15a). The comparison of the temperature profiles with various focus diameters is somewhat more difficult. As we have seen, the same beam power results in a lower increase in temperature for a larger focal spot size (Fig. 7.14). The temperature profile that results when irradiance is balanced to give the same peak temperature is displayed in Fig. 7.15b. Under this condition, larger beam diameters lead to deeper and broader effects in the retina.

7.2 Photodisruption

If the focus of a laser pulse lies in air or water, a spectacular event can occur: a so-called optical breakdown, recognizable by tiny sparks and audible snaps. Surprisingly, the process occurs even though the medium is transparent to the laser light. What is an optical breakdown? An optical breakdown happens when the irradiance exceeds a value on the order of 10^{10} W/cm^2. With a pulse duration of 1 ns, an energy of 0.1 mJ, and a focus diameter of 30 µm, e.g., this value is reached. At this point, the oscillating electrical field of the light is sufficiently strong to accelerate free electrons to an extent that, through their collision with molecules, further electrons are released, which further increases the absorption of light. An avalanche-type process begins. Finally, the large density of free electrons blocks the light completely. During the short pulse, plasma (cloud of electrons and ions) with a temperature of several thousand K is created. This produces great pressure that displaces the surrounding; thus, a bubble of approximately 1 mm in diameter is engendered (cavitation). The hot plasma expands and thereby cools down

[5] Except at larger distances from the PE, where the temperature increase is delayed.

[6] For the pulse durations considered, the equilibrium temperature is nearly reached.

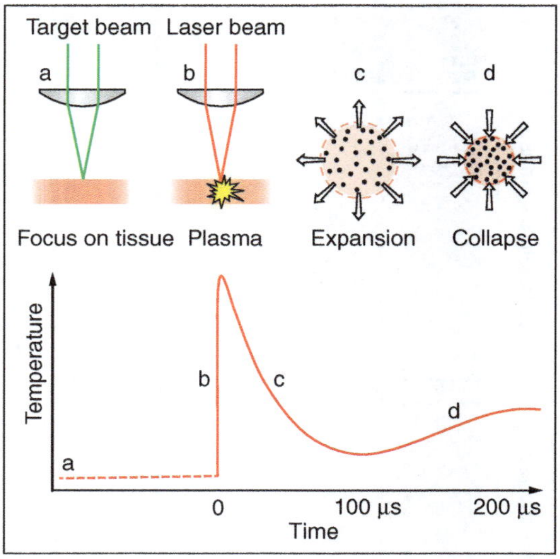

Fig. 7.16 Optical breakdown due to high irradiance in the focus of a laser beam. (*a*) Focusing on the tissue. (*b*) A nanosecond pulse with very high irradiance creates a high-temperature plasma ($\approx 10^4$ K). (*c*) Very hot plasma/gas drives the fluid away radially. A cavitation arises in a very short time. (*d*) Bubble implodes; the shockwaves are not shown

Fig. 7.17 Danièle Aron-Rosa

Fig. 7.18 Franz Fankhauser

without giving off heat to the surrounding tissue. The pressure collapses and the bubble implodes within a time of 100 µs, whereby the temperature again rises but to a lesser extent. We would like to reemphasize that the optical breakdown occurs independently of the absorption characteristics of the tissue and only when a certain threshold of irradiance is reached. One can easily imagine that this sort of event and its vehemence – mainly due to the local movements caused by the cavitation – can tear up the tissue (Fig. 7.16).

Around 1980, Danièle Aron-Rosa and Franz Fankhauser, independently and almost simultaneously, were able to disrupt the clouded capsule of an after-cataract with short pulses of a high-power Nd:YAG laser, inducing optical breakdown. This solid-state laser emits light in the near infrared (1,064 nm). The Q-switched method introduced by Fankhauser for producing short, powerful pulses is, meanwhile, the gold standard (Figs. 7.17 and 7.18). In a laser's Q-switched mode, the light energy is built up and stored inside the laser medium until it is suddenly released by an electro-optic switch. Besides photodisruption of the capsule (Fig. 7.19), iridotomies (Fig. 7.20) represent a further application.

By selecting the focus location, the operator can place the destructive event at any place within the eye. However, the critical irradiance threshold can be reached before the focus. Therefore, the focus is set slightly behind the capsule to avoid damage to the artificial lens. To simplify this procedure for the physician, instruments allow a slight separation of the foci of the aiming beam and of the therapeutic laser beam. This allows focusing of the aiming beam on the capsule.

During the duration of the pulse, the plasma can also migrate back into the light cone, especially for small cone angles, which means that the optical breakdown can occur even further ahead of the aiming focus. For this reason, the opening through which the beam enters the eye should be as large as possible. Positive contact lenses guarantee a wide cone angle of the beams within the eye.

The effects on the tissue are due mainly to the expanding bubble that can tear up directly impacted tissue. In addition, there are shockwaves: rapidly moving wave fronts with strong pressure differences on both sides. The retina is

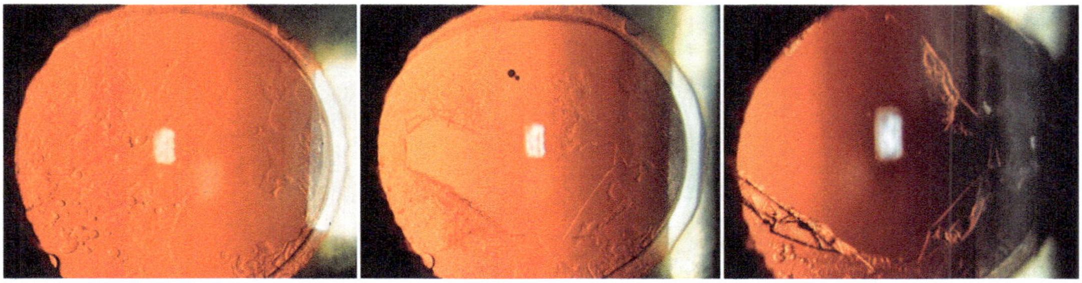

Fig. 7.19 Photodisruption of after-cataract. *Left*: before treatment. *Middle*: after the first few impacts. *Right*: after several impacts

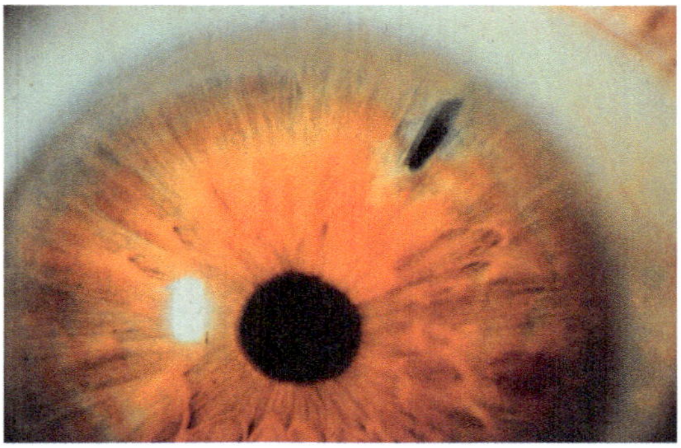

Fig. 7.20 Iris after peripheral iridotomy

protected in two ways: on the one hand, the retina is not in focus, and on the other hand, in the case of a breakdown, the energy is absorbed. Typical parameters in photodisruption of the capsule are a wavelength of 1,064 nm, a pulse duration of a few ns, a focus diameter on the order of 10 μm, a cone angle 15°, and pulse energy on the order of 1 mJ.

7.3 Photoablation

Due to its curvature, the exterior surface of the cornea contributes approximately three-quarters of the eye's total focusing power. Refractive errors can be corrected by relatively small alterations of the corneal curvature. Photorefractive keratectomies (PRK) are procedures in which stromal tissue is removed directly from the surface. During the LASIK[7] procedure, an area inside the cornea is exposed by folding back a corneal lamella; the removal of tissue takes place at this surface (Fig. 7.21).

For photoablation, lasers with wavelengths in the ultraviolet range are used, particularly the ArF excimer laser with a wavelength of 193 nm (Sect. 3.4.4). The special suitability of this short wavelength light lies in the high energies of the photons. At this wavelength, they amount to 6.4 eV, i.e., roughly twice to three times the energies of visible photons. With their high energies, the photons of these lasers can disrupt the chemical bonds typically found in tissues such as C–C bonds (binding energy 3.6 eV) or O–H bonds (4.8 eV). When a 6.4 eV photon disrupts a 4.8 eV bond, the remaining energy is transferred to the two fragments as kinetic energies initiate them to fly apart at high speed. Due to the very strong absorption, these processes take place immediately as the beam penetrates into the tissue, practically at the affected surface. At the same time, electrons are also set free (ionization). This mix of ions and electrons – plasma – is heated

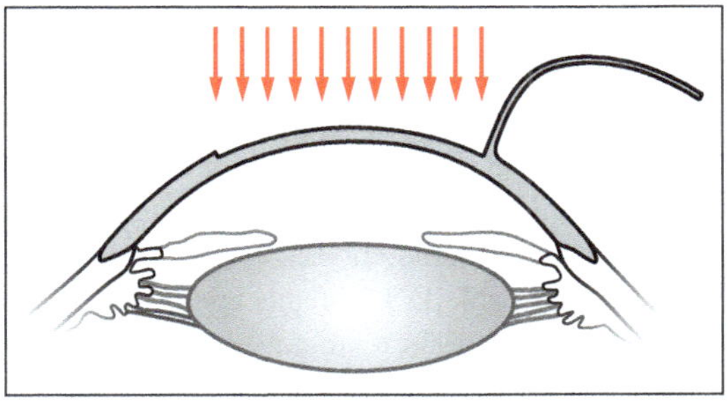

Fig. 7.21 LASIK procedure. Exposure of an inner stromal area. A corresponding flap can be cut mechanically with a microkeratome or with a femtolaser

[7]Laser In Situ Keratomileusis (Laser-Assisted In Situ Keratomileusis).

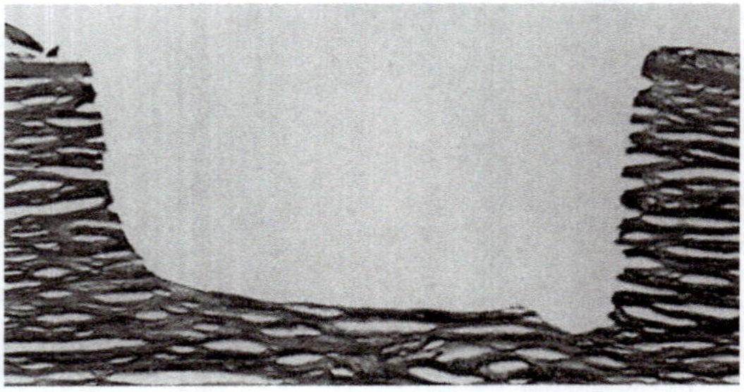

Fig. 7.22 Photoablation of the corneal tissue achieved with an ArF excimer laser (pulse duration: 14 ns; energy density: 180 mJ/cm^2) (Courtesy of T. Bende, Institute for Ophthalmic Research, Tübingen)

through the absorption of light, which is a second reason that the molecule fragments are blasted away. Depending on the energy of the pulse, the corneal tissue is removed to a precise depth of 0.1 to 1 μm, with almost no significant heating of the neighboring tissue. Figure 7.22 illustrates the precision of the operation. This process represents the basis for kerato-refractive interventions.

In contrast to photodisruption (Sect. 7.2), in this case, the treated tissue is impermeable to the light that is used. Strong absorption occurs for every level of beam irradiance. An optical breakdown, as described in Sect. 7.2, is not required and does not occur. The pulse duration amounts to approximately 10 ns. For a fluence of 1 J/cm^2, tissue removal of approximately 1 μm in depth takes place. Up to this energy density, the depth of the removal is proportional to the pulse energy. At higher levels, the resulting plasma blocks the deeper penetration of the light. The cross-section of the beam that is guided to the surface being operated on amounts to 1 mm^2 or more.

Why use the wavelength below 200 nm if photon energies of 240 or even 300 nm light would also be sufficient to break the bonds? Interestingly, it turned out that the higher-energy photons from the 193 nm light produce less heat damage. For this reason, the wavelength of the ArF excimer laser represents a favorable choice. However, as a side effect, unavoidable DNA breaks arise. Whether or not their repair leads to mutations that could be potentially carcinogenic is not yet known.

The removal of corneal tissue with the excimer laser was described in 1983 by Trokel, Srinivasan, and Braren. Their ideas were based on experiences with cow eyes. They explained the photochemical processes and discovered that the procedure left no traces in the surrounding tissues. Seiler is also among those who pioneered applications on human corneas beginning in the mid-1980s and developed PRK (Figs. 7.23 and 7.24).

Fig. 7.23 Stephen Trokel

Fig. 7.24 Theo Seiler

7.4 Cutting with the Femtosecond Laser

The successful femtosecond pulse[8] that has its application in detaching corneal lamellae in preparation for a refractive surgery is also based on optical breakdowns at high energy densities at the focal point. However, it uses considerably

[8] 1 femtosecond = 1 fs = 10^{-15} s. Typical pulse durations in this procedure are on the order of 100 fs.

different values of pulse duration and pulse energies compared to applications with nanosecond pulses (Sect. 7.2). The main difference lies in the fact that the threshold energy for the optical breakdown declines with decreasing pulse durations. A reduction in the pulse duration from nanoseconds to picoseconds lowers the threshold energy at a given spot diameter by roughly a factor of 10 to 20. With sufficiently short pulses, it is possible to trigger the phenomenon of the optical breakdown in transparent media – e.g., in the cornea – with far lower energies to keep the effect within well-dosed limits. Thus, cutting the corneal lamellae is possible by applying a large number of pulses focused on the intended plane. Each focal point creates a tiny bubble and these lie close together, permitting a lifting of the lamella with almost no disturbing tissue bridges (Fig. 7.25).

The typical parameters used to create an optical breakdown are as follows (orders of magnitude): spot size of 1–2 μm, pulse duration of 100 fs, and pulse energy of 0.1 μJ. The pressure of the plasma briefly displaces the surrounding tissue. Due to the cooling caused by its expansion, the bubble collapses rapidly, whereas a small volume filled with a gas such as O_2 and H_2O remains and is not immediately reabsorbed. This residual cavitation exists for minutes. At its maximum extent, the radius of the bubble amounts to approximately 25 μm; the diameter of the residual cavitation is on the order of 1 μm. For example, to detach a surface of 8 × 8 mm^2, a nearly continuous carpet consisting of millions of residual cavitations must be set.

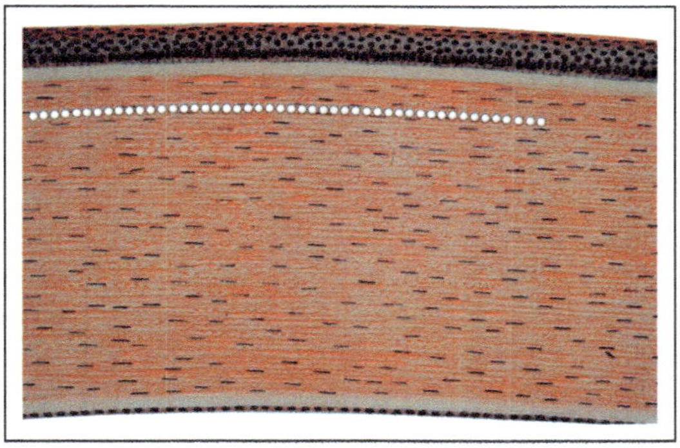

Fig. 7.25 Schematic drawing of the impacts of femtolaser pulses. Optical breakdowns create numerous tiny bubbles with diameters of a few μm. They form a two-dimensional carpet of adjacent vesicles. In reality, the bubbles are smaller than drawn here

With femtosecond pulses, the triggering mechanism of the optical breakdown is not the same as in nanosecond pulses (Sect. 7.2). With extremely short pulses, the energy per volume and, thus, the spatial density of photons is so large that several photons can contribute their energies simultaneously for the ionization of a single molecule (multiphoton absorption). The high irradiance that is attained only with extremely brief pulses is crucial. A femtosecond laser works at a wavelength of approximately 1,040 nm, corresponding to a photon energy of somewhat more than 1 eV. The energies of a few photons together are sufficient for typical ionization processes.

8 Some History of Chemistry

Related sciences such as physics, chemistry, and biology merge seamlessly into each other, which makes it difficult for us to clearly distinguish among these sciences. The terms "biophysics," "physical chemistry," or "biochemistry" imply the flowing connectivity between scientific fields. The topics in this book are, therefore, assigned to the various scientific chapters somewhat arbitrarily.

What is the link between ophthalmology and chemistry? Chemistry is the basis of biology, which, in turn, provides information about the function of the eye. Chemistry is the science of the composition, structure, properties, and reactions of matter. We shall begin by describing some of the first steps toward modern chemistry.

8.1 First Steps Toward Modern Chemistry

The "father of modern chemistry" was the French chemist Lavoisier,[1] the son of a prominent advocate, born to a wealthy family in Paris. At Lavoisier's time, chemistry was so underdeveloped it could hardly be called a science. The main view of combustion, or burning, was that of the "Phlogiston Theory," which stated that certain materials contain a fire-like element called "Phlogiston," which was liberated by burning; conversely, when those materials were heated, the "phlogiston" entered the material. One major problem with this theory was that, when some metals such as magnesium (which were considered to be rich in phlogiston) were oxidized, the resulting oxidized metal was heavier than the initial metal even though it was supposed to have lost weight. Lavoisier disproved the phlogiston theory by showing that combustion required a gas, oxygen, which had a weight. In a paper titled "Memoir on Combustion in General," he presented his theory that combustion was a reaction in which oxygen combines with other elements. A simple example is the combustion reaction between hydrogen and oxygen (Fig. 8.1).

Lavoisier also discovered that, in a chemical reaction, matter is neither created nor destroyed, known as the "law of conservation of matter" (the mass of the reactants equals the mass of the products). For the first time, he formulated chemical reactions in the form of chemical equations based on the conservation of mass. Lavoisier was among the first to have a clear concept of a chemical element and the first to list the known elements. He was also the first to develop a rational system for naming chemical compounds. For these reasons, he is known as the father of modern chemistry.

$$2H_{2(g)} + O_{2(g)} \longrightarrow 2H_2O_{(l)} + \text{heat}$$

Fig. 8.1 Combustion reaction. Hydrogen reacts with oxygen in a combustion reaction to produce water and heat

[1] Antoine Lavoisier (1743–1794), French chemist who disproved the "Phlogiston theory."

J. Flammer et al., *Basic Sciences in Ophthalmology*,
DOI 10.1007/978-3-642-32261-7_8, © Springer-Verlag Berlin Heidelberg 2013

Due to his prominence in the pre-revolutionary government in France, this famous scientist was guillotined during the revolution. Today, statues of Lavoisier can be found in the city hall or the Louvre in Paris (Fig. 8.2).

Chemistry is the study of matter and its changes and interactions. Matter is anything that has mass and takes up space. All matter is composed of atoms. An element is defined as matter consisting of atoms that cannot be broken down by further chemical means. Elements can be arranged according to their atomic numbers in a tabular display organized on the basis of their properties – the "periodic table" as depicted in Fig. 8.3. The atomic number corresponds to the number of protons in the nucleus, which, in turn, corresponds to the number of electrons in the non-ionized state.

Isotopes are elements with the same atomic number and, therefore, lie in the same location in the periodic table but with a different number of neutrons and, thus, a different atomic mass. Figure 8.4 shows three isotopes of hydrogen (according to their abundance in nature: protium with 1 proton and 1 electron, deuterium, and tritium).

Can elements be transformed into other elements? This was a common perception of

Fig. 8.2 Statue of Antoine Lavoisier in the Louvre in Paris

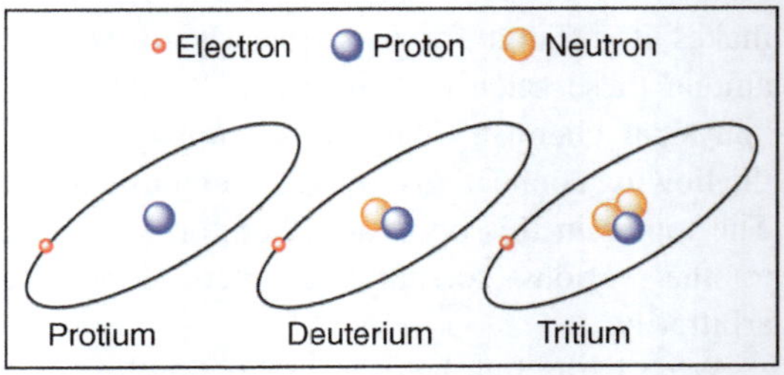

Fig. 8.4 Isotopes of hydrogen, which has three naturally occurring isotopes. The most common isotope is protium, which consists of one proton and one electron. Deuterium contains one proton and one neutron in its nucleus. Tritium contains one proton and two neutrons in its nucleus

H																	He
Li	Be											B	C	N	O	F	Ne
Na	Mg											Al	Si	P	S	Cl	Ar
K	Ca	Sc	Ti	V	Cr	Mn	Fe	Co	Ni	Cu	Zn	Ga	Ge	As	Se	Br	Kr
Rb	Sr	Y	Zr	Nb	Mo	Tc	Ru	Rh	Pd	Ag	Cd	In	Sn	Sb	Te	I	Xe
Cs	Ba	La-Lu	Hf	Ta	W	Re	Os	Ir	Pt	Au	Hg	Tl	Pb	Bi	Po	At	Rn
Fr	Ra	Ac-Lr	Rf	Db	Sg	Bh	Hs	Mt	Ds	Rg	Cn						

Fig. 8.3 The periodic table. The colors indicate groups of elements in following manner: *Lighter green*: Alkali metals. *Orange*: Alkali earth metals. *Yellow*: Transition metals. *Green*: Lanthanides and actinides. *Violet*: Other metals. *Pink*: Metalloids. *Grey*: Non-metals. *Light blue*: Halogens. *Dark blue*: Noble gases

Fig. 8.5 Oil painting of an alchemist by Josef Wright of Derby in 1771

Fig. 8.6 Antoine Becquerel and Marie Curie

$$^{14}_{7}N + ^{4}_{2}\alpha \longrightarrow ^{17}_{8}O + ^{1}_{1}H$$

Fig. 8.7 Transmutation. Nitrogen exposed to alpha radiation can change into oxygen

alchemists who believed in converting inexpensive metals, such as iron, into the more valuable gold and silver. The transformation of elements was believed to be achieved by using the "Philosopher's Stone." This stone apparently had a component that was supposedly capable of turning base metals, such as lead, into gold. This stone was also thought to contain a magical component that cured diseases and made humans younger. Figure 8.5 shows an alchemist searching in vain for the secret of transforming base metals into gold.

The spontaneous transformation of one element into another is known as radioactive decay. This happens by changing the number of protons of an atom in the nucleus. In the nineteenth century, Becquerel[2] and later Curie[3] (Fig. 8.6) discovered that certain atoms have radioactive properties.

Transmutation, or the change of one element into another, involves a change in the nucleus of an atom and is, therefore, a nuclear reaction. When the number of protons in an atom is changed, the atom is transmuted into an atom of another element. Transmutation may either occur spontaneously or be induced. A few years after the discovery of Curie, Rutherford,[4] in 1919, showed that nitrogen exposed to alpha radiation changed into oxygen (Fig. 8.7).

8.2 The Birth of Elements

But how did elements arise in the first place? The earliest phases of the "birth" of our universe, "the Big Bang," are subject to much speculation. Before the Big Bang theory, the universe was believed to be essentially eternal and unchanging. One of the first indications that the universe might change as time passes came in 1917 when Einstein[5] (Fig. 8.8) developed his general theory of relativity. From his equations, it was realized that the universe could either be expanding or contracting. Nevertheless, Einstein tried to stick

[2]Antoine Henri Becquerel (1852–1908).

[3]Marie Skłodowska-Curie (1867–1934).

Antoine Becquerel first discovered that uranium had radioactive properties. Marie Curie and her husband Pierre Curie later discovered that the elements polonium and radium also had radioactive properties. The Nobel Prize for physics 1903 was divided with one half awarded to Antoine Becquerel the other half jointly to Marie Curie and her husband Pierre Curie.

[4]Ernest Rutherford (1871–1937) showed that nitrogen exposed to alpha radiation changed into oxygen.

[5]Albert Einstein (1879–1955) received the Nobel Prize for physics for the discovery of the law of the photoelectric effect.

Fig. 8.8 Albert Einstein

to static solutions. In fact, only in the late 1920s, the expansion of the universe was observed by the astronomer Edwin Hubble.

The Big Bang created the universe. The standard model of the Big Bang theory proposes that the universe was once an extremely hot and dense state that expanded rapidly.

Within the famous first two minutes, neutrons, protons, electrons, and some light nuclei such as helium, lithium, and beryllium were created as particles in very hot plasma. This is the "big bang nucleosynthesis." When the free electrons recombined with the nuclei, the light neutral atoms were formed, in the first place hydrogen and helium. Today, hydrogen is estimated to make up more than 90% of all the atoms or three-quarters of the mass of the universe. However, most elements were formed during fusion processes in stars. This is true up to iron. Everything heavier was created during supernova explosions.

In the next chapter, we shall describe some of the important elements and molecules and their chemical properties with particular relevance to ophthalmology.

9 Oxygen

9.1 The Oxygen Atom

In the universe, oxygen is the third most abundant element after hydrogen and helium. Oxygen is synthesized at the end of the life of massive stars, when helium is fused into carbon, nitrogen, and oxygen nuclei. Stars burn out, explode, and expel the heavier elements into interstellar space. Later, oxygen plays a crucial role in the emergence of life. Oxygen is not always reactive to the same extent. The oxygen atom (O) is more reactive than the oxygen molecule (O_2). To understand this, we will review some of the basics of oxygen. The oxygen atom is depicted in Fig. 9.1.

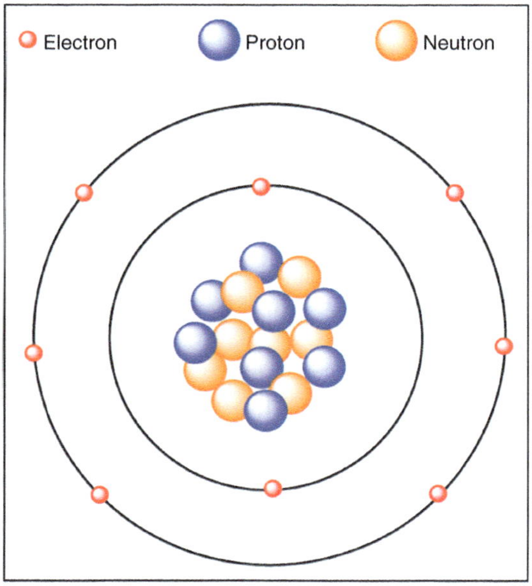

Fig. 9.1 Oxygen atom, which has eight protons, eight neutrons, and eight electrons

The eight electrons of the oxygen atom fill the "s" and "p" orbitals. The names "s" and "p" indicate the orbital shape and are used to describe the electron configurations. S orbitals are spherically symmetric around the nucleus, whereas p orbitals are rather like two identical balloons tied at the nucleus. The electron configuration for the oxygen atom reads as follows: $1s^2\ 2s^2\ 2p^4$. There are two electrons in the first shell and six in the second (Fig. 9.2). In the second shell, two electrons occupy an s-type orbital and four occupy p-type orbitals. Given that a p-type orbital has a capacity of six electrons, the oxygen atom falls short by

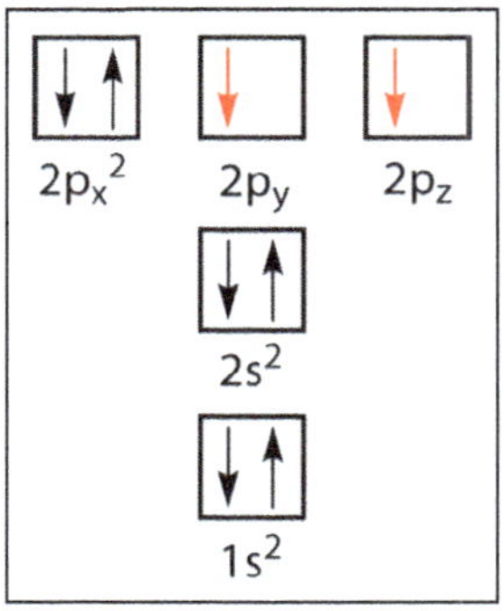

Fig. 9.2 Electron configuration of the oxygen atom. Two electrons occupy the first shell of the oxygen atom and six electrons occupy the second shell (electron configuration: $1s^2\ 2s^2\ 2p^4$)

J. Flammer et al., *Basic Sciences in Ophthalmology*,
DOI 10.1007/978-3-642-32261-7_9,

two electrons of its "wanting" to fill its outermost shell to its full natural capacity.

This explains the high reactivity of the oxygen atom. Oxygen is, after fluorine, the most electronegative element. The electronegativity of an element describes its "electron hunger." Atoms or molecules with an unpaired electron in their outer shell are called free radicals. The oxygen atom is a free radical. Since the two 2p orbitals (each containing a lone electron) are not full, the oxygen atom tries to become stable by reacting with other atoms and trying to add the electron of the other atom to its own shell. This makes the oxygen atom highly reactive. In nature, an oxygen atom typically steals away an electron from one or two other atoms to form a molecule, such as water (H_2O). To form an oxygen molecule, each oxygen atom donates two electrons to the other oxygen atom. In the case of the formation of water, each hydrogen atom "donates" one electron to the oxygen (Fig. 9.3).

This is an example of a redox reaction where hydrogen atom is oxidized ("loses electrons") and oxygen is reduced ("gains electrons"). This electron transfer sets energy free; in other words, it releases heat, which is why this reaction is called exothermic. We can, therefore, also say that water is formed when hydrogen is "burned" by oxygen.

Similar to the oxygen atom, molecular oxygen also has two unpaired electrons in its last orbital that have the same spin (Fig. 9.4). Interestingly, however, the oxygen molecule, although a biradical, is only minimally reactive because the unpaired electrons in the oxygen molecule have the same spin. Thus, for the oxygen molecule to be able to react, it would need another molecule or atom with two unpaired electrons with a parallel spin opposite to that of the oxygen molecule. The latter will only rarely occur. Hence, the oxygen molecule is minimally reactive (spin restriction).

As oxygen is almost always needed in biological energy metabolism, we shall discuss the role of oxygen in more detail.

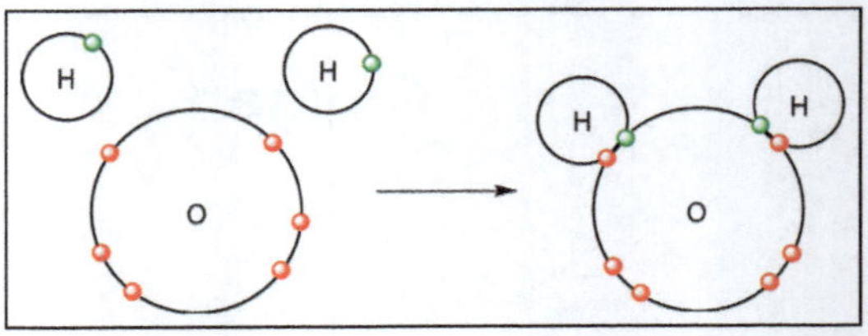

Fig. 9.3 Example of a redox reaction. When hydrogen and oxygen bind, the electron of the hydrogen is "donated" to the oxygen. Hydrogen is, therefore, oxidized, whereas oxygen is reduced

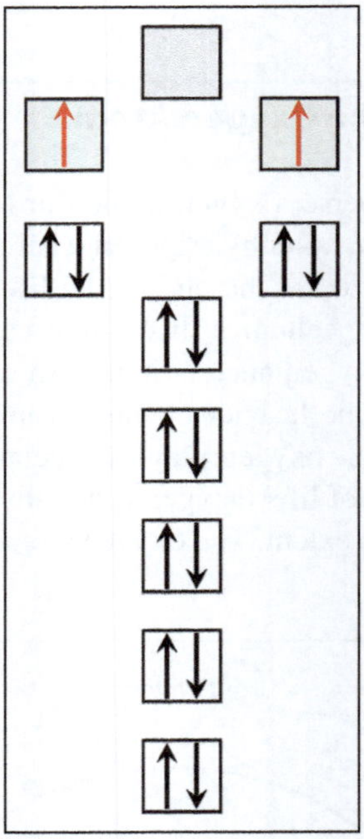

Fig. 9.4 Molecular oxygen, which also has two unpaired electrons in its last orbital that have the same spin direction. To bind with another molecule, it would need another molecule or atom with two unpaired electrons with a parallel spin opposite to that of the oxygen molecule. This will rarely occur, explaining why molecular oxygen is minimally reactive

9.2 Oxygen and Energy Production

Oxygen is one of the most important molecules required for human life. One of the key ways for a cell to gain useful energy is by aerobic respiration (in other words, by oxidizing high-energy

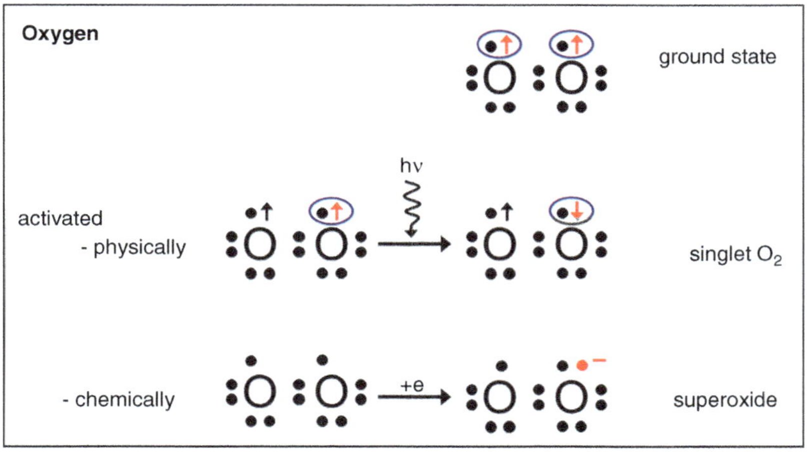

Fig. 9.5 Activation of ground-state oxygen, which can be either activated physically (direct excitation by light) to form the singlet oxygen or it can be activated chemically by a reduction reaction (one electron reduction of dioxygen) to form the superoxide anion. (From Mozaffarieh M, Flammer J (2009) *Ocular blood flow and glaucomatous optic neuropathy*. Springer Publ, Berlin. With permission)

molecules). In environments, however, where oxygen is not sufficiently available, both primitive organisms and cells in our body can metabolize glucose by fermentation without using oxygen. This process also yields energy, though in lower amounts, and produces byproducts such as lactic acid.

9.3 Biochemical Reactions of Oxygen

Combustible material (e.g., sugar or wood) can be found anywhere on earth. Fortunately, most atmospheric oxygen (molecular oxygen) is inert, meaning that it reacts very sluggishly with other molecules. The oxygen in the air we breathe is normally in its "ground" (not energetically excited) state. However, for use in mitochondria (see Sect. 14.3), it does not need to be activated since mitochondria have special machinery that donates electrons to the oxygen molecule. If, however, oxygen reacts outside the oxidative phosphorylation pathway, it needs to be activated. This activation of oxygen may occur by two different mechanisms: either through the absorption of sufficient energy to reverse the spin on one of the unpaired electrons or through monovalent reduction (Fig. 9.5).

Thus, if ground-state oxygen absorbs sufficient energy to reverse the spin of one of its unpaired electrons, the two unpaired electrons now have opposite spins. In this more reactive form of oxygen, namely singlet oxygen ($^1O_2$), one of these unpaired electrons has a changed spin direction (Fig. 9.6).

Singlet oxygen, being much more reactive than ground-state oxygen, reacts destructively with molecules with double bonds.

The second mechanism of activation is by the stepwise monovalent reduction of oxygen that gives rise to superoxide anion radical ($O_2^{\bullet-}$), hydrogen peroxide (H_2O_2), and, finally, water (H_2O). The situation is similar to that of a burning fire. To make a fire, we first need to transfer heat to a combustible material because the fire contains a cloud of electrons that facilitates the transfer of electrons to oxygen.

During combustion (burning), electrons are transferred from, for example, hydrogen or carbon to oxygen, resulting in the final product of water or carbon dioxide, respectively. We inhale oxygen and exhale carbon dioxide (the carbon comes from sugar or fat). To close the circuit and reach the "steady state" of these molecules in the atmosphere, oxygen must be regenerated by plants with the help of photosynthesis.

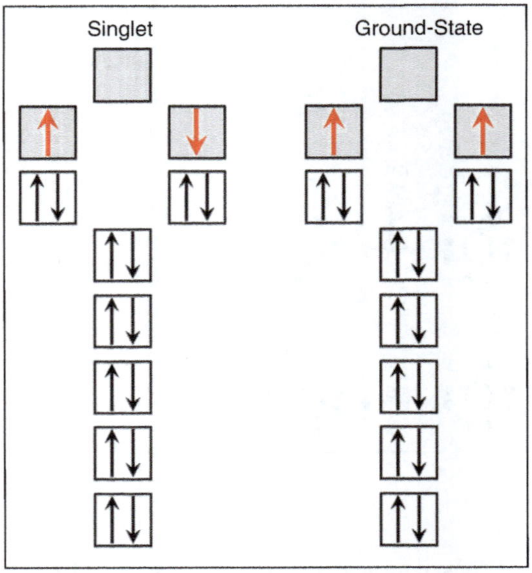

Fig. 9.6 Singlet oxygen. If ground-state oxygen absorbs sufficient energy to reverse the spin of one of its unpaired electrons, the two unpaired electrons now have opposite spins. This more reactive form of oxygen is called singlet oxygen ($^1O_2$)

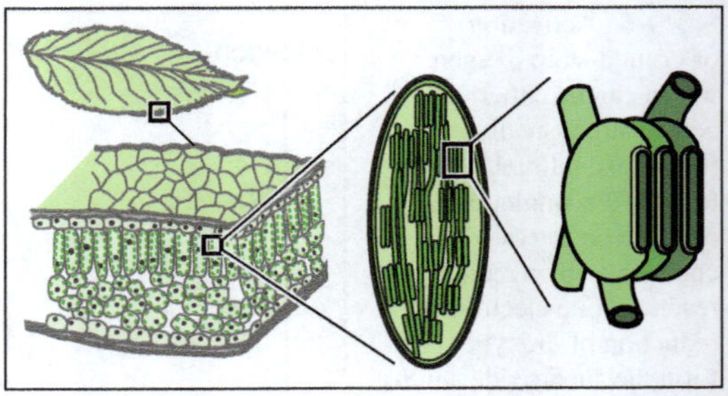

Fig. 9.7 Structure of chloroplast. Green plants are green because they contain the pigment chlorophyll, found in the thylakoid sacs of the chloroplast. *Left*: Cross-section of a leaf. *Middle*: Thylakoids within the stroma of the chloroplast. *Right*: Thylakoids are arranged in stacks called grana

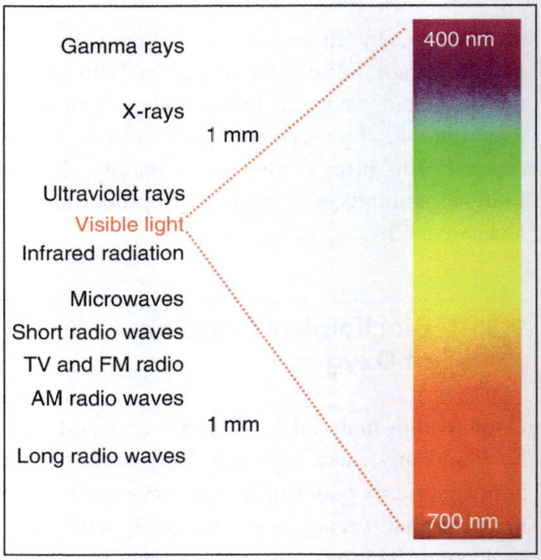

Fig. 9.8 Visible light. From among the broad spectrum of electromagnetic waves, only a small portion is perceived as light

The term "photosynthesis" simply implies "synthesis" with the necessary energy coming from light ("photo"). The energy from sunlight is absorbed by green plants with the aid of the green pigment (chlorophyll) (Fig. 9.7).

Light absorbed by the chlorophyll is in the visible spectrum, which is a small part of the electromagnetic spectrum, as shown in Fig. 9.8.

The longer the wavelength of visible light is the more red the color will be. Likewise, the shorter wavelengths are toward the violet side of the spectrum. Since not all wavelengths of visible light are absorbed by chlorophyll, the leaves of plants appear green rather than black. Chlorophyll plant pigments are capable of absorbing light. Their light absorption spectrum is shown in the illustration in Fig. 9.9.

To regenerate oxygen, the chlorophyll molecules in plants need to take away an electron from the oxygen atom bound within a molecule, which in the case of photosynthesis is water. Prior to this, however, the chlorophyll molecule itself needs to lose an electron, as only the oxidized chlorophyll can take an electron away from oxygen. Chlorophyll is oxidized with the help of sunlight energy. However, one photon of light does not give the chlorophyll molecule enough energy to lose an electron. For this reason, the energy of several photons is summed up in the so-called "antenna" of chlorophyll molecules (Fig. 9.10).

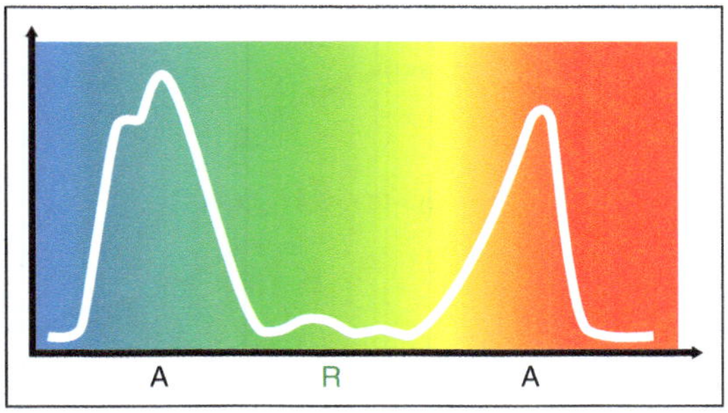

Fig. 9.9 Chlorophyll light absorption. Chlorophyll pigments absorb light in the blue (short-wavelength) and the red (long-wavelength) regions of the visible light spectrum, as marked by *A*. The wavelengths of light that are not absorbed include the remaining green-yellow colors, between 500 and 600 nm (marked by *R*), which explains why plants appear green. The wavelengths of light that are not absorbed include the remaining green-yellow colors, between 500 and 600 nm

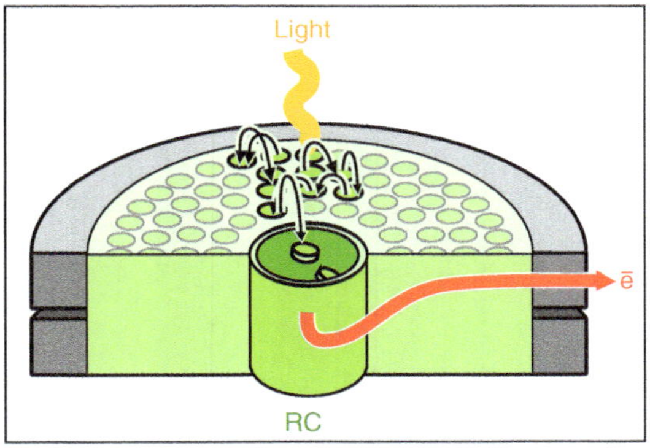

Fig. 9.10 Accumulation of energy by chlorophyll. The different chlorophyll molecules are arranged to form a light-harvesting complex also known as the "antenna" of chlorophyll molecules

This energy is then transferred from molecules absorbing short wavelengths of light to those absorbing longer wavelengths of light. In other words, this energy is transferred to a single "burning point" (reaction center, RC). This accumulated energy is enough to transfer an electron to another molecule (Fig. 9.11).

The chlorophyll, having lost an electron, now has great "electron hunger." In fact, it hungers for an electron so much that it readily scavenges an electron from the oxygen atom in the molecule

Fig. 9.11 Oxidation of chlorophyll. The light energy absorbed by the individual molecules is transferred from one molecule to another and finally concentrated in the reaction center (*RC*), rendering enough energy to liberate an electron

(H_2O in the case of photosynthesis). This reaction closes the circuit of energy transfer: plants take an electron away from the oxygen molecule and, in the process of energy metabolism, oxygen receives this electron back. Thus, the concentration of oxygen and carbon dioxide would theoretically remain constant in the atmosphere (Fig. 9.12).

Today's massive combustion of oils and gases (fossil fuels), however, disturbs this balance. Far more carbon dioxide is produced by burning than is used by plants, with the net effect that the concentration of carbon dioxide in the atmosphere is increasing. Sunlight reaches the earth and a part of it is reflected and scattered back to the universe, while another part is transferred into heat. The emission of heat (longer wavelength) is reduced by carbon dioxide. Therefore, an increased concentration of carbon dioxide leads to so-called global warming (Fig. 9.13).

It is important for the cells in our body to have a permanent supply of oxygen because oxygen, unlike sugar or carbohydrates, cannot be stored. The high-energy electrons coming from the sugar and fatty acids via NADH (nicotinamid-adenine dinucleotide) and $FADH_2$ (flavin-adenine dinucleotide) running from one complex to the other within the respiratory chain deliver the energy to build up the proton gradient (Fig. 9.14). The final recipient of the electron from the respiratory chain is oxygen. This is why oxygen is a prerequisite for aerobic respiration.

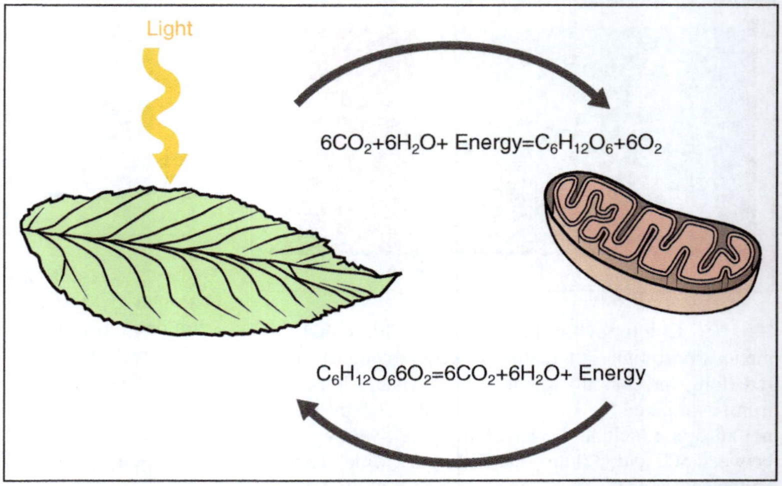

Fig. 9.12 Photosynthesis and respiration create a balance. Photosynthesis uses energy from sunlight to liberate oxygen (O_2), whereas respiration gains energy by using oxygen

Fig. 9.13 Global warming. The outgoing light has, on average, a longer wavelength than the incoming light. The carbon dioxide concentration in the atmosphere has a greater impact on the outgoing light

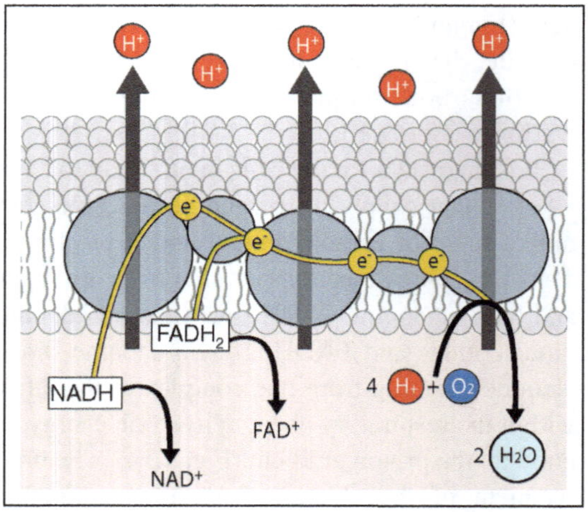

Fig. 9.14 The respiratory chain. The complexes involved in the respiratory chain are embedded in the inner membrane of the mitochondria

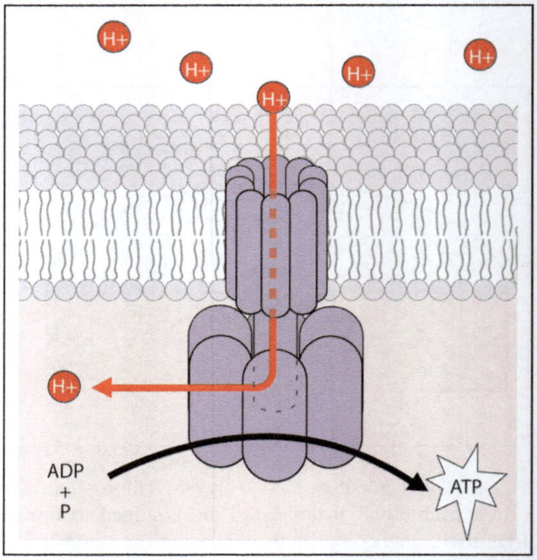

Fig. 9.15 ATP production. The proton gradient is used as energy to drive the ATPase enzyme, which transfers phosphate to ADP to produce ATP

The important carrier of energy within our cells is ATP (adenosine-triphosphate), which is produced by the transfer of phosphate to ADP (adenosine-diphosphate) with the aid of the enzyme F0F1-ATPase (Fig. 9.15). This requires energy. The energy driving the ATPase is the proton gradient in the mitochondria.

In the following section, we shall learn how oxygen from the atmosphere is delivered to the tissues.

9.4 Oxygen Delivery to Biological Tissues

How does oxygen from the ambient air reach the mitochondria of the different cells? The gaseous oxygen in the alveoli of the lungs diffuses down the gradient in the capillaries (Fig. 18.5). However, since the water solubility of oxygen is relatively low (Table 18.1), the transport capacity of oxygen is increased with the help of hemoglobin. In the end, organ oxygen diffuses from the hemoglobin through the vessel wall into the tissue, partially assisted by local carriers such as myoglobin or by locally produced hemoglobin (Figs. 9.16 and 9.17).

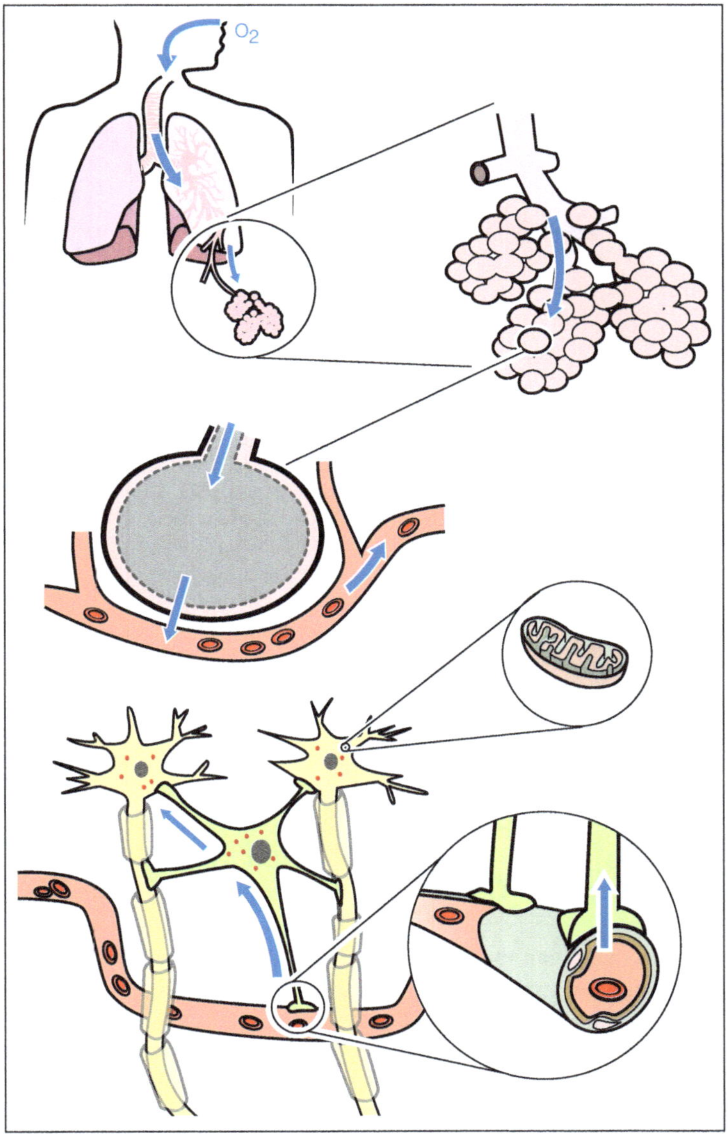

Fig. 9.16 Oxygen transport. The gaseous oxygen in the alveoli of the lungs diffuses into the capillaries, where it is physically dissolved in water. Since the water solubility of oxygen is relatively low, the transport capacity of oxygen to the tissues is increased with the help of hemoglobin. In the capillaries, oxygen diffuses from the blood into the neighboring tissue. In case of the brain and retina, the diffusion from the capillaries to the neural cells occurs mainly through the astrocytes. To facilitate this, both astrocytes and neurons also contain some hemoglobin. Once in the cell, the oxygen diffuses into the mitochondria

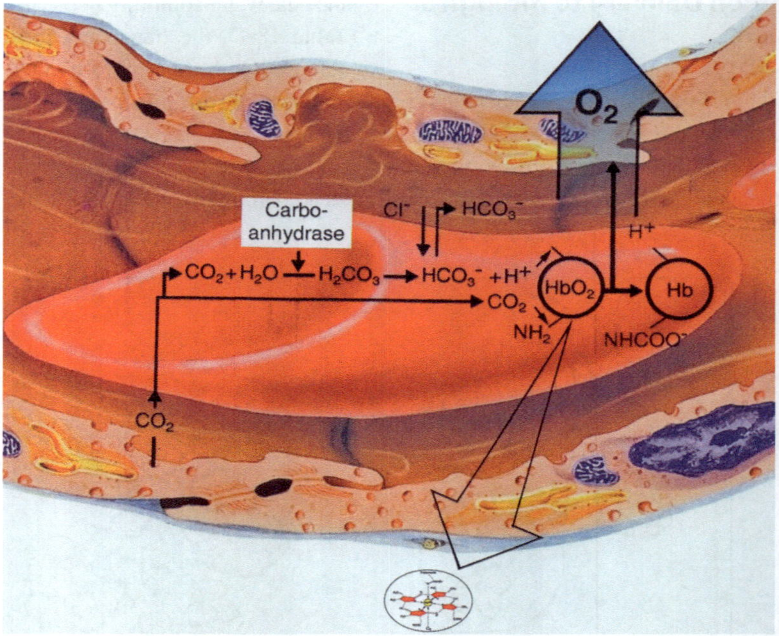

Fig. 9.17 Transport of oxygen and carbon dioxide by hemoglobin. Hemoglobin (Hb) is not only important in the transport of oxygen to the tissues but also in the transport of carbon dioxide and in the buffering of hydrogen ions. In the tissues, dissolved CO_2 passes into the red blood cells, where it combines with water to form carbonic acid. This reaction is catalyzed by the enzyme carbonic anhydrase. Carbonic acid then dissociates into bicarbonate and hydrogen ions. The hydrogen ions bind to reduced hemoglobin. The bicarbonate ions (HCO_3^-) generated by this process pass back into the plasma in exchange for chloride ions (Cl^-). (Courtesy of MSD Sharp & Dohme GMBH, Haar, Germany)

Let's consider the oxygen transport per minute. As a rule of thumb, in a resting state, we breathe about 5 L of air (which corresponds to about 1 L of oxygen), a quarter of which is finally taken up by the blood. This 250 mL of oxygen is transported by 5 L of blood.

9.5 Oxygen Deficiency in Tissues

Oxygen deficiency (hypoxia) can be mild or ample, transient or chronic. For example, atherosclerosis normally leads to mild chronic ischemia and thereby to a mild and most often asymptomatic hypoxia. This hypoxia, however, increases if the demand for oxygen increases. If the oxygen demand in muscles increases, such as by walking, it can lead to symptoms of intermittent claudication (stumbling). The symptoms of claudication occur because the demand for oxygen is greater than the delivering capacity. Another example of a cause for a transient hypoxia is reversible vasoconstriction (vasospasms).

Yet how do our tissues respond to hypoxia? Under normal conditions, the constitutively produced factor hypoxia-inducible factor-1 alpha (HIF-1 α) is oxidized, ubiquitinated, and degraded by the proteosomes. However, if the oxygen concentration in a cell is lowered, less HIF-1 α is oxidized and degraded. The stabilized HIF-1 α moves into the nucleus of the cell, where it acts as a transcription factor, stimulating the production of vascular endothelial growth factor (VEGF), Endothelin (ET), erythropoietin (EPO), von Willebrand factor, and other molecules (Fig. 9.18).

Similar changes occur during Von Hippel Lindau (VHL) disease. The Von Hippel Lindau tumor suppressor, also known as the pVHL, is a protein that is encoded by the VHL gene.

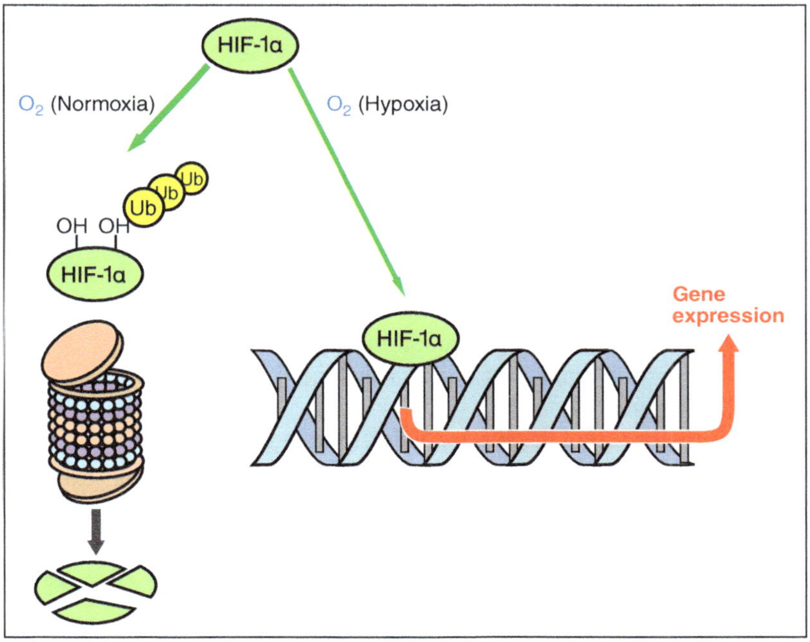

Fig. 9.18 Hypoxia-inducible factor-1 alpha (HIF-1 α). If oxygen concentration in a cell is lowered, less HIF-1 α is oxidized and degraded. Thus, stabilized HIF-1 α moves into the nucleus of the cell, where it acts as a transcription factor, inducing gene expression

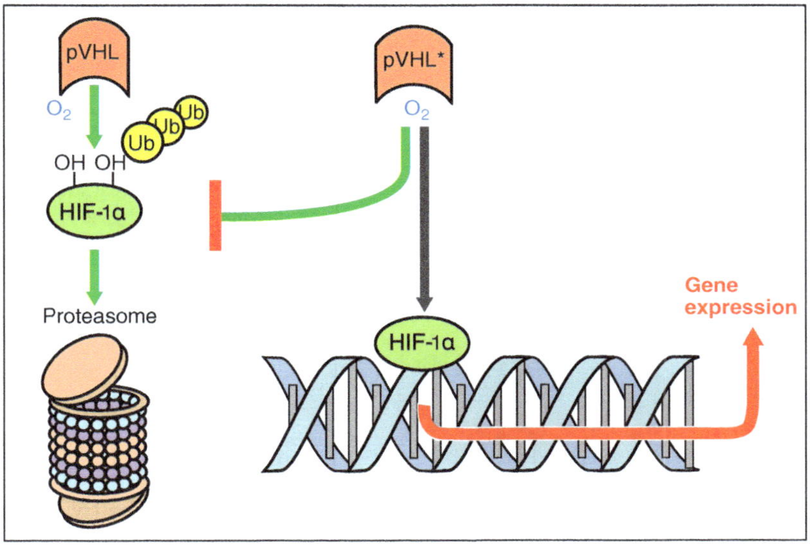

Fig. 9.19 Von Hippel Lindau gene mutation. *Left*: Normal cell with active VHL protein. *Right*: Cell with mutated VHL gene leading to VHL disease

Mutations in the VHL gene in humans are associated with VHL disease. In a normal cell with active VHL protein, HIF-1 α is oxidized, ubiquitinated, and degraded by the proteosomes. Cells with abnormal pVHL, however, behave as if they were in a hypoxic environment, so less HIF-1 α is oxidized and degraded. Thus, HIF-1 α moves into the nucleus of the cell, where it acts as a transcription factor and stimulates gene expression (Fig. 9.19).

During hypoxia, but particularly when the oxygen concentration rises again, the amount of reactive oxygen species (ROS) increases. A mild increase is beneficial, as it leads to preconditioning and this, in turn, makes the cells more resistant to future hypoxic episodes. If the increase in

ROS is more extensive, the resulting oxidative stress damages the tissue (see Sect. 13.3). This is known as reperfusion injury.

9.6 Oxygen in the Eye

The amount of oxygen consumption in the eyes differs in the various parts. We have all the extremes in the eye, from the lens that needs extremely little oxygen to the retina that has a very high consumption of oxygen. But where does the oxygen come from?

The anterior part of the eye, including the cornea, aqueous humor, and lens, receive their oxygen directly from the air by diffusion through the tear film and cornea. There is an oxygen gradient with falling oxygen concentration from the tear film to the cornea, to the anterior chamber, and, finally, to the lens (Fig. 9.20).

Oxygen diffuses corresponding to this gradient. This gradient also changes after a cataract and/or a vitrectomy surgery. The reason for this change in gradient is, on the one hand, because a pseudophakic lens does not consume oxygen and, on the other hand, because removal of the lens or the vitreous body facilitates oxygen diffusion. Oxygen supply to the cornea at night (closed eyes) is explained in the following section dealing with the consequences of hypoxia.

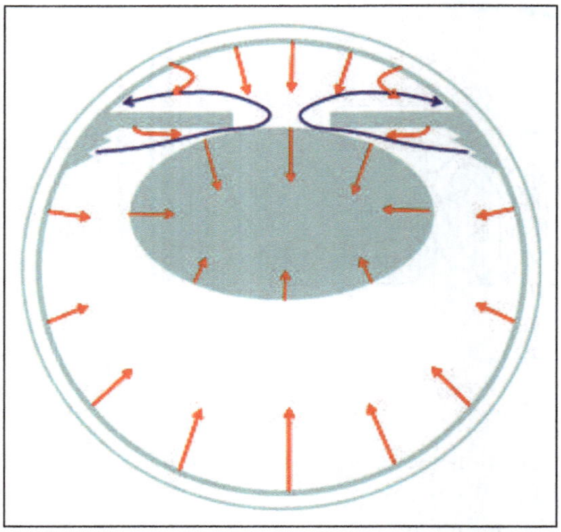

Fig. 9.20 Oxygen supply to the eye. The anterior parts of the eye receive oxygen directly from the air by diffusion through the tear film and cornea. The other parts of the eye receive oxygen through the ocular vasculature. (From Shui YB, et al. (2006) Invest Ophthalmol Vis Sci, 47. With permission)

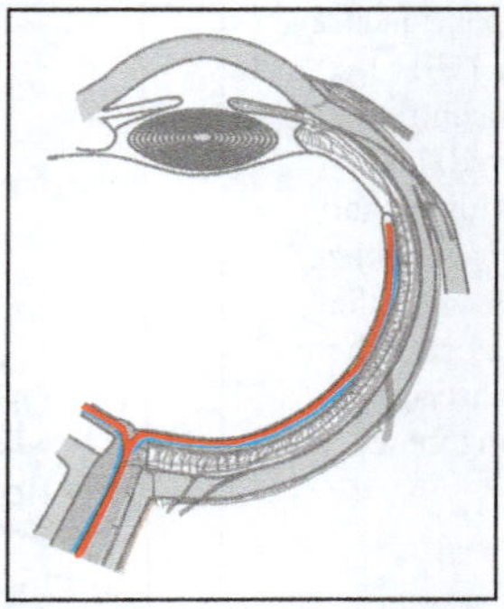

Fig. 9.21 Retinal vasculature. The retinal blood flow is supplied by the central retinal artery and drained by the central retinal vein

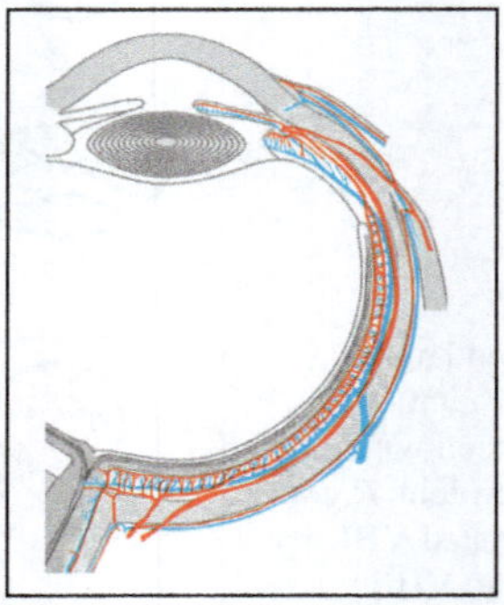

Fig. 9.22 Uveal vasculature. The uveal blood flow is supplied mainly by the long ciliary arteries and by the short ciliary arteries. Uveal circulation is drained by the vortex veins

The other parts of the eye receive oxygen through the ocular vasculature. For this reason, we shall shortly discuss ocular blood flow. We basically have two separate vascular systems in the eye: the retinal blood flow (Fig. 9.21) and the uveal blood flow (Fig. 9.22).

The retinal blood flow is supplied by the central retinal artery and drained by the central retinal vein. The uveal blood flow consists of iris and ciliary body circulation (anterior uvea), supplied by the long ciliary arteries and the choroidal circulation (posterior uvea) supplied by the short ciliary arteries. Both anterior and posterior uveal circulations are drained by the vortex veins. The

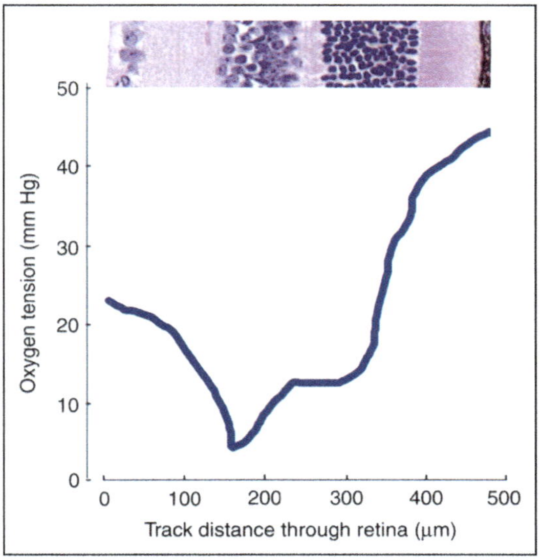

Fig. 9.23 Oxygen gradient in the retina. The oxygen concentration is highest at the level of the rods and cones

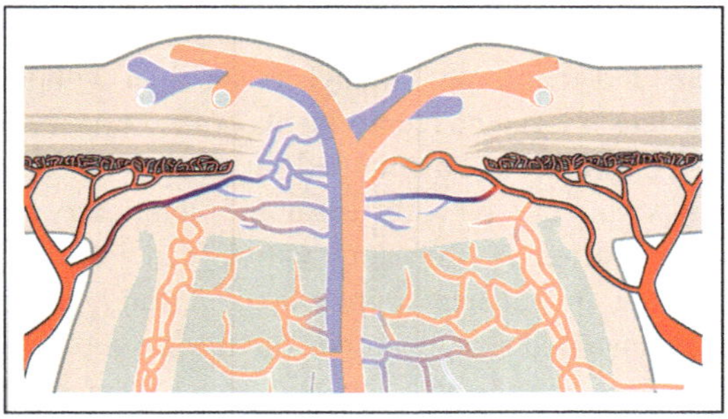

Fig. 9.24 Vascular supply to the preliminary part of the optic nerve head. The preliminary area of the optic nerve head receives the arterial supply by the ciliary (choroidal) arteries, but its venous drainage occurs through the central retinal veins

posterior part of the retina (the rods and cones, which have a particular high demand for oxygen) gets its oxygen supply by diffusion through the choroid. Because oxygen does not diffuse easily and because the rods and cones consume huge amounts of oxygen, a steep oxygen gradient results in the retina (Fig. 9.23).

The optic nerve head has a special status. The preliminary area receives the arterial supply by the ciliary (choroidal) arteries, but its venous drainage occurs through the retinal veins (Fig. 9.24). In the inner retina and optic nerve head, oxygen transport is facilitated by intracellular hemoglobin, which is also expressed in the glial cells and in the retinal ganglion cells.

9.7 Consequences of Hypoxia in the Eye

Although the eyes are small compared to most of the body's other organs, they require a lot of oxygen to function well. In the following section, we give a few examples of consequences of hypoxia in the anterior and posterior segments of the eye.

As described in the previous section, the cornea receives most of its oxygen directly from the air. This can, of course, happen only when the eye is open. At night, when the eyes are closed, the cornea receives oxygen from the capillaries of the conjunctiva of the lids. Thus, at night, the partial pressure of oxygen in the tear film is lower.

The behavior of the cornea is interesting. The cornea is thin and this dimension is relatively constant over time. How is this achieved? If we isolate a cornea and put it into water, it will swell because the macromolecules such as collagen fibers and hyaluronic acid in the stroma attract water. Once swollen, the cornea loses its transparency (see Sect. 2.3) because the distance between the collagen fibers is larger than half of the wavelength of light that ought to pass through. Water diffuses passively into the stroma. To avoid swelling, the endothelial cells of the cornea constantly need to pump water out. This pumping mechanism requires energy, which, in turn, requires oxygen. If the oxygen tension is reduced, such as during the night, a light swelling of the cornea occurs, even under physiological conditions.

By the way, the most frequent cause of stromal corneal edema is not hypoxia but, rather, damage to the endothelial cells such as that which occurs, for example, in Fuchs' endothelial dystrophy or after complicated surgery.

We also have to distinguish between corneal stromal edema and corneal epithelial edema. Why do we sometimes see stromal and, at other times, epithelial edema (Fig. 9.25)? If either the influx of water into the cornea is increased or if

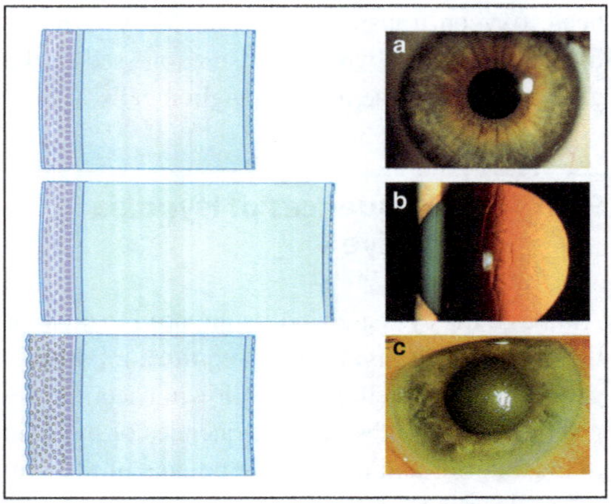

Fig. 9.25 Corneal thickness depending on water content. *Left*: schematic cross-section through a cornea. *Right*: clinical picture: (**a**) healthy condition, (**b**) stromal edema, and (**c**) epithelial edema

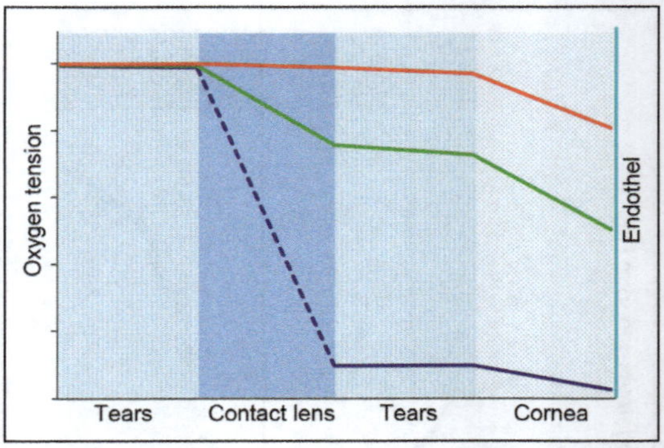

Fig. 9.26 Relative oxygen tension during sleep (closed eyes). *Red*: without contact lenses. *Green*: with permeable contact lenses. *Blue*: with impermeable lenses (gradient from the tears to the cornea)

the pumping effect of the corneal endothelial cells is decreased, the cornea swells. However, the stroma can never increase in thickness by swelling anteriorly because this would imply an elongation of the corneal collagen fibers. Therefore, the thickening of the cornea occurs by swelling posteriorly. However, in the case of an acute glaucoma attack, this posterior swelling is also inhibited by the backpressure; therefore, the only way for the additional fluid to flow is through the epithelium, which leads to corneal epithelial edema.

This information is important for contact lens wearers (Fig. 9.26). Basically, two types of lenses are available: oxygen-permeable and oxygen-impermeable lenses. When wearing oxygen-impermeable lenses, we have high oxygen tension in the tear film in front of the contact lens but lower oxygen tension in the tear film between the contact lens and the cornea. This oxygen tension in the tear film between the contact lens and the cornea is not zero because the tear film is constantly renewed by the blinking mechanism. At night, on the one hand, we do not blink and, on the other hand, the oxygen tension in the tear film in front of the contact lens is lower. The resulting oxygen tension between the contact lens and cornea is particularly low. Therefore, the cornea would not receive enough oxygen at night with oxygen-impermeable lenses.

This situation is quite different when wearing gas-permeable contact lenses. Here, the oxygen diffuses through the contact lens; nevertheless, it builds up a certain diffusion barrier, resulting in an increased gradient. The partial pressure of oxygen in the tear film between the gas-permeable lens and the cornea, though lower than without the lens, meets the requirements of the cornea.

As previously mentioned, oxygen deficiency (hypoxia) can be mild or ample, temporary or long-lasting. A transient hypoxia can be mild, leading to transient impairment of function as demonstrated, for example, in the case of the brain as a transient ischemic attack. Transient hypoxia, however, can also be so severe that it leads to an infarction, as demonstrated in a cerebrovascular insult in the case of a transient cardiac arrest. Long-lasting but mild hypoxia leads to adaptation. As mentioned before, a relatively low oxygen supply may still be sufficient for baseline conditions, but it would be insufficient when the consumption is increased. This explains why severe atherosclerosis in the leg leads to pain only when walking (claudicatio intermittens).

One of the consequences of hypoxia in the retina is the interruption of the axoplasmic flow within the nerve fiber layer, resulting in the characteristic appearance of cotton wool spots (Fig. 9.27). More extensive hypoxia results from a branch retinal artery occlusion (Fig. 9.28). Severe hypoxia in the optic nerve may lead to a stroke (anterior ischemic optic neuropathy) (Fig. 9.29).

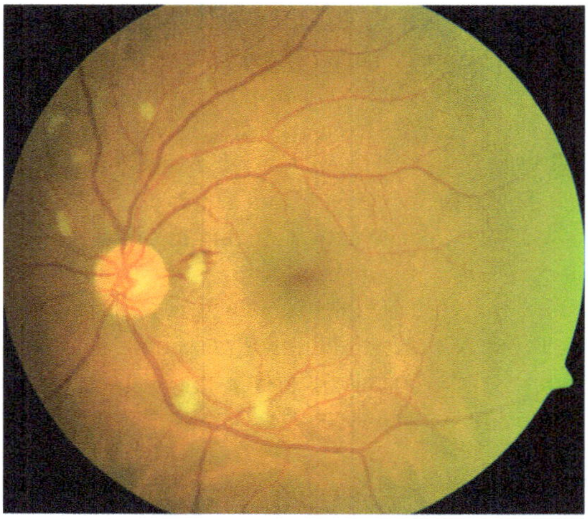

Fig. 9.27 Cotton wool spots. Hypoxia leads to interruption of the axoplasmic flow and thereby to swelling of the corresponding fibers, resulting in a decrease in the transparency of the retina

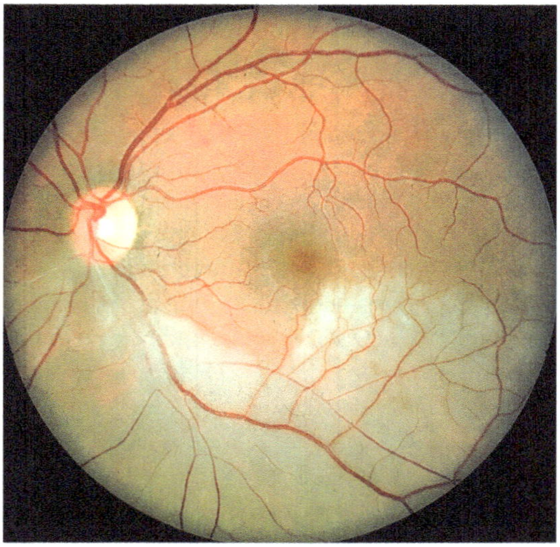

Fig. 9.28 Branch retinal artery occlusion. Severe hypoxia leads to necrosis of the retina. The necrotic retina is less transparent; therefore, the red color of the choroid is obscured

The most important causes of hypoxia are reduced blood flow, which, in turn, can be a consequence of systemic causes such as heart failure or a drop in blood pressure. However, reduced blood flow may also be the consequence of local causes, which may either be structural, such as atherosclerosis, or functional, such as vascular dysregulation (for more information, refer to Josef Flammer's (2006) book *Glaucoma*).

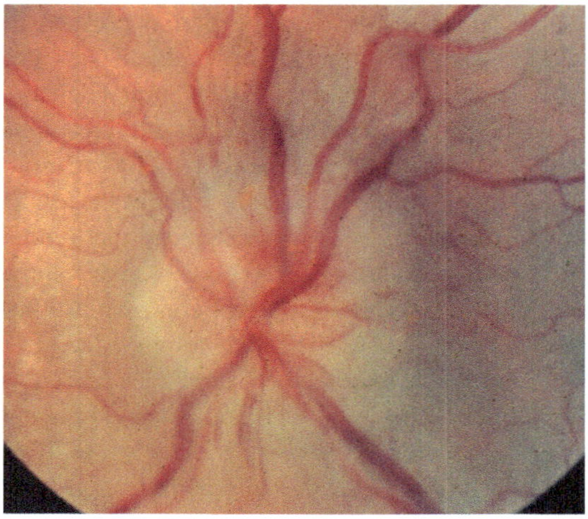

Fig. 9.29 Anterior ischemic optic neuropathy. Severe hypoxia in the preliminary portion of the optic nerve head leads to swelling and hemorrhages and, in the late stage, to an atrophic, pale optic nerve head (not shown)

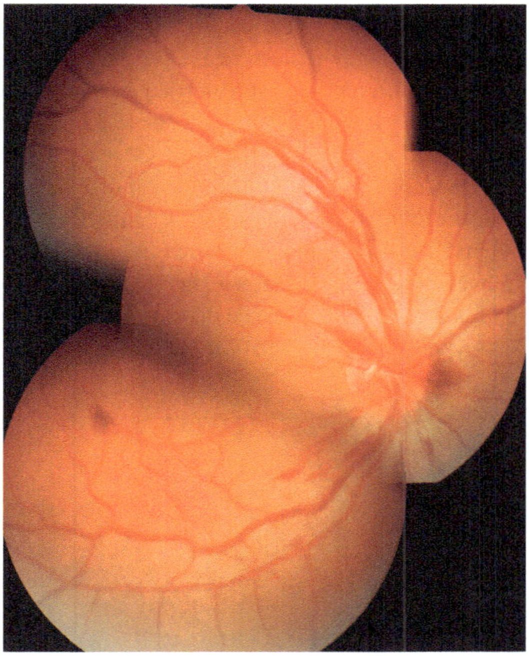

Fig. 9.30 High-altitude retinopathy. The retinal veins are dilated and hemorrhages occur in the retina and vitreous. The optic nerve head is swollen. (From Schnetzler G, et al. (2006) Forum Med Suisse, 6. With permission)

Another example where moderate hypoxia leads to eye disorder is high-altitude retinopathy (Fig. 9.30). The partial pressure of oxygen decreases with an increase in altitude over sea level. At high altitudes, the partial pressure is lower, which is normally no problem provided that there is enough time for adaptation. High altitude, however, causes retinopathy in predisposed people. In these patients, the Endothelin in the circulating blood is increased. This also leads to, among other changes, an increase in retinal venous pressure and dilated veins, a decrease in the blood-retinal barrier. This all together contributes to the clinical picture of high-altitude retinopathy.

As previously described, hypoxia leads to an increase in HIF-1 α, which, in turn, stimulates the production of various molecules such as VEGF (vascular endothelial growth factor), ET-1 (Endothelin-1), EPO (erythropoietin), von Willebrand factor, MMPs (matrix metalloproteins), and others. An increase in ET-1 constricts the veins at the level of the optic nerve head, leading to high venous pressure and to dilated veins. An increase in VEGF and MMPs weakens the blood–brain barrier, leading to edema and, in extreme cases, to hemorrhages.

Quite different are the consequences of an unstable oxygen supply. Tissues such as the optic nerve head can also be damaged by an unstable oxygen supply. As mentioned in the previous section, the return to a normal blood flow after a period of impaired circulation is known as reperfusion. Why does reperfusion damage living tissue? When perfusion to the optic nerve decreases, a shortage of oxygen, particularly in the mitochondria, results. Mitochondria are the cell's power plants. As previously described, the inner mitochondrial membrane contains the electron transport chain. The high-energy electrons from the Krebs cycle are transported via NADH and $FADH_2$ to the respiratory chain, where they travel from a high-energy state to a low-energy state and are finally received by oxygen. As they travel from a high- to a low-energy state, they release energy, which is used to transport protons across the inner mitochondrial membrane. This generates a proton gradient, which, in turn, drives the enzyme ATPase to generate ATP (Fig. 9.31).

Under conditions of hypoxia (due to reduced perfusion), not enough oxygen is available to receive the electrons from the last complex of the respiratory chain. The electrons are, thus, congested in the respiratory chain, and although they are pushed, they shear out only minimally, as there is not enough oxygen available to receive them. After the blood flow returns to normal, the oxygen tension increases again. This leads to the recovery of the normal transfer of electrons (always two together with an opposite spin). However, as long as a congestion of electrons is present, individual electrons jump out falsely (before they reach the last complex) and react with oxygen to form oxygen free radicals.

This, in turn, leads to a number of changes that will be discussed in a later chapter on oxidative stress.

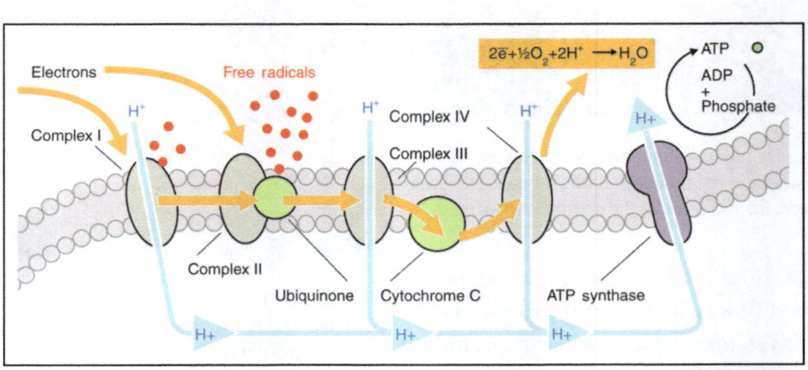

Fig. 9.31 Formation of free radicals in the mitochondria. Under normal circumstances, the electrons travel down from one complex to another until they are finally accepted in pairs by the oxygen molecule. However, if oxygen concentration is reduced, the electrons congest and electron flow is impaired. As soon as the oxygen concentration normalizes, the electron flow is again normalized, but because the electrons were congested and some complexes are damaged, some of these electrons go astray and react with nearby oxygen molecules to form free radicals

10 Water

10.1 What Is Water?

Besides oxygen, water is essential to sustain life on earth. This molecule, which we often take for granted, is quite phenomenal (see Chapter 18). H_2O can change its physical state from ice to liquid (water) to steam within a temperature range of 100 °C. The important feature of water is its polar nature. Since the oxygen atom in the water molecule has a higher electronegativity than hydrogen, the electrons are not homogeneously distributed but are, rather, closer to the oxygen than the hydrogen atom. This creates a weak electrical field, with a negative pole at the oxygen and a positive pole at the hydrogen (Fig. 10.1).

If the kinetic energy of the molecules falls below a certain limit (at 0 °C), the forces resulting from the kinetic energy are lower than the mutual attractions; therefore, the liquidity disappears (ice formation). More detailed information can be found in Chapter 18. On the other hand, if the kinetic energy exceeds a certain limit (at 100 °C), the resulting forces are large enough to disrupt the hydrogen bonds (Fig. 10.2), allowing the individual molecules to escape (steam formation). Water not only has a high fluidity but is also transparent and colorless because the molecules are small and have no double bonds, so they allow light to travel through without scattering or absorption. The high solubility and weak polarity make water an ideal solvent for ions (e.g., Na^+, Cl^-) or any molecules with a certain polarity (e.g., glucose). In crystal form (ice), the water molecules are less densely packed due to the formation of hexagonal crystals and the increase in hydrogen bonding. Therefore, the specific weight of ice is lower than that of water. This is the reason that ice floats on water. This may seem trivial, but it is actually quite important as the ice would otherwise fall into deep seas and would not melt by the following summer. This would be

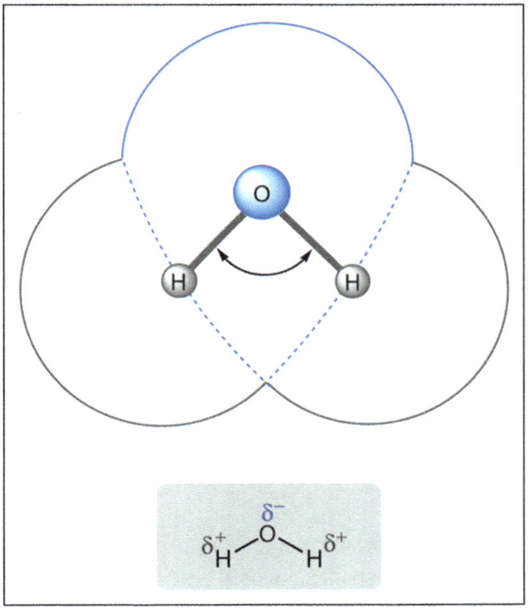

Fig. 10.1 Dipole nature of water. An important feature of water is its polar nature. The water molecule forms an angle, with hydrogen atoms at the tips and oxygen at the vertex. The side of the molecule with the oxygen atom has a partial negative charge, whereas the side with the hydrogen atom is partially positive. This charge difference (or dipole) causes water molecules to be attracted to each other and to other polar molecules

J. Flammer et al., *Basic Sciences in Ophthalmology*,
DOI 10.1007/978-3-642-32261-7_10, © Springer-Verlag Berlin Heidelberg 2013

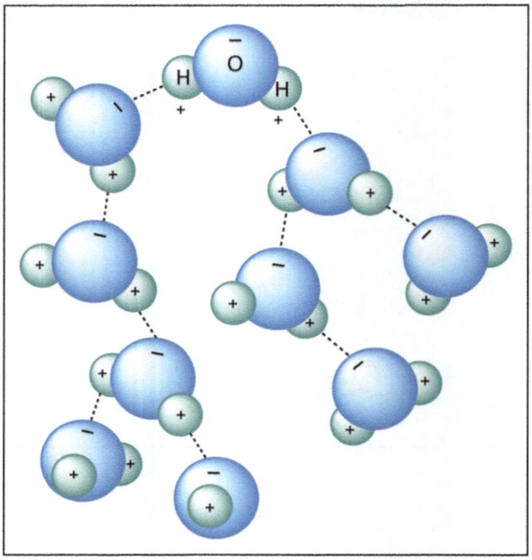

Fig. 10.2 Hydrogen bonding, a relatively weak attraction between the water molecules, responsible for a number of water's physical properties

disastrous in the long run, as most of the sea would consist of ice. The fact that ice requires greater volume than water also explains why a water pipe can break open in freezing temperatures.

10.2 Water in the Universe

How much water exists? The universe has huge intergalactic spaces, which are more or less empty. These spaces are, however, not totally empty but contain some "dust particles."

These "dust particles" contain, among other atoms and molecules, mostly water (Fig. 10.3). This is why the universe is said to be "wet." Although the number of particles per volume is extremely low, the total mass of these particles is greater than the mass of all the stars together simply because the intergalactic space is incredibly huge. Thus, in terms of quantity, water is the most important molecule in the universe.

Fig. 10.3 Water in the universe. Water is the most abundant molecule in the universe. Photograph of dust particles in the universe consisting mainly of water

10.3 Water on Earth

What about the water on our planet? Water on our planet rained down from space billions of years ago. It covers more than two-thirds of our planet (Fig. 10.4).

Our earth functions the way it does only as a result of the continuous cycling of water (Fig. 10.5). The energy of our sun leads to evaporation and the resulting clouds are transported by the wind. Once cooled, water droplets form, finally precipitate, and then rain down on the earth. This water supply is a prerequisite for land-based plants. Not only do a number of animals and plants live in water but all other plants and animals both contain and require water for their survival: without water, there is no life.

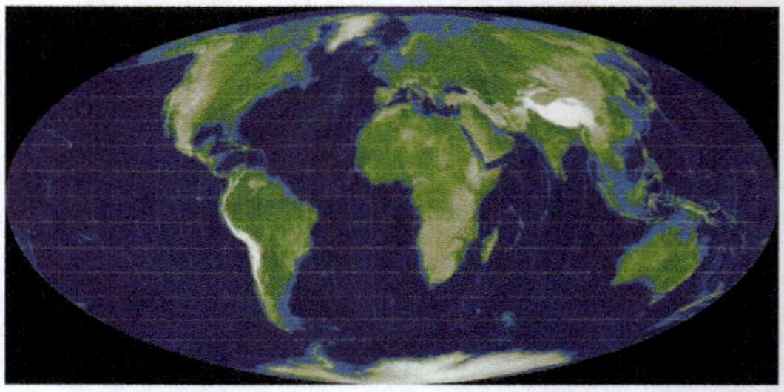

Fig. 10.4 Water on earth. There is clearly more water on earth than on other planets of our solar system. Water covers about 70 % of the Earth's surface (copyright Lars H. Rohwedder)

Fig. 10.5 The water cycle, which describes the continuous movement of water on, above, and below the surface of the Earth

10.4 Water in Biology

Why is water so essential in biology? The simple answer is that all biologically important chemical reactions take place in aqueous solutions. Not only do individual cells depend on water for survival but the functioning of the entire body also has an equal dependence on water, as illustrated, for example, by the blood circulation.

10.5 Water in Medicine

What role does water play in medicine? Because water plays such an important role in biology, the regulation of water intake, transportation, and excretion is essential (as demonstrated by the importance of a well-functioning kidney). In medicine, we not only strive to keep all these regulatory processes intact but we can also purposely interrupt the physiological regulation of water. If we withdraw water from tissues and cells (such as by cooling down the surroundings), cellular metabolism ceases but, under ideal conditions, the cellular structure is preserved. This feature is used as a way of preserving tissues and cells such as sperm (cryopreservation). However, if we cool down a tissue (or cell) so rapidly that water freezes before it can escape the cell, water crystals form within the cell and destroy it. This is called cryocoagulation and it is used to destroy tissues such as tumors. Although the term "coagulation" is commonly used in this context, it is not strictly correct.

10.6 Water in the Eye

As part of its role in biology, water also serves a vital function in the eye. Water regulation is an important topic for ophthalmologists. On one hand, there are spaces that are filled with water (the anterior or posterior chamber) or that consist predominantly of water (the vitreous humor). On the other hand, there are structures that do not function properly if they contain too much water, such as the cornea or the lens, and, finally, there are areas in the eye that largely lose their function if the water content increases, such as the retina (Fig. 10.6).

For this reason, the eye has important water barriers such as the inner and outer blood retinal barrier, as well as important water transport mechanisms, as seen in the epithelium of the ciliary body (aqueous humor production) or the pumping mechanisms in the corneal endothelial cells. One mechanism by which water can selectively pass through cell membranes is by means of aquaporins (Fig. 10.7), integral membrane proteins that belong to the larger family of the so-called major intrinsic proteins (MIP). These proteins form pores in the membranes of biological cells and selectively conduct water molecules in and out of the cell while preventing the passage of ions and other solutes.

Aquaporins can be found in the eye in areas such as the corneal endothelium or in the glial cells of the retina.

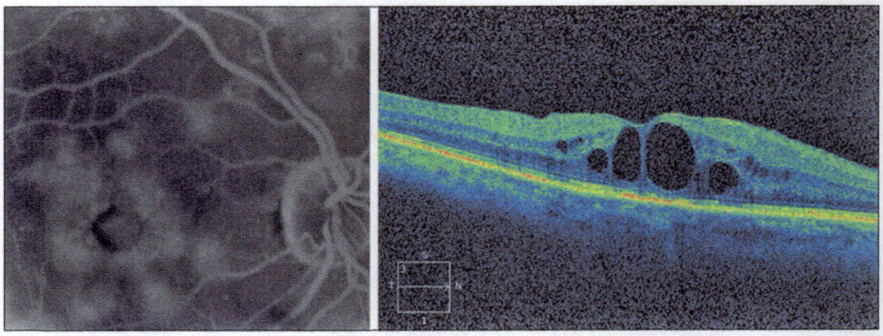

Fig. 10.6 Macular edema. Illustration of macular edema as detected by fluorescence angiography (*left*) and by optical coherence tomography (OCT) (*right*). The water accumulates within Henle's fiber layer

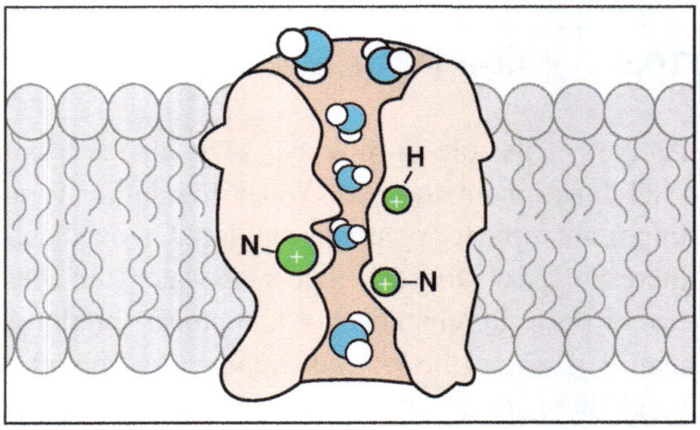

Fig. 10.7 Aquaporins, proteins that are embedded in the cell membrane and regulate the flow of water

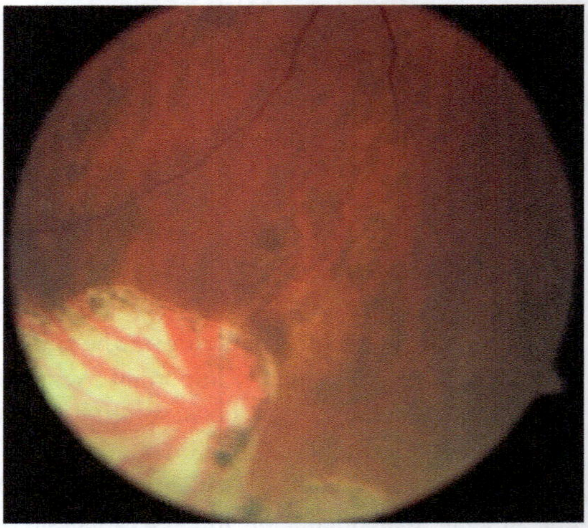

Fig. 10.8 Retinal scar after cryocoagulation of the retina

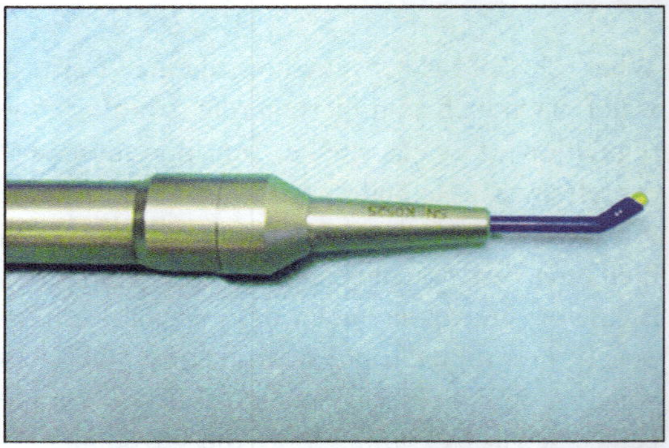

Fig. 10.9 Cryoprobe. The surgeon places a cryoprobe on the outside of the eye in the vicinity of a retinal tear

In the previous section on "water in medicine," we explained how water can be used for therapeutic purposes such as cryocoagulation, which may be used in the retina to induce a retinal scar following retinal detachment surgery (Figs. 10.8 and 10.9).

Cryocoagulation of the ciliary body reduces aqueous humor production and, thus, intraocular pressure (IOP). The ophthalmologist may be interested to know how to freeze the tissue quickly enough. In a tissue that is well perfused, the cold is transported away rapidly (or, more correctly, the heat is delivered rapidly). For this reason, vitreoretinal surgeons will push the cryoprobe onto the bulbus to cause a local reduction of choroidal blood flow and thereby increase the speed of cooling of the retina and the retinal pigment epithelium.

We shall now move on to the next molecule of interest; carbon dioxide.

Carbon Dioxide (CO_2) 11

11.1 What Is Carbon Dioxide?

Carbon dioxide is a molecule composed of two oxygen atoms covalently bonded to a single carbon atom (Fig. 11.1). It is colorless and exists in gaseous form at room temperature. In nature, carbon dioxide is used by plants during photosynthesis to make sugar, for example, and it is liberated by all organisms that depend on oxidative phosphorylation. Burning in a fire or "burning" in mitochondria also leads to carbon dioxide formation through a set of redox reactions summarized as "cellular respiration."

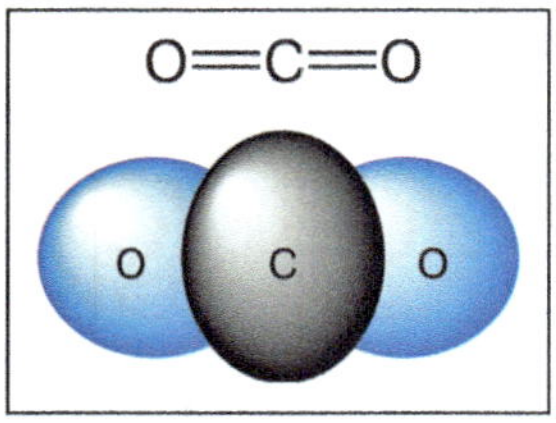

Fig. 11.1 Carbon dioxide (CO_2), produced by combustion reactions. It is a symmetrical molecule with a slight polarity, which makes it more soluble than molecular oxygen. At room temperature, unlike water, carbon dioxide is a gas

11.2 Transport of Carbon Dioxide

As a waste product of cellular metabolism, carbon dioxide needs to be constantly transported away from the cells and, ultimately, from the body. Although carbon dioxide has a greater solubility than oxygen in water, only 5 % is transported in unchanged form (physically dissolved in water). About 10 % of carbon dioxide is bound to reduced hemoglobin and the vast majority is transported as bicarbonate ions (HCO_3^-) (Fig. 11.2).

Carbon dioxide can be found in two forms: an unhydrated (CO_2) and hydrated bicarbonate form (H_2CO_3). The unhydrated (CO_2) form can easily diffuse through membranes, whereas the hydrated form (HCO_3^-) is water-soluble and is, therefore, the preferred form for transport in aqueous solutions. This implies the necessity of a fast transition between the hydrated and unhydrated forms. To facilitate this conversion in both directions, nature has developed an enzyme called carbonic anhydrase. Several isoforms of carbonic anhydrase exist and some of them are membrane-bound, as shown in Fig. 11.3.

J. Flammer et al., *Basic Sciences in Ophthalmology*,
DOI 10.1007/978-3-642-32261-7_11, © Springer-Verlag Berlin Heidelberg 2013

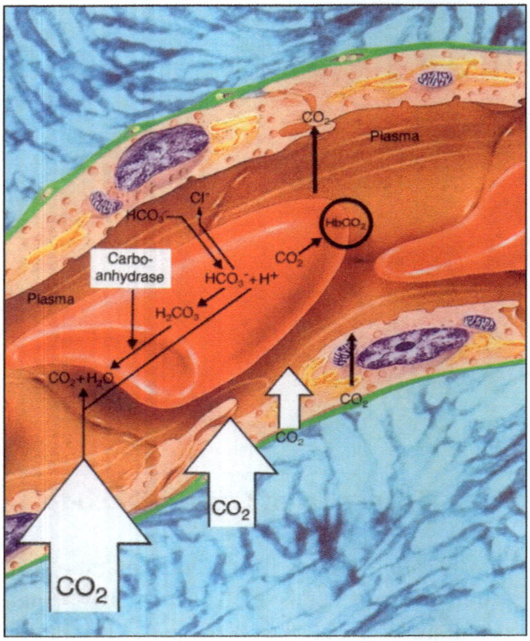

Fig. 11.2 Transport of carbon dioxide in the blood. Carbon dioxide produced in the tissue cells diffuses into the blood plasma and, from there, the majority diffuses into the red blood cells. The carbon dioxide is transported either dissolved in the plasma or bound to hemoglobin. (Courtesy of MSD Sharp & Dohme GMBH, Haar, Germany)

11.3 Carbon Dioxide in Medicine

Although efficient elimination of carbon dioxide is needed, the molecule, particularly in its hydrated form, also plays a major role in the regulation of pH. For this reason, the elimination of bicarbonate by the kidney and of carbon dioxide in the lungs is highly regulated. The inhibition of carbonic anhydrase leads to a decrease in pH in the extracellular space and increased loss of sodium (Na+) and potassium (K+) through excretion in the kidney at least transiently, as well as increased excretion of water. The inhibition of carbonic anhydrase reduces cerebrospinal fluid (CSF) formation and is, therefore, used to reduce CSF pressure.

11.4 Carbon Dioxide in the Eye

The diffusion of carbon dioxide through the cornea requires the conversion of carbon dioxide and water to bicarbonate and protons, as well as the reverse reaction. For this reason, carbonic anhydrase is also found in the cornea. The inhibition of this enzyme induces a slight swelling of the

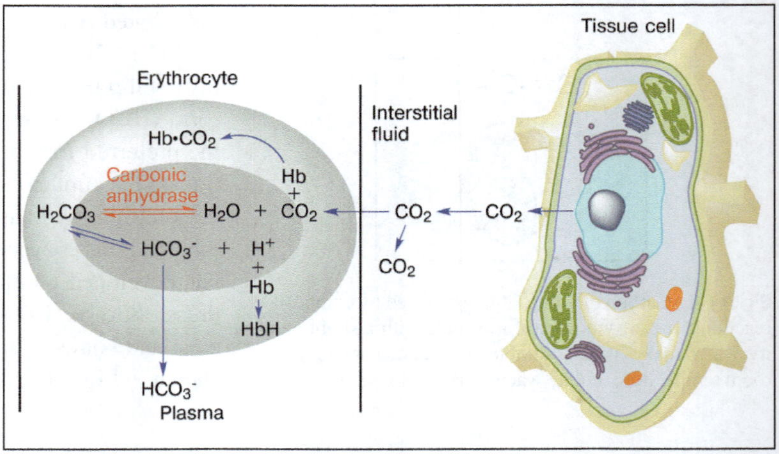

Fig. 11.3 Carbonic anhydrases are enzymes that catalyze the hydration of carbon dioxide and the dehydration of bicarbonate. This chemical reaction is crucial for the diffusion and transport of carbon dioxide

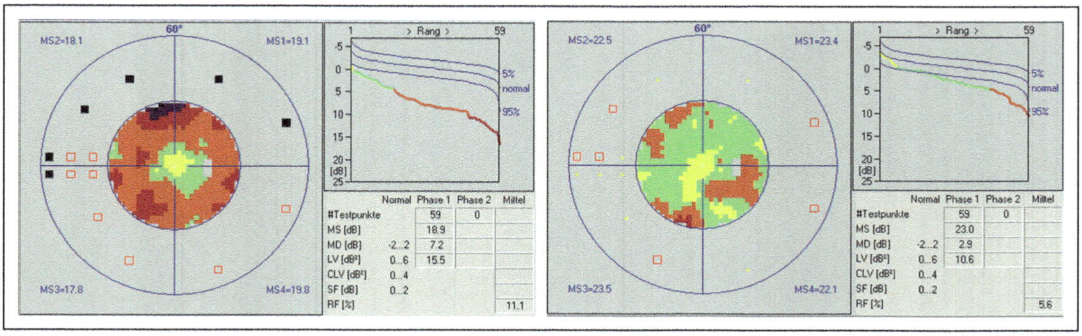

Fig. 11.4 Visual fields of a 25-year-old woman with marked primary vascular dysregulation (Octopus program G1; see Sect. 4.1.13). The results are presented with a comparison display printout, the Bebie curve, and visual field indices. *Left*: visual field before treatment. *Right*: visual field after 1 month of treatment with a systemic carbonic anhydrase inhibitor

cornea. The vast majority of carbon dioxide in the eye is transported away by the blood circulation. Carbonic anhydrase is also found in the epithelium of the ciliary body and its inhibition is of particular importance in the reduction of aqueous humor production and, consequently, of intraocular pressure (IOP). Carbonic anhydrase inhibition has additional effects. The speed of carbon dioxide elimination is reduced; therefore, more carbon dioxide accumulates in the tissue. This side effect has a positive component, as increased carbon dioxide concentration leads to vasodilation. The resulting vasodilation, in turn, leads to a better oxygen supply, which is of particular interest in glaucoma. The acute effect of the normalization of oxygen supply on visual function was discussed earlier in the chapter on oxygen. This explains why carbonic anhydrase inhibition can improve visual fields in patients with disturbed autoregulation [as a result of primary vascular dysregulation (PVD) syndrome] within hours (Fig. 11.4).

For more information on this topic, please refer to www.glaucomaresearch.ch.

We shall now discuss another molecule of interest in ophthalmology: nitric oxide.

Nitric Oxide

12

12.1 Nitric Oxide Molecule

Nitric oxide (NO) is a diatomic free radical consisting of one nitrogen and one oxygen atom (Fig. 12.1).

Nitric oxide is both lipophilic and small, so it can easily pass through cell membranes. In its natural form, it is a colorless gas. This gas is distinct from and not to be confused with nitrous oxide (N_2O), also known as laughing gas, which is used for anesthetic purposes.

This molecule has a history as one of the components of air pollution. Nevertheless, it also plays such an important role in physiological and pathophysiological processes that, in 1982, the journal *Science* named nitric oxide as its choice for "Molecule of The Year."

Before we discuss the involvement of nitric oxide in a variety of biological processes, we shall briefly discuss the discovery of nitric oxide.

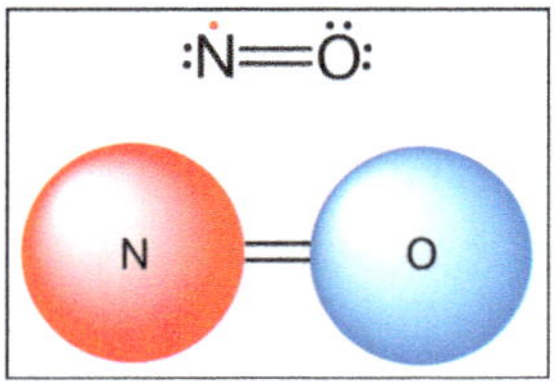

Fig. 12.1 Molecular structure of nitric oxide and its chemical structure, showing that it is a free radical (*dot*: an extra electron). NO is only slightly polar, making it soluble in both water and lipids

12.2 Nitric Oxide in History

The mechanism of blood circulation was not very well understood before the seventeenth century, when Harvey[1] established the heart as the center of the blood system and showed that blood circulates to and from the heart. While researching a cure for the heart disease angina, physicians began experimenting with amyl nitrite, which seemed to reduce both blood pressure and angina pain. As its effectiveness was short-lived, they began researching related chemicals. One related chemical compound was nitroglycerin (Fig. 12.2), a compound formed by the combination of glycerol and nitric and sulfuric acids.

Pure nitric oxide was first synthesized by the Italian chemist Sobrero.[2] In his private letters and in a journal article, Sobrero warned vigorously against the use of this compound because of its explosive properties, stating that it was extremely dangerous and impossible to handle. In fact, he was so frightened by what he had created that he kept it a secret for over a year.

[1]William Harvey (1578–1657), established the heart as the center of the blood system in the seventeenth century.

[2]Ascanio Sobrero (1812–1888), an Italian chemist, first synthesized nitric oxide.

J. Flammer et al., *Basic Sciences in Ophthalmology*,
DOI 10.1007/978-3-642-32261-7_12, © Springer-Verlag Berlin Heidelberg 2013

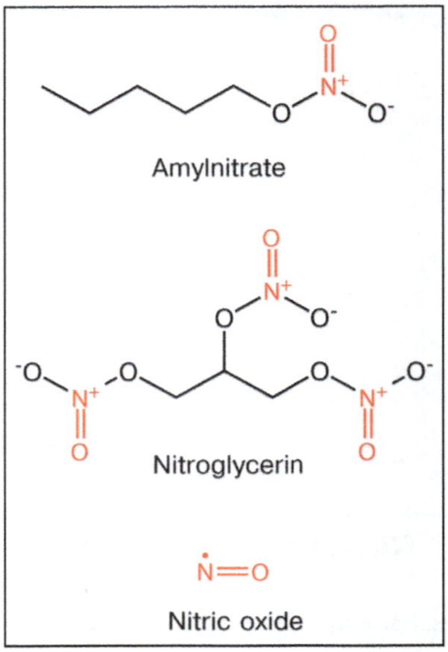

Fig. 12.2 Nitrogen oxides. The chemical structure of amyl nitrite (*top*), nitroglycerin (*middle*), and nitric oxide (*bottom*). Both amyl nitrite and nitroglycerin liberate nitric oxide (NO)

Nevertheless, the word got out. Four years after Sobrero's discovery, Nobel[3] learned about nitroglycerin's explosive properties. In 1863, after extensive experimentation, Nobel (Fig. 12.3) found that, when nitroglycerin was incorporated in an absorbent inert substance such as kieselguhr (diatomaceous earth), it became more convenient to handle, and he patented this mixture in 1867 as "dynamite."

During the development of nitroglycerin, workers discovered that they developed headaches when they came into contact with the compound (the workers called it the "nitro headache"). This suggested to the physician Murrell[4] that nitroglycerin could act as a vasodilator, much like the previously discovered amyl nitrate, and could be used as a treatment for cardiac angina.

Fig. 12.3 Alfred Nobel

In 1876, Murrell first used nitroglycerin for angina, yet because he was worried that his patients would be alarmed at the prospect of ingesting explosives, Murrell marketed his finding as "glyceryl trinitrate."

Much later, in 1978, Furchgott[5] recognized that acetylcholine induced vasorelaxation only if vascular endothelium cells (VEC) were present. Ignarro[6] found that a substance he called "endothelium-derived relaxing factor" (EDRF) was released by VEC. By 1986, Murrad[7] had found that EDRF was, in fact, nitric oxide (NO) (Fig. 12.4). Furchgott, Ignarro, and Murrad jointly received the Nobel Prize in 1998.

12.3 Nitric Oxide in Biology

Nitric oxide (NO) is a highly reactive molecule with a half-life in biological tissues of only a few seconds. Indeed, the half-life of NO is so short that it acts locally and its concentration is not homogenous even within a cell. Because it is small and liposoluble, it can easily diffuse across membranes. These attributes make nitric oxide

[3]Alfred Nobel (1833–1896), Swedish chemist, inventor, and engineer. Among others, he was also the inventor of dynamite.

[4]William Murrell (1853–1912), physician who used nitroglycerin as a remedy for angina pectoris.

[5]Robert Francis Furchgott (1916–2009) discovered the role of the endothelial cell.

[6]Louis José Ignarro found that "endothelium-derived relaxing factor" (EDRF) was released by endothelial cells.

[7]Ferid Murad discovered that endothelium-derived relaxing factor (EDRF) was nitric oxide (NO).

Fig. 12.4 Robert F. Furchgott, Louis J. Ignarro, and Ferid Murrad

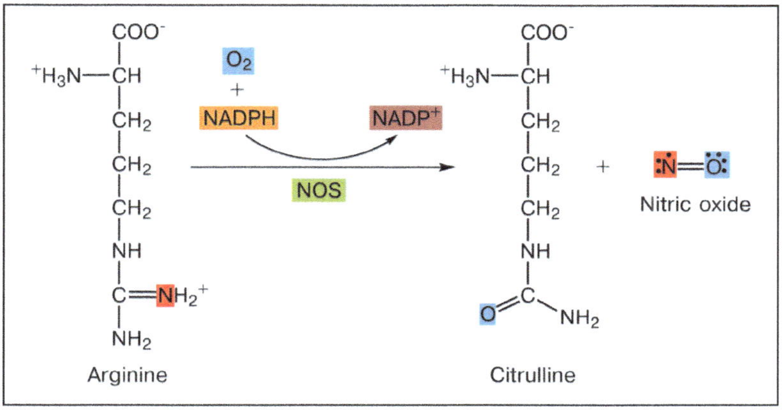

Fig. 12.5 Synthesis of nitric oxide from arginine and oxygen by various nitric oxide synthases (NOS)

ideal for a paracrine (between adjacent cells) and autocrine (within a single cell) signaling molecule. Indeed, NO is one of the few gaseous signaling molecules known. A signaling molecule is a chemical that transmits information within and/or between cells. NO is synthesized in living cells from the amino acid L-arginine, as shown in Fig. 12.5.

This reaction is catalyzed by a group of enzymes called nitric oxide synthases (NOS). There are three types: neuronal NOS (nNOS=NOS1), which produces NO in neuronal tissue, inducible NOS (iNOS=NOS2), which can mostly be found in the immune cells, and endothelial NOS (eNOS=NOS3=cNOS), which is constitutively expressed in vascular endothelial cells. NOS1 is always present in all neuronal cells, while the expression of NOS2 occurs when "induced" by a variety of factors (Table 12.1).

Table 12.1 Nitric oxide synthase enzymes – comparisons of the different human NOS enzymes

nNOS NOS1	Chromosome 12	Ca++ dependent
iNOS NOS2	Chromosome 17	Ca++ independent
eNOS NOS3	Chromosome 7	Ca++ dependent

Let's take a closer look at NOS3. In blood vessels, an equilibrium exists between endothelium-derived vasoconstriction and endothelium-derived vasodilation. If this equilibrium is disturbed by inhibition of the production of a molecule such as NO, the "net result" is vasoconstriction. Therefore, a basal production of NO is a prerequisite to keep the vessels dilated. The necessary enzyme, NOS3, is constantly expressed in vascular endothelial cells. Its activity to produce NO depends on cytosolic calcium concentration. Molecules such as acetylcholine

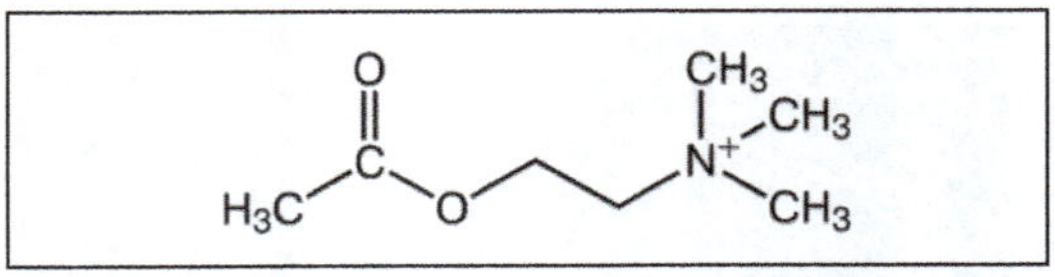

Fig. 12.6 Acetylcholine, in addition to bradykinin, is the major stimulator of NO production. Interestingly, this molecule is also a neurotransmitter and can be used to constrict the pupil intraoperatively

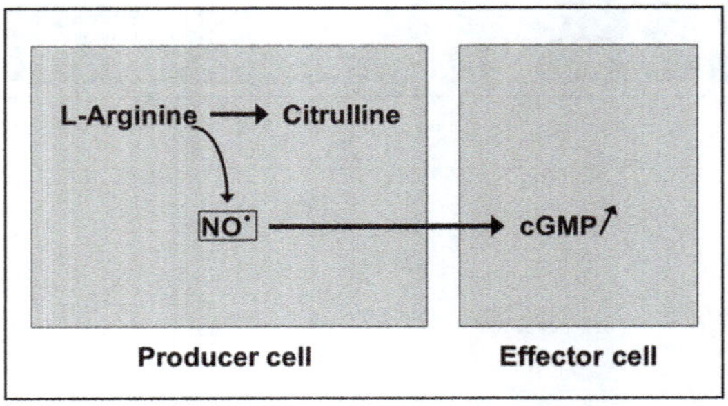

Fig. 12.7 The effect of nitric oxide diffusing into neighboring cells. Nitric oxide (NO) production leads to an increase in cyclic guanosine monophosphate (cGMP) in neighboring cells. The effect of cGMP depends upon the cell type. In smooth muscle cells, it promotes relaxation (From Mozaffarieh M, Flammer J (2009) *Ocular blood flow and glaucomatous optic neuropathy*. Springer Publ, Berlin. With permission)

(Fig. 12.6) or bradykinin stimulate NOS and thereby increase NO production.

If NO diffuses into neighboring smooth muscle cells or pericytes, it stimulates the enzyme guanylate cyclase, thereby leading to increased production of cyclic guanine monophosphate (cGMP) (Fig. 12.7). This leads to a variety of effects depending on the cells. In the case of smooth muscle cells and pericytes, NO promotes relaxation and, therefore, vasodilation.

Is NO a good or a bad molecule? On its own, it is vasodilatory and neuroprotective. However, if it is present in the vicinity of free oxygen radicals, it has a detrimental effect.

12.4 Nitric Oxide in Medicine

As mentioned above, NO, by itself, is actually a good molecule and is, therefore, used therapeutically. However, it reacts quite rapidly with many

$$\bullet NO + \bullet O_2^- \longrightarrow ONOO^-$$

Fig. 12.8 Formation of peroxynitrite ($ONOO^-$). Nitric oxide reacts with the superoxide anion to form the highly damaging peroxynitrite

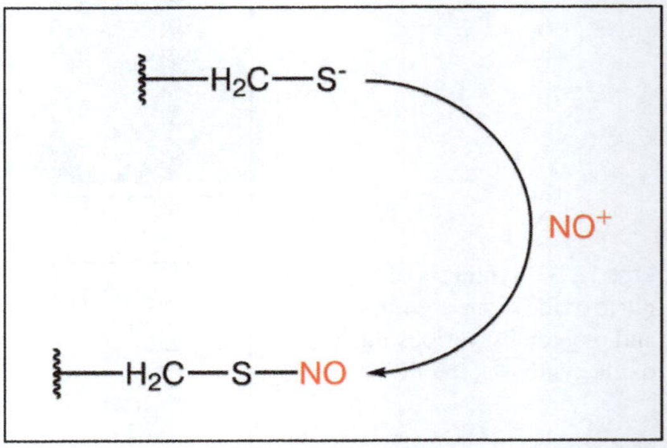

Fig. 12.9 Nitrosylation reaction. Peroxynitrite leads to the nitrosylation of protein, particularly of the SH groups

other radicals, including the oxygen radical superoxide anion ($O_2^{\bullet -}$), generating harmful peroxynitrite ($ONOO^-$) (Fig. 12.8). In fact, in a variety of nitrosylation disorders of SH-groups frequently occur in diseased tissues. SH-groups react either with NO or (more damaging) with peroxynitrite (Fig. 12.9). This has been shown in neurodegenerative diseases, atherosclerosis, rheumatoid arthritis, adult respiratory distress syndrome, cancer, and others.

The measurement of NO production is used by some for diagnostic purposes in various disorders. The half-life of NO is much longer in air than in tissues. NO in air can be measured directly with a microsensor. An example in which the direct measurement of NO is used for diagnostic purposes is the measurement of exhaled NO as an index for airway inflammation (e.g., chronic lung inflammation). However, since NO has only a very short half-life in tissues, it is difficult to quantify under clinical conditions. Therefore, the metabolites of NO – such as the inorganic anions nitrate (NO_3^-) and nitrite (NO_2^-) – are measured as a surrogate for NO. Incidentally, NO can also be recycled from these metabolites. The multiple roles of NO in the eye will be covered later.

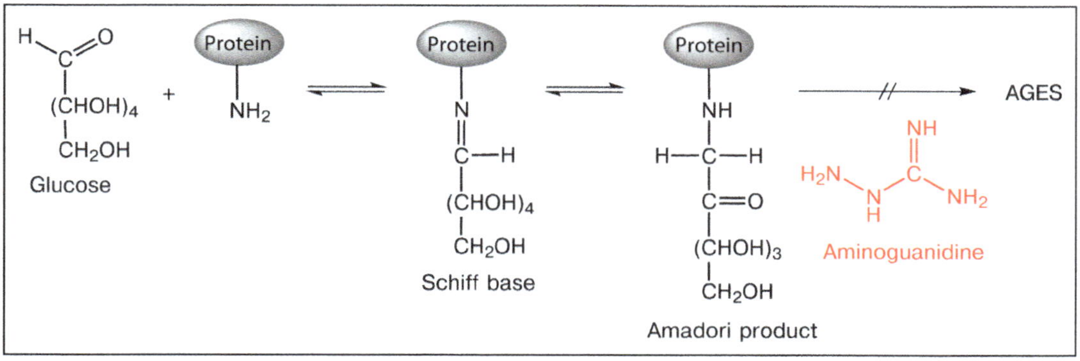

Fig. 12.10 Prevention of the formation of advanced glycation end products (AGEs) by aminoguanidine. AGEs are unwanted products of the reaction between glucose and proteins. This reaction can be interrupted by the molecule aminoguanidine

Reducing NO concentration by inhibiting the NOS enzyme is also therapeutically promising. Since the inhibition of NOS decreases peroxynitrite production, it has therapeutic value in neurodegenerative diseases. An example of a selective inhibitor of NOS2 is the drug aminoguanidine. This drug has been used for a long time to treat type 2 diabetes mellitus because it stimulates the production of insulin. Later, aminoguanidine was used as an NOS2 inhibitor in experimental studies to prevent neurodegenerative diseases. Correspondingly, in experimental animals, aminoguanidine prevented or reduced the development of GON (glaucomatous optic neuropathy).

Aminoguanidine also has additional independent effects, such as in the prevention of the formation of advanced glycation end products (AGEs). AGEs are the end products of glycation reactions, where a sugar molecule bonds to either a protein or lipid molecule without an enzyme to control the reaction. The initial product of this reaction is called a Schiff base, which spontaneously rearranges itself into an Amadori product (Fig. 12.10). A lowered glucose concentration will unhook the sugars from the amino groups to which they are attached; conversely, high glucose concentrations will have the opposite effect if they are persistent. A key characteristic of certain reactive or precursor AGEs is their ability to form covalent cross-links between proteins, which alters protein structure and function. The formation and accumulation of AGEs has been implicated in the progression of age-related diseases.

It may sound paradoxical that both NOS inhibitors and NO donors have certain beneficial effects for clinical use. NO donors, for example, have been used for more than a century to treat angina pectoris.

We shall now discuss the role of NO with particular reference to the eye.

12.5 Nitric Oxide in the Eye

As discussed, the effect of NO is somewhat like a double-edged sword. On one hand, NO exerts beneficial effects (e.g., in the regulation of vascular tone or aqueous humor production). On the other hand, it contributes to pathologies such as GON. The reason for the latter is the ability of NO to fuse with free oxygen radicals, which leads to the production of toxic molecules. Thus, the ultimate effect of NO, as we shall see later on, depends on its location and concentration, and on the presence of other molecules in the vicinity.

12.5.1 NO and Aqueous Humor Dynamics

The presence of NOS, as well as the demonstration of NO production in the ciliary body, suggests that NO plays a role in the physiology of

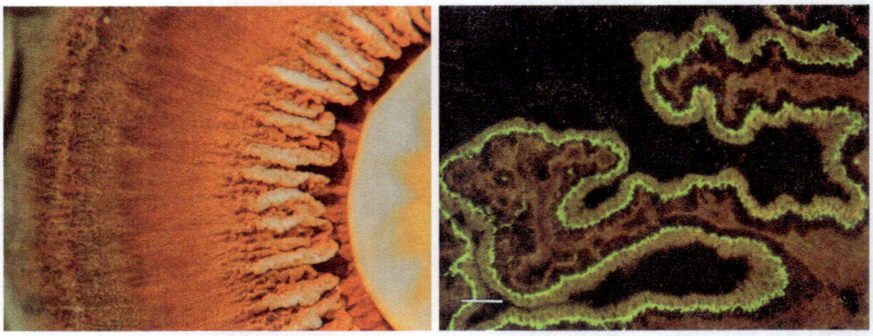

Fig. 12.11 Nitric oxide synthase in the ciliary body. *Left*: ciliary body processes in a cadaver eye. *Right*: immunostaining of nitric oxide synthase in a histological section of the ciliary body (green staining indicates staining of NOS2). (*Left*: from Flammer J (2006) *Glaucoma*. Hogrefe&Huber, Bern. With permission. *Right*: from Meyer P, et al (1999) Curr Eye Res, 18. With permission)

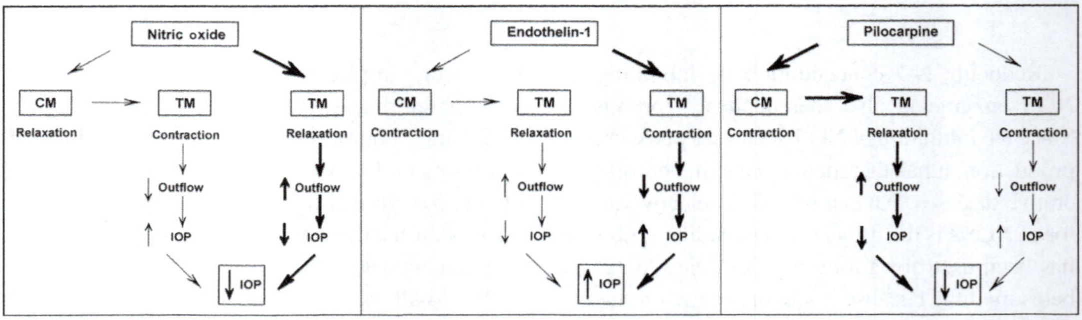

Fig. 12.12 Nitric oxide and Endothelin as important modulators of IOP. *Left*: NO causes relaxation of the contractile elements of the trabecular meshwork (TM), leading to increased outflow and, thus, a lower IOP. The direct effect on the TM is stronger than the indirect opposite effect via the ciliary muscle. *Middle*: Endothelin-1 (ET-1) causes a contraction of the contractile elements of the trabecular meshwork (TM), leading to decreased outflow and, thus, a higher IOP. *Right*: Pilocarpin leads to a contraction of the ciliary muscle and thereby to a relaxation of the contractile elements of the trabecular meshwork (TM), leading to increased outflow and, thus, a lower IOP. This indirect effect is stronger than the direct effect on the TM. (From Wiederholt M. In: Haefliger IO, Flammer J (1998) *Nitric oxide and Endothelin in the pathogenesis of glaucoma*. Lippincott-Raven, Philadelphia. With permission)

aqueous humor production. In the ciliary body, NO synthesis occurs via the activity of NOS2 and NOS3, depending on the particular species. Immunohistochemical findings are presented in Fig. 12.11. Interestingly, NOS2 (as COX-2) in the ciliary body is expressed constitutively, whereas in other tissues, it is expressed only by induction.

The trabecular meshwork also contains NOS3 in its outer, intermediate, and juxtacanalicular areas. The trabecular meshwork contains contractile elements, which are relaxed by NO. This relaxation leads to decreased outflow resistivity and, thus, a lower IOP. As side information, Endothelin (ET) (see Chap. 16) causes exactly the opposite – a contraction of the TM contractile elements, increased outflow resistivity, and, thus, a higher IOP. Obviously, NO (and ET) are important modulators of IOP. We would like to emphasize that NO and ET both have an effect on ciliary muscles as well as on the trabecular meshwork. While contraction of the trabecular meshwork increases IOP, contraction of the ciliary muscle decreases the IOP. The net effect, however, is what is clinically relevant. In the case of pilocarpin, the effect on the ciliary muscle is stronger than that on the trabecular meshwork. For this reason, the resulting effect is a decrease in IOP (Fig. 12.12).

Incidentally, the main cause of increased IOP is chronic changes in the trabecular meshwork

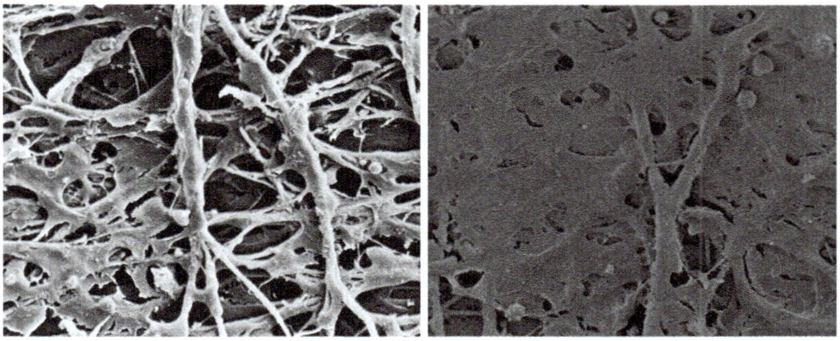

Fig. 12.13 Changes in the trabecular meshwork (TM). *Left*: normal TM. *Right*: pathologically changed TM showing depositions reducing the outflow facility. (From Flammer J (2006) *Glaucoma*. Hogrefe&Huber, Bern. With permission)

(Fig. 12.13), which lead to decreased aqueous outflow. These changes are brought about by oxidative stress.

12.5.2 NO and Ocular Blood Flow

The endothelial cells release a number of vasoactive factors, both intraluminally, influencing the rheology, as well as abluminally, influencing the vessel size (Fig. 12.14).

One of the major players with a vasodilatory effect is NO, produced under normal conditions by the constitutively expressed e-NOS. Molecules such as bradykinin increase the production of NO.

The basal production of NO is a prerequisite for physiological ocular blood flow (OBF). Ex vivo studies in isolated ciliary vessels have shown that inhibition of NO production by L-NMMA (L-NG-monomethyl argenine) leads to marked vasoconstriction, as shown in Fig. 12.15.

Similarly, when NO production is inhibited in a perfused eye model, the perfusion of the eye drops by 40–50% under the condition that the perfusion pressure is kept constant (Fig. 12.16). A reduction of OBF can also be observed in healthy subjects when NOS inhibitors are infused intravenously.

NO also plays a role in neurovascular coupling. Neurovascular coupling refers to the vascular response to increased neuronal activity (Fig. 12.17).

If, for example, we expose the retina to flickering light, the retinal vessels widen. Neurovascular coupling can be measured by means of a retinal vessel analyzer (RVA), an

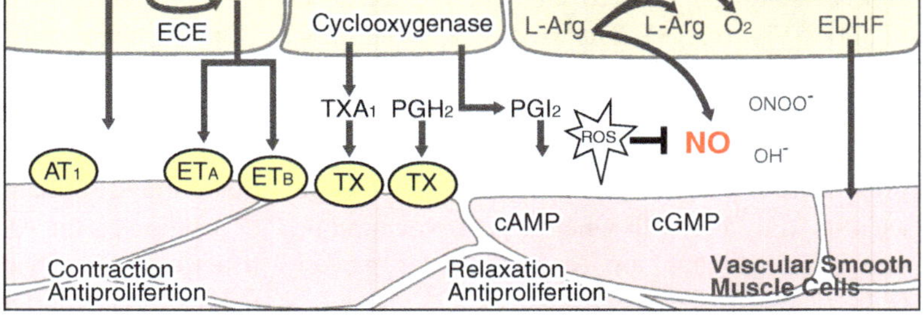

Fig. 12.14 Vasoactive factors released by endothelial cells. Different vasoactive factors are released both intraluminally and abluminally by the endothelial cells. One important factor is nitric oxide (NO), which has a vasodilatory effect. (Modified after Flammer AJ, et al (2010) Pflugers Arch)

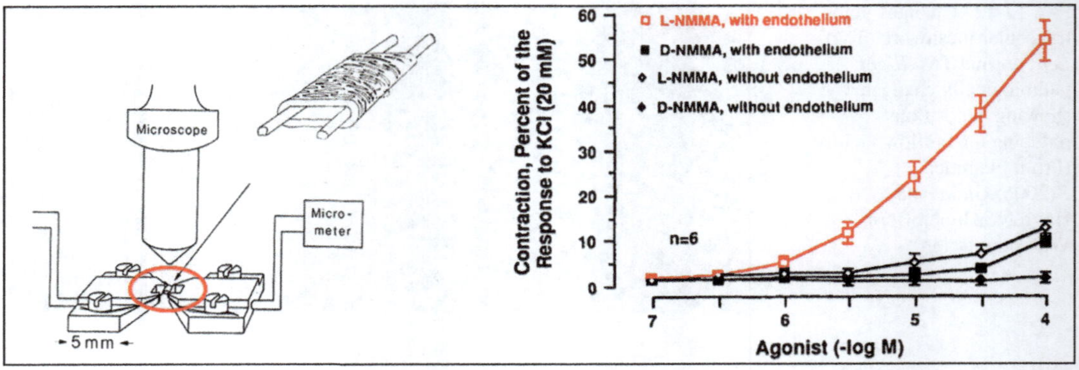

Fig. 12.15 Inhibition of NO production leads to marked vasoconstriction. *Left*: Small rings of vessels mounted on a myograph system. *Right*: If L-NMMA is added, this vessel ring constricts in a dose-dependent manner. (From Yao K, et al (1991) Invest Ophthalmol Vis Sci, 32. With permission)

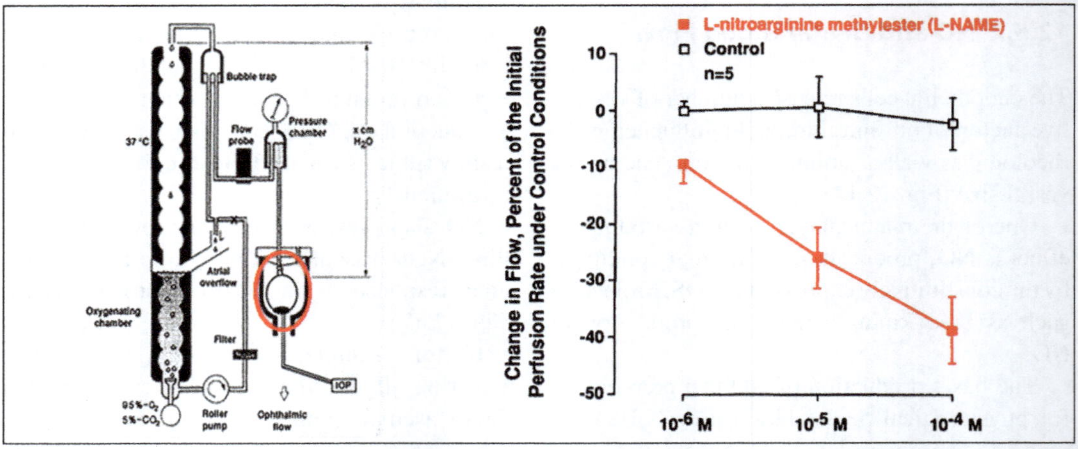

Fig. 12.16 The perfused eye model. *Left*: an enucleated eye is perfused under constant perfusion pressure. *Right*: if the production of NO is inhibited by L-NAME, the perfusion drops. (From Meyer P, et al (1993) Invest Ophthalmol Vis Sci, 34. With permission)

instrument that quantifies the size of the arteries and veins along a selected segment over a period of time (Fig. 12.18). The vessel diameter provides only indirect information about ocular blood flow. Nevertheless, the size of the arteries and veins and their spatial and temporal variation provides very useful information that can be used to study in depth whether a provocation or treatment dilates or constricts a vessel (Fig. 12.19).

12.5.3 NO in Eye Disease

Both the over- and underproduction of NO can lead to pathological conditions in the eye. The concentration of NO may be quite inhomogeneous in a certain organ, such as the eye, meaning that it may simultaneously be too high or too low depending on the particular area. This even holds true for the distribution of NO within an individual cell. Decreased levels of NO in vascular

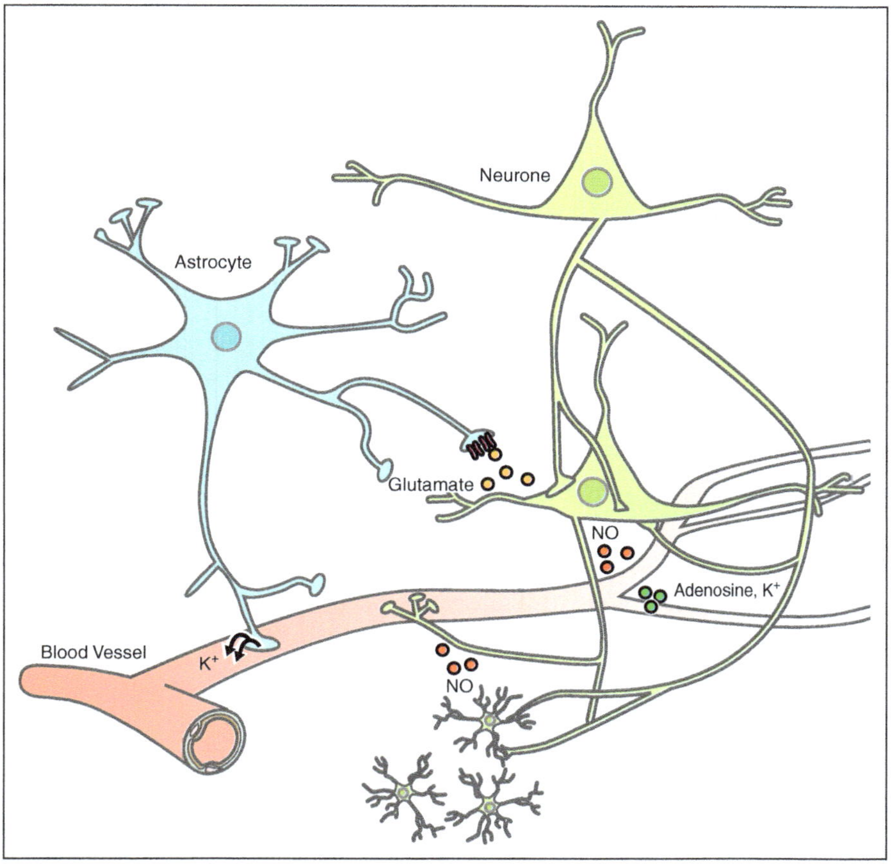

Fig. 12.17 Neurovascular coupling. Neuronal activation causes the release of glutamate, nitric oxide (NO), potassium (K+), and adenosine. In this context, NO is the main vasodilator (Modified after D'Esposito M, et al (2003) Nature Reviews Neuroscience)

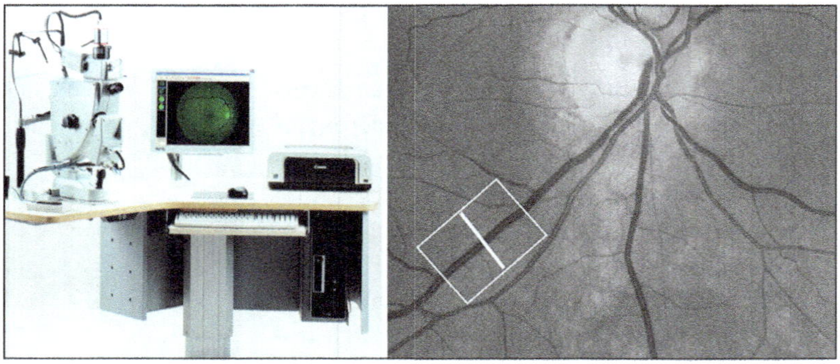

Fig. 12.18 Measurement of neurovascular coupling by means of a retinal vessel analyzer. *Left*: photograph of a "retinal vessel analyzer" used to measure neurovascular coupling. (Courtesy of Imedos Systems UG, Jena, Germany). *Right*: fundus showing a segment of a vessel monitored during a session

smooth muscle cells are unfavorable, as this leads to decreased perfusion, as in the case of diseases with an endothelial dysfunction, such as atherosclerosis. However, other conditions are recognized where the enhanced production of NO leads to damage. Next, we shall address this latter condition with the help of two examples.

A high concentration of NO in and around the neuronal axons can be damaging and lead to glaucomatous optic neuropathy (GON). The

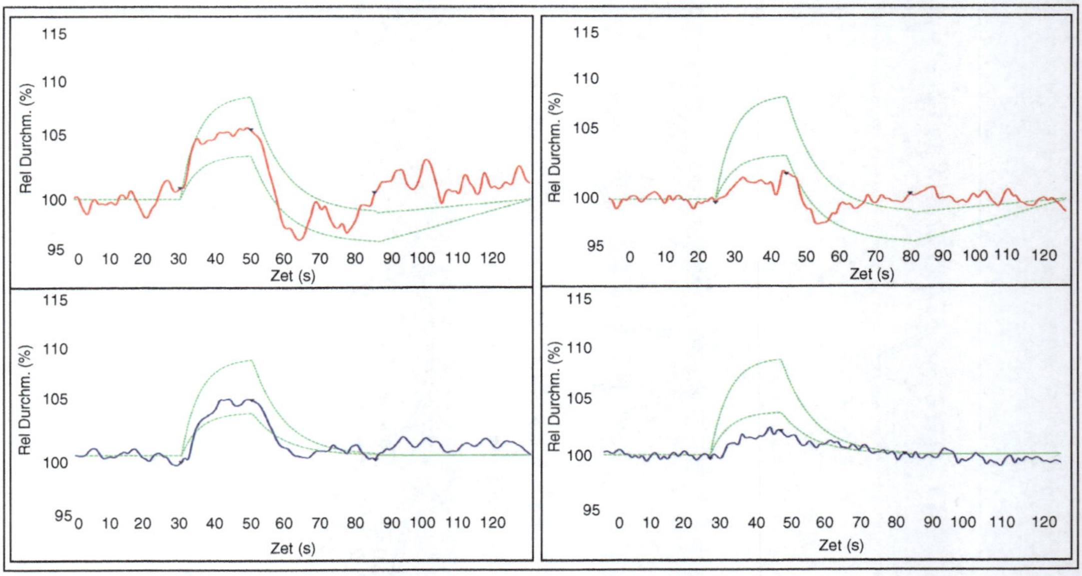

Fig. 12.19 Retinal vessel analyzer (RVA). *Left*: reaction of the artery (*top*) and vein (*bottom*) to flickering light of a healthy volunteer. *Right*: reaction of the artery (*top*) and vein (*bottom*) to flickering light of a glaucoma patient

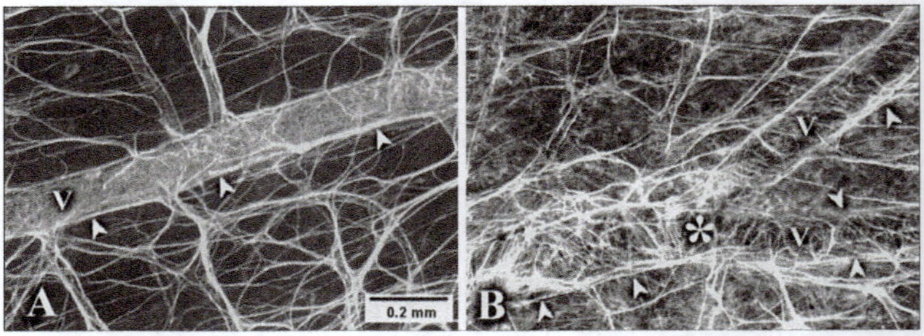

Fig. 12.20 Activation of glial cells in the retina. *Left*: the processes of astrocytes connect vessels with neurons. *Right*: the regular pattern is lost if astrocytes are activated. (From Wang L, et al (2002) Invest Ophthalmol Vis Sci, 43. With permission)

reason is that NO fuses with oxygen free radicals such as superoxide anion ($O_2^{\bullet-}$) to form peroxynitrite ($ONOO^-$), which kills cells. The question arises as to why NO is produced in excess. A lot of NO is produced when astrocytes are activated, either mechanically or by Endothelin. Thus, NO may be produced by the activation of retinal astrocytes (Fig. 12.20) or optic nerve head astrocytes, as well as by Müller cells (Fig. 12.21).

Incidentally, when astrocytes are activated, they change their morphology, which increases light scattering. This can be observed clinically as glinting spots ("gliosis-like alterations").

Superoxide anion ($O_2^{\bullet-}$), on the other hand, is overproduced in the mitochondria if the oxygen

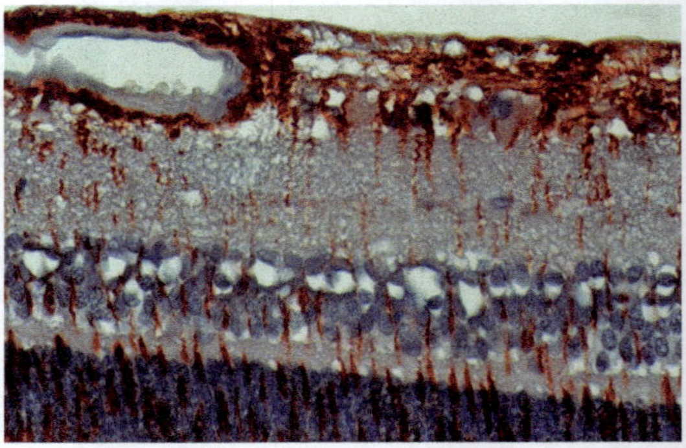

Fig. 12.21 Activation of Müller cells in the retina. Müller cells, similar to astrocytes, can be activated as demonstrated in this cross-section of the retina by glial fibrillary acid protein (GFAP) staining (*brown color*). (Courtesy of P. Meyer, University of Basel)

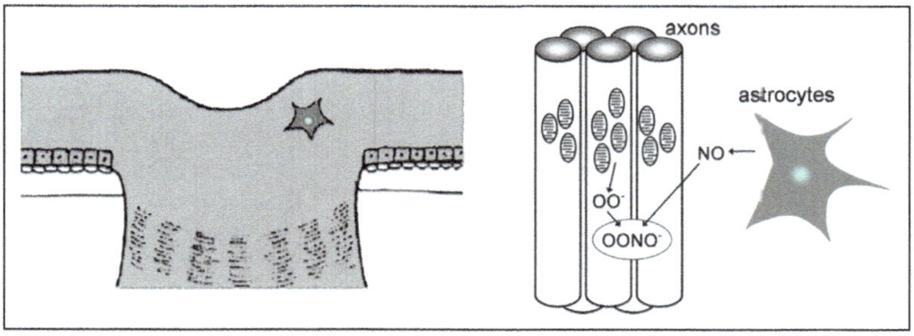

Fig. 12.22 Formation of peroxynitrite in the optic nerve head. *Left*: schematic drawing of astrocytes in the optic nerve head. *Middle*: axons of the optic nerve head crowded with mitochondria. *Right*: astrocyte capable of producing NO (modified after Neufeld 1999)

supply is unstable. This is particularly the case in the axons of the optic nerve head, where mitochondria are crowded due to the lack of myelin sheaths. In this area, the oxygen supply is frequently unstable (see Sect. 9.7). As depicted in Fig. 12.22, NO can easily diffuse from the astrocytes into the axons and react with the superoxide anion ($O_2^{\bullet-}$) to form the damaging toxic peroxynitrite ($ONOO^-$). In contrast, the superoxide anion ($O_2^{\bullet-}$) and peroxynitrite ($ONOO^-$) cross the cell membrane very poorly. Peroxynitrite, however, can diffuse within the axons in both directions (toward the retina as well as toward the lateral geniculate ganglion) and induces apoptosis (cell death).

A high amount of NO is also produced in uveitis. Here, immune cells involved in inflammatory processes produce additional NO. Large amounts of reactive oxygen species (ROS) are produced in addition to NO. Similar to what we discussed for the pathogenesis of GON, NO can fuse with the superoxide anion ($O_2^{\bullet-}$) to produce peroxynitrite ($ONOO^-$), which is particularly damaging to tissues. This explains why the inhibition of NOS2 has certain therapeutic effects. We shall now discuss the role of NO in therapy.

12.5.4 NO in Therapy

Since both the enhanced production of NO as well as the underproduction of NO are involved in various pathologies, both NO donors and inhibitors are used for therapeutic purposes.

NO donors are a group of nitrogen-containing compounds that are able to release NO in tissues. An example of an NO donor is nitroglycerin (Fig. 12.2), which has been used for many years to treat angina pectoris. It works by decreasing the oxygen demand of the heart and by dilatating vessels or reversing vasospasm respectively. By the way, an attempt has been made to use an NO-donating latanoprost derivative to deliver NO locally to the eye.

The inhibition of NO production is a promising approach that has already proven to be effective on an experimental level. As described before, in glaucoma, the activated astrocytes produce high amounts of NO through the increased expression of NOS2. Several possibilities can reverse this condition. The first possibility is to inhibit NOS2 using aminoguanidine. A second possibility is to prevent the activation of astrocytes either by blocking the epidermal growth factor receptor (EGFR) using a tyrosine kinase inhibitor or by inhibiting the ET effect. Indeed, in experimental animals, these approaches can help to prevent GON.

The inhibition of NO can also be counterproductive. Retrobulbar anesthesia reduces ocular blood flow more than what can be expected by the mechanical compression of adding volume to the orbit. This is due to the fact that local anesthetics inhibit NOS3 in retroocular vessels, in part directly and in part by reducing the stimulating effect of bradykinin or acetylcholine (Fig. 12.23). This is one of the reasons that surgeons prefer to give local anesthetics subconjunctivally instead of retrobulbarly, particularly in the case of advanced glaucoma patients.

The vasoconstrictive effect of three local anesthetics, lidocaine, bupivacaine, and mepivacaine, is illustrated in Fig. 12.24.

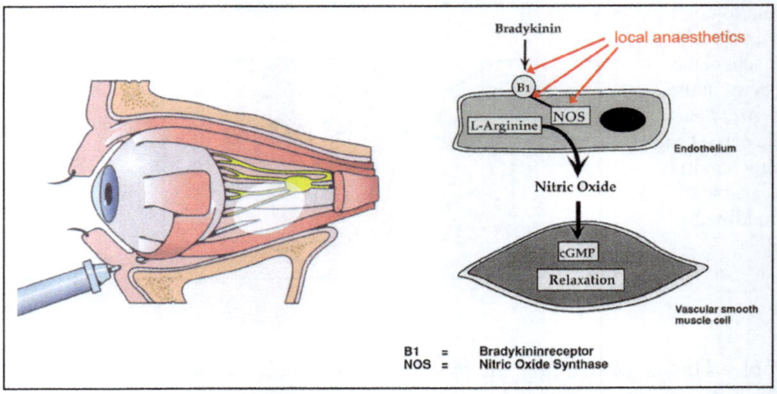

Fig. 12.23 Effect of local anesthesia. *Left*: schematic illustration of retrobulbar injection of anesthesia. *Right*: local anesthetics inhibit NOS3 in retroocular vessels and, therefore, the production of NO, leading to vasoconstriction and to a reduction in ocular blood flow. (*Left*: From Flammer J (2006) *Glaucoma*. Hogrefe&Huber, Bern. With permission. *Right*: From Grieshaber MC, et al (2007) Surv Ophthalmol, 52 Suppl 2. With permission)

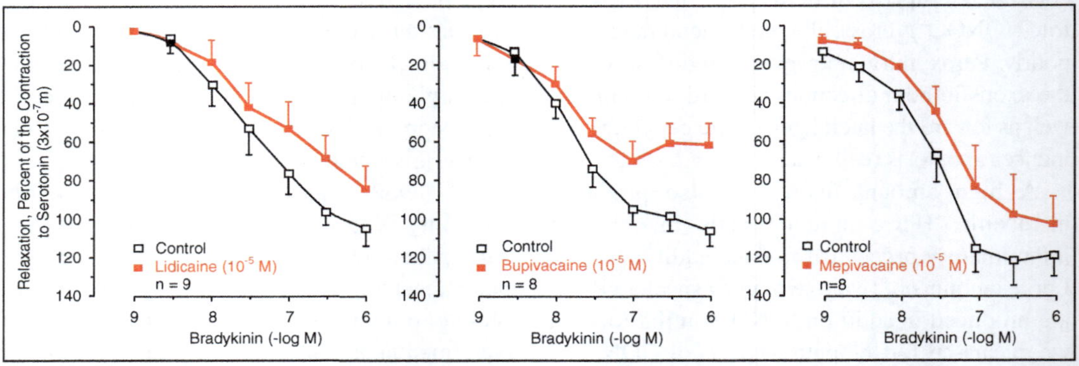

Fig. 12.24 Nitric oxide synthase (NOS) inhibition by local anesthetics. The inhibitory effect on NOS is similar for the different types of anesthetics (lidocaine, bupivacaine, mepivacaine) used. (From Meyer P, et al (1993) Invest Ophthalmol Vis Sci, 34. With permission)

Redox Reactions

13

13.1 Redox Chemistry and Terminology

The term "oxidation" refers to the loss of electrons and "reduction" refers to the gain of electrons, as shown in Fig. 13.1.

The oxidation of iron is an example of a redox reaction. When iron is oxidized by oxygen, iron oxide results, which is known as "rust." An example is the formation of rust in the cornea when a foreign body sticks in the cornea (Fig. 13.2).

Besides the formation of iron oxide, iron can also be oxidized by transferring electrons to hydrogen peroxide in the so-called Fenton reaction, thereby creating reactive oxygen species (ROS) (Fig. 13.3).

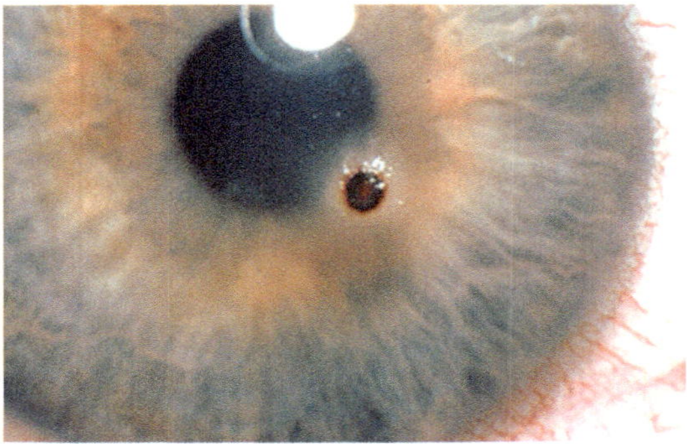

Fig. 13.2 Corneal foreign body. Iron in the cornea forms a rust ring quite quickly. This rust ring, in turn, damages the cornea by facilitating redox reactions

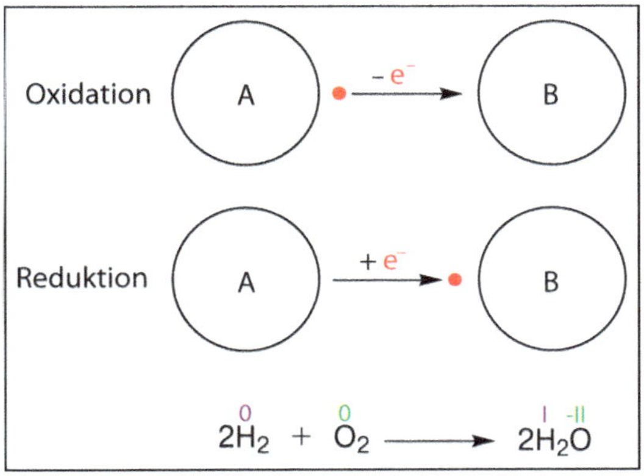

Fig. 13.1 Redox reactions. Oxidation is the loss of electrons by a molecule, atom, or ion. Reduction is the gain of electrons by a molecule, atom, or ion. Substances that have the ability to oxidize other substances are known as oxidizing agents. Substances that have the ability to reduce other substances are known as reducing agents

$$Fe^{2+} + H_2O_2 \xrightarrow{-e^-} Fe^{3+} + \bullet OH + OH^-$$

Fig. 13.3 Fenton reaction, which occurs between iron II and hydrogen peroxide in aqueous solutions. Hydrogen peroxide oxidizes iron II (Fe^{2+}) and, in the process, generates the highly reactive hydroxyl radical (OH)

Pro-oxidants are molecules that can oxidize other molecules; in other words, they have an oxidizing potential that is stronger than the oxidizing potential of the molecule they react with. A small but pathophysiologically important subgroup of prooxidants is reactive oxygen species (ROS). ROS shall subsequently be discussed in more detail.

J. Flammer et al., *Basic Sciences in Ophthalmology*,
DOI 10.1007/978-3-642-32261-7_13, © Springer-Verlag Berlin Heidelberg 2013

13.2 Production of ROS

ROS are oxygen-containing molecules that are highly reactive. ROS can occur in the form of free radicals (e.g., superoxide anion) or non-radical species (e.g., singlet oxygen). Free radicals are molecules that contain one or more unpaired electrons. The presence of an unpaired electron makes free radicals highly reactive because of the need to pair up the unpaired electron.

The normally occurring oxygen molecule in the atmosphere is relatively inert and called "ground-state oxygen" (Fig. 13.4) due to the fact that the unpaired electrons in the oxygen molecule have the same spin (refer to Chap. 9). Thus, for the oxygen molecule to react, it would need another molecule or atom with two unpaired electrons with a parallel spin opposite to that of the oxygen molecule. This rarely occurs, which explains why oxygen in the ground state is only minimally reactive.

The activation of ground-state oxygen may occur by two different mechanisms: either through the absorption of sufficient energy to reverse the spin on one of the unpaired electrons or through monovalent reduction. If ground-state oxygen absorbs sufficient energy to reverse the spin of one of its unpaired electrons, the two unpaired electrons now have opposite spins (see Sect. 9.5). Due to a change in spin direction, this oxygen molecule, called singlet oxygen ($^1O_2$), is much more reactive than ground-state oxygen and reacts destructively, particularly with molecules with double bonds.

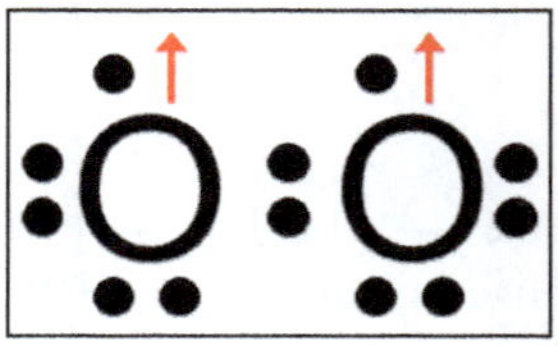

Fig. 13.4 Ground-state oxygen, in which the last two electrons of the oxygen molecule are located in a different π* antibonding orbital. These two unpaired electrons have the same quantum spin number (they have parallel spins) and qualify ground-state oxygen as a diradical

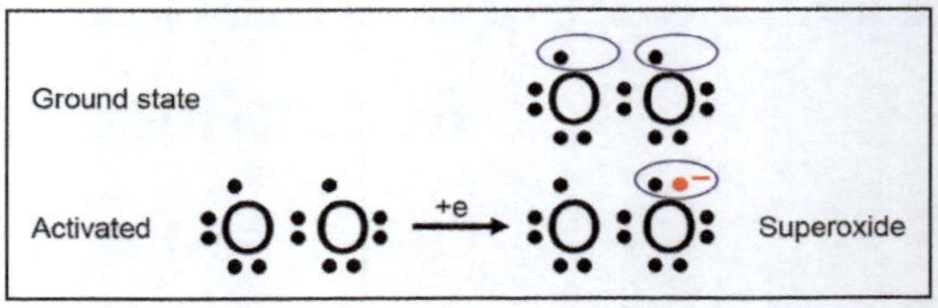

Fig. 13.5 Activation of ground-state oxygen, which can occur by the monovalent reduction of ground-state oxygen to form the more reactive superoxide anion

The second mechanism of activation is by the stepwise monovalent reduction of oxygen, which gives rise to the superoxide anion radical (O_2^-) (Fig. 13.5), hydrogen peroxide (H_2O_2), and, finally, to water (H_2O).

Both of these mechanisms also occur in our bodies. The production of singlet oxygen is of particular importance in organs exposed to light, such as the eye and the skin, whereas the production of superoxide anion occurs essentially in any tissue. In aerobic cells, energy is gained by oxidizing molecules such as sugar. During this energy metabolism, a certain amount of ROS is produced physiologically as a byproduct. This occurs mainly during electron transfer in the respiratory chain. The tetravalent reduction of molecular oxygen (together with hydrogen) by the mitochondrial electron–transport chain produces water. This system, however, does not always work perfectly and, as a result of electron leakage, the univalent reduction of oxygen molecule forms superoxide anions (O_2^-). ROS can also be built as intermediates in enzymatic reactions. Certain cell types such as macrophages can produce high amounts of ROS as a part of their defense mechanism (Fig. 13.6).

ROS production is obviously not only detrimental but can also be beneficial. It is, for example, a prerequisite to achieve the optimal physical training effect. However, an unwanted increase in ROS occurs in a number of pathologic conditions such as those resulting from an unstable oxygen supply. This is particularly relevant in elderly people, as the capacity to eliminate free radicals decreases with age (Fig. 13.7). Under optimal conditions, the magnitude of ROS formation is

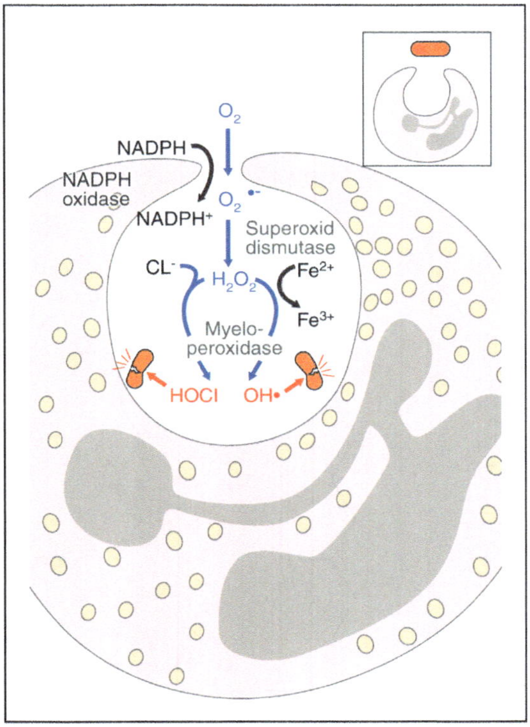

Fig. 13.6 Respiratory burst. The rapid release of reactive oxidative species (ROS) is referred to as "respiratory burst." Neutrophils have oxygen-dependent mechanisms (myeloperoxidase system) for killing bacteria. After phagocytosis, NADPH oxidase, located in the leukocyte membrane, converts ground-state oxygen into ROS, which, in turn, attacks bacteria

balanced by the rate of ROS elimination through the available antioxidants. If the production of ROS exceeds the capacity of its elimination, the amount of ROS increases to a level that leads to oxidative stress. This condition is described in greater detail below.

13.3 Oxidative Stress

Nature has provided us with mechanisms to cope with ROS with the help of antioxidants. An antioxidant, by definition, is any substance that reduces oxidative damage, such as that caused by free radicals. The antioxidant

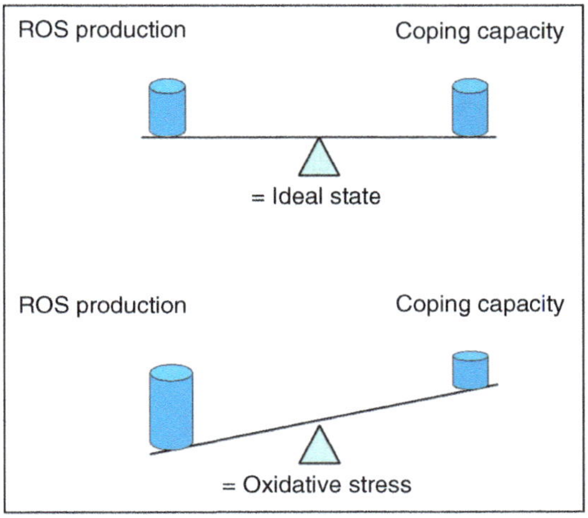

Fig. 13.7 Coping with free radicals. Under optimal conditions, the rate and magnitude of oxidant formation is balanced by the rate of oxidant elimination through the action of antioxidants. If ROS production, however, exceeds this capacity, oxidative stress damages different molecules

defense system is comprised of a variety of molecules. These include proteins such as enzymes (e.g., superoxide dismutase and catalase) and low-molecular mass molecules (e.g., some vitamins).

Oxidative stress can damage molecules. One example is lipid degradation initiated by lipid peroxidation (Fig. 13.8). This is a process by which molecular oxygen (O_2) is inserted into an unsaturated fatty acid (called a lipid peroxide). This process is initiated by free radicals. The formation of peroxide again leads to the formation of free radicals, inducing a chain reaction.

As long as nature is capable of repairing damaged molecules (e.g., DNA) or eliminating them (e.g., proteins via proteosomes), no major structural damage will occur. However, if oxidative stress exceeds the capacity of antioxidants and the molecular damage exceeds the capacity of repair or elimination mechanisms, structural damage will occur. This results in damage that is ultimately clinically relevant as a basis of diseases (Fig. 13.9). This can occur in any organ, including the eye. In this context, we will focus on eye diseases.

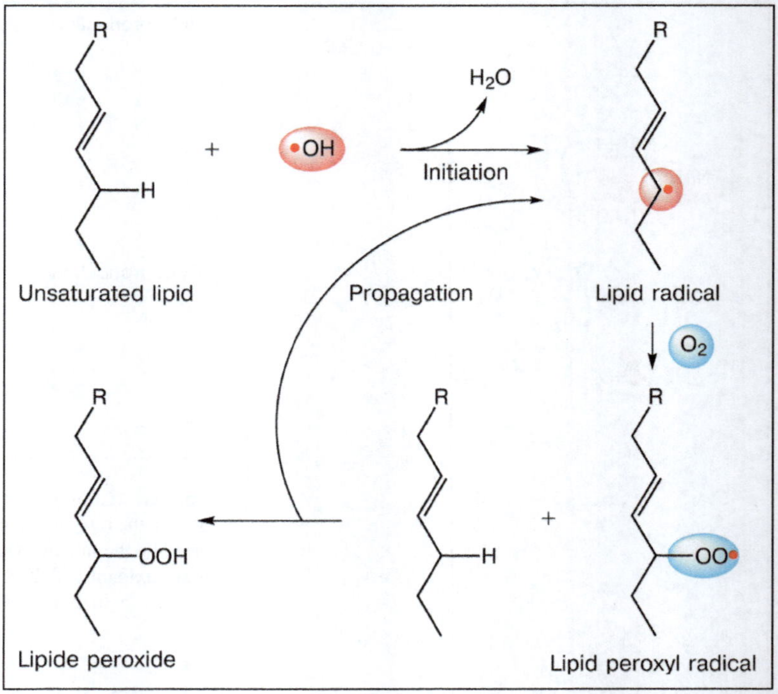

Fig. 13.8 Lipid peroxidation, the oxidative degradation of lipids. It is the process through which free radicals "steal" electrons from the lipids, forming a lipid radical. This lipid radical can then further react with molecular oxygen to form a lipid peroxyl radical, which, in turn, can oxidize other lipid molecules

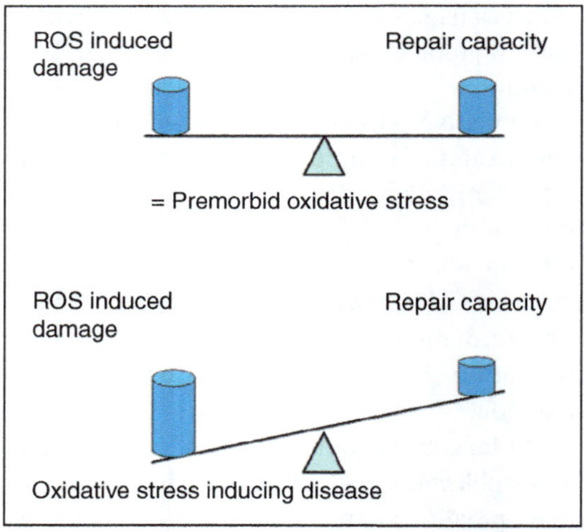

Fig. 13.9 Repair of molecular damage. *Top*: balance between induced damage and repair capacity. *Bottom*: imbalance between induced damage and repair capacity

13.4 Oxidative Stress in the Eye

The cornea and the conjunctiva are in direct contact with the ambient air. This has a positive effect on the cornea, as it can receive its oxygen supply from the air, but it also has a negative effect, as the cornea and the conjunctiva are exposed to toxic substances such as ozone in the air. Likewise, the retina requires light for vision but light can also contribute to the production of ROS via photodynamic procedures. We will now discuss some of the consequences of oxidative stress in the eye.

Oxidative stress plays a role in the pathogenesis of cataract formation (Fig. 13.10).

The lens of our eye contains water-soluble proteins, so-called crystallins. These proteins compose 90 % of the lens. For the lens to remain transparent, crystallins have to be regularly organized and tightly packed (see Sect. 2.3). The lens loses its transparency when the distance between the lens fibers is larger than half the wavelength of light that ought to pass through or if these fibers lose their regular arrangement. Both can be the result of oxidative stress. Indeed, proteins isolated from cataractous lenses reveal oxidative damage.

Oxidative stress may also affect the vitreous humor. With age, "floaters" (also called "mouches volantes") are often perceived (Fig. 13.11). The cause of these is the depo-

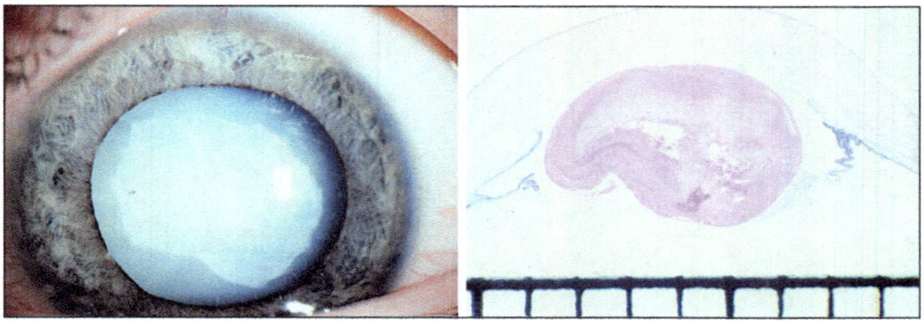

Fig. 13.10 Cataract. *Left*: clinical picture. *Right*: histological section. Cataract can be caused by oxidative stress (Courtesy of P. Meyer, University of Basel)

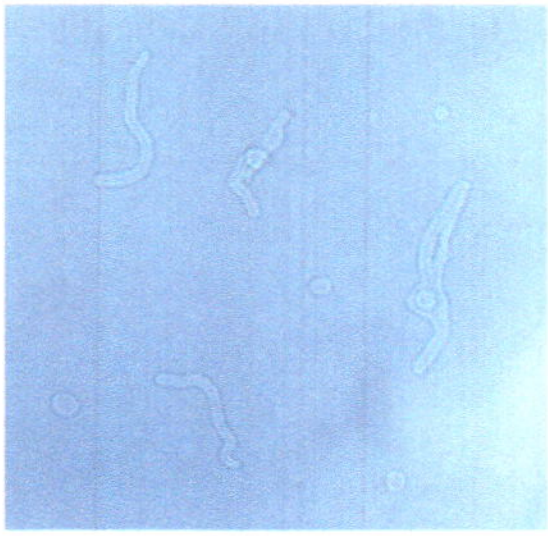

Fig. 13.11 Mouches volantes. Packed hyaluronic acids throw moving shadows on the retina

lymerization of hyaluronic acid, which is due in part to oxidative stress. The depolymerization leads to the clumping of the collagen fibrous skeleton.

Oxidative stress in the trabecular meshwork increases IOP. The human trabecular meshwork is composed of collagen lamellae lined by endothelial cells. The space between the collagen beams of the trabecular meshwork is filled with extracellular matrix, composed mostly of glycoproteins and proteoglycans, where the aqueous humor filters through (Fig. 12.13). Oxidative stress leads to an overexpression and alteration of glycoproteins as well as the loss of trabecular endothelial cells. This, in turn, leads to impaired aqueous humor outflow and causes an increase in IOP. The role of oxidative stress in the pathogenesis of GON was described in Sect. 12.5.3.

Oxidative stress also plays a role in age-related macular degeneration (AMD). In the retina, a particularly high production of ROS results from the interplay of high light exposure, high concentration of oxygen, and lipofuscin (pigment granules composed of lipid-containing residues of lysosomal digestion). Oxidative stress is quite damaging to the retina due to its high concentration of polyunsaturated fatty acids (Fig. 13.8), which are readily peroxidized as they contain a high number of double bonds (see Chap. 17).

The chronic inflammatory component of AMD is at least partially due to the oxidative modified molecules. These modified molecules (e.g., malonaldehyde) activate the alternative pathway of the complement system. This is regulated by, among others, the inhibitory factor complement *H* (which also binds to the epitopes of malonaldehyde) (Fig. 13.12). If this factor *H* is

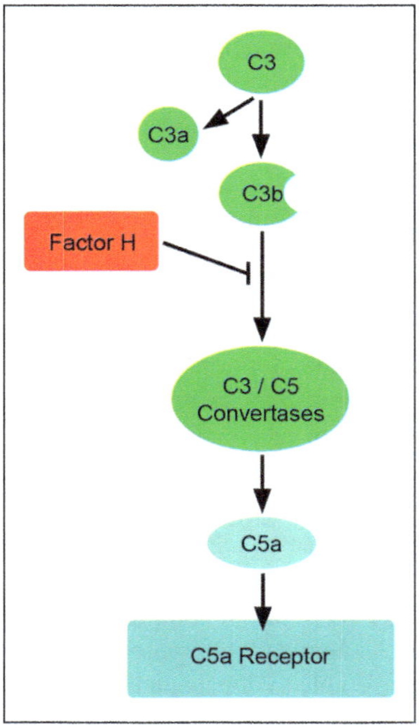

Fig. 13.12 Complement factor *H*, which inhibits the activation of the alternative pathway, which is activated in AMD

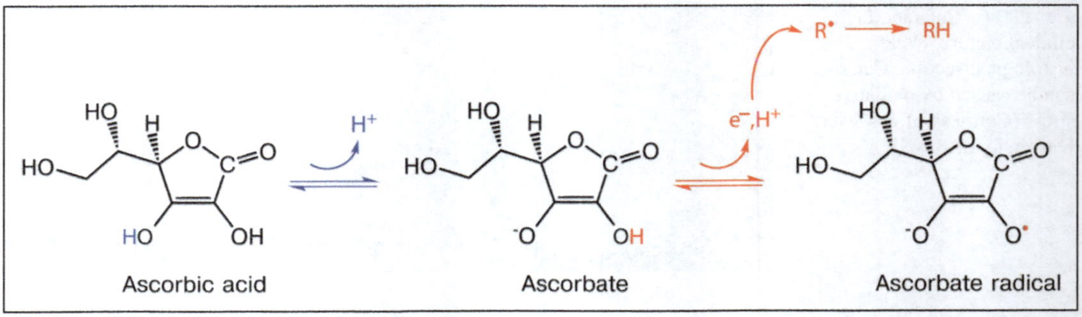

Fig. 13.13 Antioxidative effect of ascorbic acid. Ascorbic acid donates an electron to a free radical to neutralize it

Fig. 13.14 Citrus fruits contain vitamin C

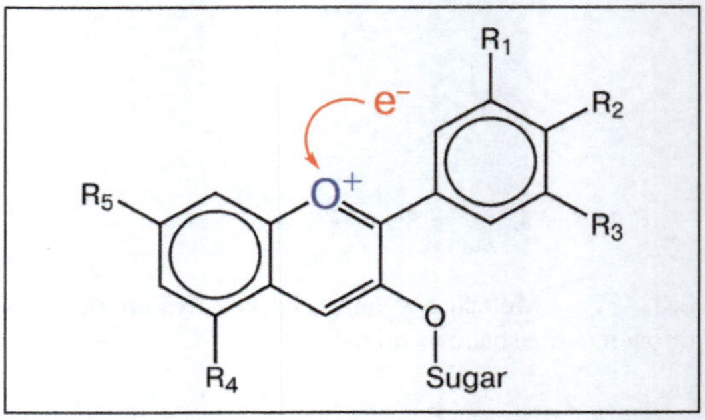

Fig. 13.15 Antioxidative effect of anthocyanins, which accept an electron from a free radical to neutralize it

Fig. 13.16 Blueberries contain anthocyanins

altered (polymorphism), the inhibitory effect is decreased, resulting in increased inflammation.

Because oxidative stress is involved in a variety of eye diseases, there is much interest in counteracting it through the regulation of antioxidant balance. This topic will be discussed in the following section.

13.5 Antioxidants

Nature has provided us with mechanisms to cope with ROS through the use of antioxidants. An antioxidant, by definition, is any substance that, when present at a lower concentration than that of an oxidizable substrate, significantly delays or prevents the oxidation of that substrate. The antioxidant defense system is composed of a variety of molecules. These include enzymes (e.g., superoxide dismutase) that are capable of the catalytic removal of ROS, and low-molecular mass molecules. An example of a low-molecular mass molecule that is able to donate an electron is vitamin C (Fig. 13.13). Vitamin C occurs particularly in citrus fruits (Fig. 13.4). Antioxidants also act by accepting an electron such as anthocyanin (Fig. 13.15), which occurs especially in blueberries and grapes (Fig. 13.16).

Another way of eliminating ROS is by the catalytic removal of ROS by enzymes such as catalase. Indirect protective mechanisms also exist, explained here by means of the example of iron (Fe). Iron (Fe^{2+}) supports the production of ROS through the Fenton reaction (Fig. 13.3). To avoid this reaction, iron is transported and stored by complex proteins such as transferrin and ferritin. Among other molecules, ROSs also attack other proteins, disrupting their tertiary structure. This is averted by heat shock proteins, whose expression is increased when cells are exposed to elevated temperatures or other stress. Heat shock proteins (HSPs) function as chaperones (Fig. 13.17). They play an important role in helping proteins to fold and keep their proper conformation and to prevent unwanted protein aggregation.

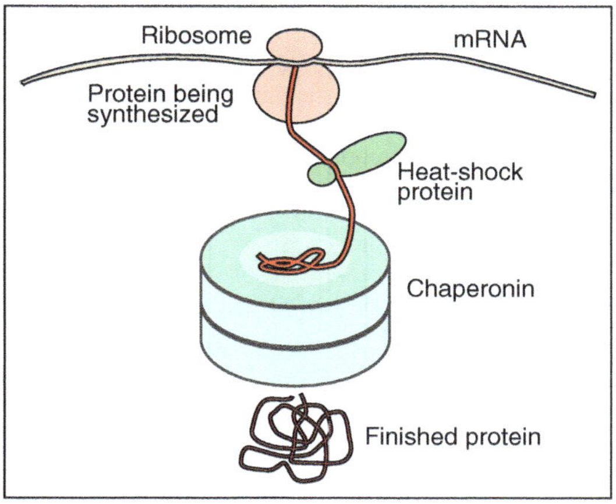

Fig. 13.17 Chaperones, which fold and unfold other proteins. Some chaperonins are heat shock proteins that are particularly expressed when a cell is under stress

Certain nutritional components, such as vitamin C or E, are known to have beneficial antioxidative properties. Vitamins C and E, for example, can prevent lipid peroxidation by working synergistically (Fig. 13.18).

The "nutraceutical" industry has, therefore, isolated and synthesized these molecules to deliver them in pure and highly concentrated forms. Although no doubt exists that antioxidative nutrition is beneficial, the question arises as to whether these highly concentrated, high-dose "drugs" in the form of tablets are meaningful. While some studies have shown beneficial effects, most have shown no effects and some even adverse effects. Although vitamin C deficiency is known to promote cataract formation, too high a level of vitamin C concentration, paradoxically, also leads to cataract formation. Too high a level of vitamin E (>400 IU per day) even increases mortality.

The question now arises as to why a high dose has an opposite and, therefore, an adverse effect. When antioxidants are given in high doses, the chances of their giving or taking an electron to non-radical species, thereby forming ROS, becomes higher. In addition, a

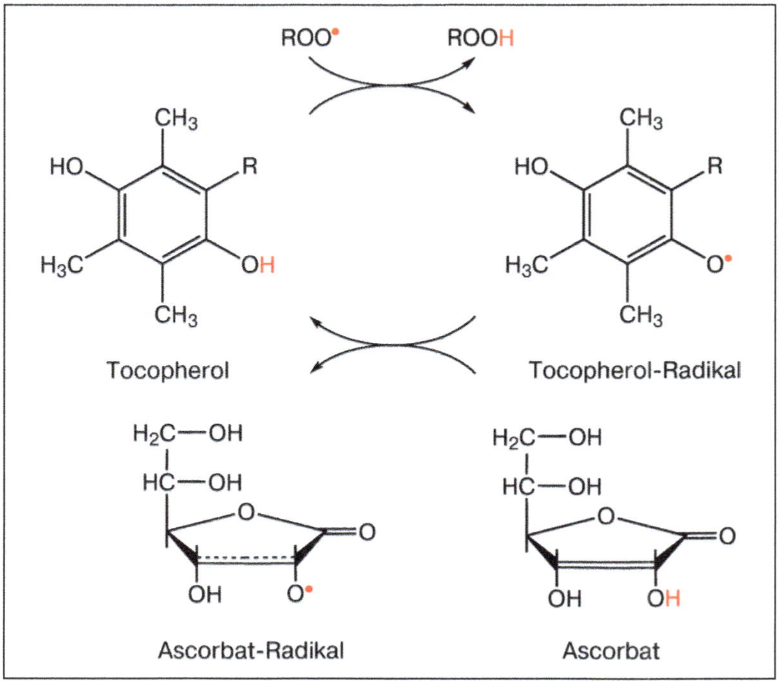

Fig. 13.18 Synergistic antioxidant activity of vitamin C and E. Vitamin E is a lipid soluble vitamin that can readily donate an electron to free lipid radicals to neutralize them. Oxidized vitamin E can, in turn, be reduced to the active state by the water-soluble vitamin C. In a final reaction, oxidized vitamin C can be converted back to the active reduced state by glutathione (not depicted)

certain amount of oxidative stress may be required for a limited time for certain functions such as, for example, exercise training. Here, the resulting effects of physical exercise are reduced through the intake of these antioxidants prior to the exercise. The question also arises as to why, in contrast to intake in the form of tablets, antioxidant intake in natural nutritional form is beneficial. Plants and fruits are also exposed to ROS, particularly due to exposure to sunlight and oxygen. This is why, during evolution, these plants adapted by producing antioxidants. Through natural selection, those fruits and vegetables survived best that had the best combination and dosage of antioxidants. Thus, the most effective antioxidants are those found naturally in foods.

We shall now discuss additional classes of compounds with antioxidant activity that can be found in natural sources.

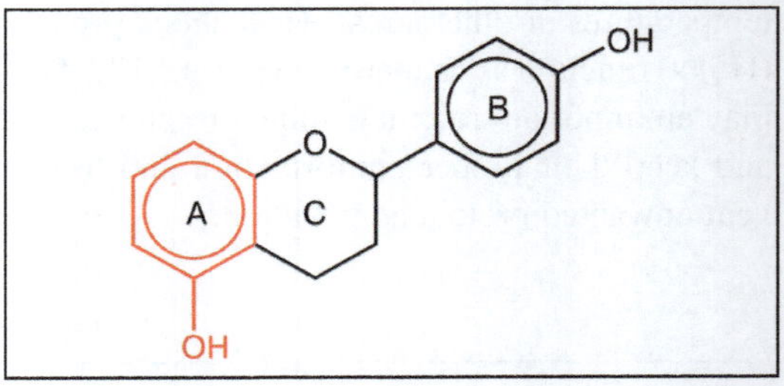

Fig. 13.19 General chemical structure of polyphenolic flavonoids, compounds containing more than one phenolic ring (one of which is colored *red*). These phenols consist of a hydroxyl group (–OH) attached to an aromatic hydrocarbon group

13.6 Further Antioxidants in Nutrition

Flavonoids are a group of naturally occurring compounds that are widely distributed in nature and ubiquitous in food sources such as vegetables, berries, fruits, chocolate, and plant-derived beverages such as tea or red wine. More than 5,000 different flavonoids have been identified.

Flavonoids are an important class of phenolic compounds. One thing that all of the different phenolic compounds have in common is that their molecular structure includes a phenol ring, a ring of six carbon atoms with at least one hydroxyl (OH) group attached (Fig. 13.19).

When a compound has more than one phenolic group, it is referred to as a polyphenol. Polyphenols provide much of the flavor and color of fruits and vegetables due to the fact that the electron cloud around the aromatic ring of a phenol has the capacity to absorb light at visible wavelengths. Moreover, a phenol is very easily oxidized. It is this ease of oxidation (polyphenols act as reducing agents) that gives polyphenols their antioxidant properties. The hydroxyl group of a phenol enables these compounds to readily donate an electron to free radicals to neutralize them.

Polyphenol antioxidants are found in a wide array of phytonutrients. Ginkgo biloba leaf extract is an example of the most widely sold phytomedicine in Europe, where it is used to treat the symptoms of early-stage Alzheimer's disease, vascular dementia, peripheral claudication, and tinnitus of vascular origin. It also is one of the 10 best-selling herbal medications in the United States. The ginkgo tree is one of the oldest living tree species, growing on earth for 150–200 million years. Ginkgo is indigenous to China, Japan, and Korea, where it exists in remote mountainous areas. The active components of ginkgo are extracted from the leaves and the seeds of the ginkgo fruit, which contain flavonoids and terpenoids (Fig. 13.20).

Recently, the effect of Ginkgo biloba extract as a potential antiglaucoma therapy has raised great interest. Unlike vitamins (e.g., vitamin C or E) which cannot penetrate into mitochondrial cell membranes, gingko biloba compounds can penetrate into the mitochondrial cell membrane and thereby protect the mitochondria from oxidative stress. Thus, it also protects retinal ganglion cells. Ginkgo has also been shown to improve the visual fields of glaucoma patients. A daily dose of 120 mg is sufficient and acceptable.

Polyphenols are found in a variety of other naturally occurring substances. Both green and black teas are rich sources of the flavonoids

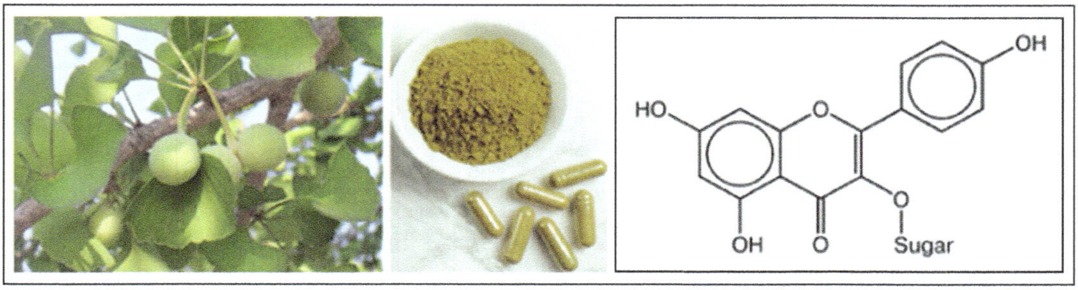

Fig. 13.20 Ginkgo biloba. *Left*: the medicinal components of ginkgo are the extracts of the leaves and the seeds of the ginkgo fruit, which contain flavonoids and terpenoids. *Right*: the chemical structure of one of the 26 flavonol glycosides from Ginkgo biloba leaves. This molecule can donate an electron to a free radical to neutralize it. The sugar moiety increases the water solubility

Fig. 13.21 Tea. Both green and black tea are rich sources of flavonoids such as catechin (C), epicatechin (EC), and epigallocatechin (EGC). However, the concentration of these flavonoids is higher in green tea (*left*). *Right*: the chemical structure of epicatechin

catechin (C), epicatechin (EC), and epigallocatechin (EGC) (Fig. 13.21).

Green tea has one of the highest levels of phenolic compounds among all foods, about 35 % by dry weight. The difference between black and green tea is how the leaves are processed after picking. In black tea, catechins are converted to complex fermentation products, namely theaflavins (TFs) and thearubigins (TGs), which give black tea its characteristic color.

Coffee also has good antioxidant properties due to the presence of polyphenolic compounds (Fig. 13.22). In addition, coffee contains the molecule 3-methyl-cyclopentane-1,2-dione (MCP), which is recognized as a selective "scavenger" of peroxynitrite ($ONOO^-$). MCP neutralizes the reactivity of peroxynitrite ($ONOO^-$) via electron donation.

Other foods containing naturally occurring polyphenols include dark chocolate and red wine. In 1819, the Irish physician Samuel Black first noted that the French suffer from a relatively low incidence of coronary heart disease, despite having a diet rich in saturated fats. The drinking of red wine was attributed to the low incidence of coronary heart disease in France, known as "The French Paradox". When news of this paradox was aired in the United States in 1991, the consumption of red wine increased by 44 % and some wineries began lobbying for the right to label their products as "health food."

Fig. 13.22 Coffee. *Right*: the resonance chemical structure of MCP (3-methyl-1, 2-cyclopentanedione) from coffee extract. MCP is a selective scavenger of peroxynitrite

Fig. 13.23 Red wines exhibit a stronger antioxidant capacity due to their phenolic content of resveratrol

Red wines exhibit a stronger antioxidant capacity than white wines due to their phenolic content. In addition, red wines strongly inhibit the synthesis of Endothelin-1 (ET-1). Resveratrol, a polyphenol in the skin of grapes, is found in red wine (Fig. 13.23). Resveratrol also reduces extracellular levels of vascular-derived endothelial growth factor (VEGF).

One of the mechanisms by which polyphenols in red wine improve endothelial function is via their ability to stimulate the production of endothelial nitric oxide synthase (NOS3) and thereby promote the production of nitric oxide (NO), which induces vasodilation.

Polyphenols are also found in chocolate, which comes from the seeds of the tree *Theobroma cacao*, a tropical rainforest tree (Fig. 13.24). The chocolate tree was given the Latin name *Theobroma cacao*, which means "food of the gods" by the ancient Mesoamericans (including the Mayans and Aztecs), who believed that chocolate had a source of power. Dark chocolate generally contains at least twice as much cacao and, therefore, twice as much polyphenol as milk chocolate; in addition, the milk in milk chocolate reduces the effective resorption of cacao. The antioxidative capacity of cacao has been demonstrated to be higher than that of wine or green tea due to much higher levels of polyphenols. Various studies have shown that the consumption of dark chocolate is associated with a reduced risk for vascular disease. The mechanism is due to the action of flavan-3-ols, which – among other effects – also augment endothelial NOS and thereby NO to improve endothelium-dependant vasorelaxation.

Lycopene ($C_{40}H_{56}$) is a member of another class of antioxidant phytochemicals, the carotenoids, and is a bright red pigment found mainly in tomatoes (Fig. 13.25).

The color of lycopene is due to its many conjugated carbon double bonds. Each double bond reduces the energy required for electrons to

Fig. 13.24 Dark chocolate is made from the seed of *Theobroma cacao*. Cacao is particularly rich in polyphenols, namely flavan-3-ols and procyanidin oligomers

Fig. 13.25 Lycopene is a bright red carotenoid pigment ($C_{40}H_{56}$) found in tomatoes, which acts as an antioxidant by donating electrons to free radicals

transition to higher energy states, allowing the molecule to absorb visible light of progressively longer wavelengths. Lycopene absorbs most of the visible light (except red light), which is why it appears red. Lycopene acts as an antioxidant by donating electrons to free radicals. When lycopene is oxidized (loses an electron), the double bonds between carbon atoms will be broken, cleaving the molecule into smaller molecules.

Anthocyanins are found in different types of fruits and berries, such as blueberries. Blueberries (*Vaccinium myrtillus*) are yet another class of edible substances with antioxidant properties. The use of blueberry for the treatment of conditions affecting eyesight dates back to World War II, when British Royal Air Force pilots first reported that nighttime vision increases after ingesting blueberries. In addition to polyphenolic rings, anthocyanins possess a positively charged oxygen atom in their central ring, which enables them to readily scavenge electrons (Fig. 13.26).

Whereas the compounds discussed thus far neutralize free radicals by donating an electron, anthocyanosides neutralize free radicals by accepting an electron.

We shall briefly discuss two further compounds with antioxidant activity. The first, ubiquinone (Coenzyme Q10), belongs to a class of organic compounds called quinones, which are compounds that have a fully conjugated cyclic

Fig. 13.26 Anthocyanins not only contain phenolic rings but also consist of a central ring with a positively charged oxygen atom. This positively charged oxygen atom can readily steal away or "scavenge" electrons. This particular structure gives foods such as blueberries their antioxidant properties

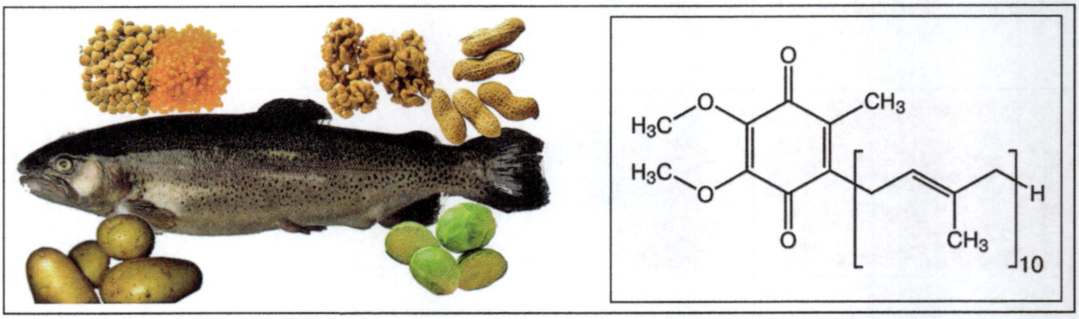

Fig. 13.27 Coenzyme Q10. Shown here is the chemical structure of oxidized ubiquinone. Dietary sources of ubiquinone include various nutritional sources, mainly certain types of fish

dione structure, such as that of benzoquinones. Coenzyme Q10 is a benzoquinone, where Q refers to the quinone chemical group and 10 refers to the number of isoprenoid units (Fig. 13.27). This compound is present in all human cells, where it acts as a coenzyme for the mitochondrial enzyme complexes involved in energy production. In other words, in each human cell, food energy is converted into energy in the mitochondria with the aid of CoQ10. It is, therefore, often used in the form of supplements as treatment for some very rare and serious mitochondrial disorders. Ubiquinone also has strong antioxidant properties. Coenzyme Q10 has been shown to prevent lipid peroxidation and DNA damage induced by oxidative stress. Ubiquinone has been studied well in dermatology but less in ophthalmology. Coenzyme Q10 (CoQ10) exists in three redox states, fully oxidized (ubiquinone), partially reduced (semiquinone anion), and fully reduced (ubiquinol) (Fig. 13.28).

Finally, we shall briefly discuss the antioxidant properties of the compound melatonin (Fig. 13.29). Melatonin (N-acetyl-5-methoxytryptamine) is secreted by the pineal gland as well as by the retina, lens, and gastrointestinal tract. Melatonin is naturally synthesized from the amino acid tryptophan (via synthesis of serotonin) by the enzyme 5-hydroxyindole-O-methyltransferase. Melatonin is found in all living creatures from algae to humans, at levels that vary in a diurnal cycle. A basic biological function of melatonin is that of an antioxidant. In fact, it serves this purpose in many lower life forms. Melatonin has been shown not only to neutralize free radicals but also to stimulate a number of antioxidative enzymes.

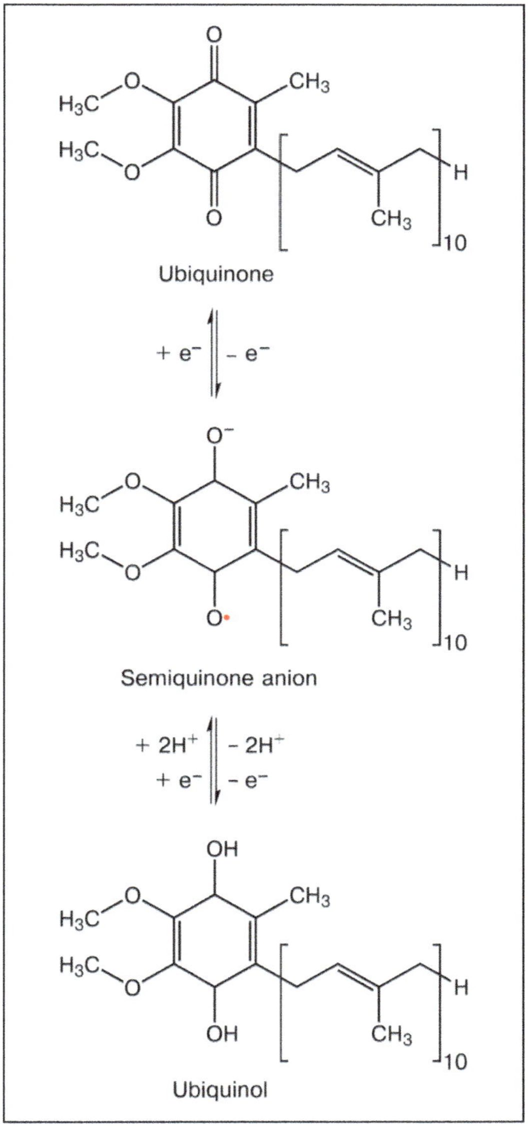

Fig. 13.28 Reduced and oxidized forms of ubiquinone

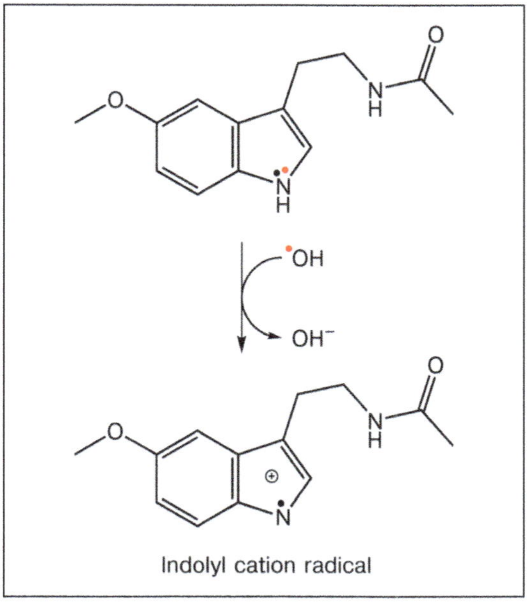

Fig. 13.29 Melatonin (*top*) can neutralize free radicals by changing its structure to an indolyl cation radical (*bottom*), which is relatively stable

14 DNA

14.1 DNA as the Hard Disk of the Cell

Deoxyribonucleic acid (DNA) can be compared with a CD-ROM compact disc. This means that information is stored that can be repeatedly read without changing the information. The DNA codes the information that finally guides the metabolism and structure of cells. Let's explain this with an analogy.

In your clinic, you have a library with a number of books. Within these books, you're searching for one with a description of a particular operation. Since the book is valuable, you cannot take it out of the library. Instead, you make a copy of the book and intend to take this copy to the operating theater. The individual chapters in the books correspond to the genes written in DNA, whereas the copies correspond to the RNA. Fortunately, you have a photocopy machine in your library. The photocopy machine corresponds to the transcriptional machinery. As you have only copied individual chapters, these copies are smaller than the books and can be carried easily. Likewise, the mRNAs are much shorter than the DNA. Correspondingly, the mRNA moves out of the nucleus to the ribosomes. With the help of this information, you carry out your operation. This corresponds to the translation of RNA to one protein.

14.2 Discovery of DNA

Since early times, man has witnessed how children inherit characteristics from their parents. However, it was not until 1866 that the most important fundamental rules of genetic inheritance were recognized. Mendel[1] conducted a series of breeding trials with plants and, thus, discovered classical genetics. Taking what he learned from pea plants, he demonstrated that the form and color of the peas were determined by "factors," now known as genes. Mendel was completely correct in his assumption that these "factors" occur as pairs, i.e., one from the mother and one from the father.

The first step in the discovery of DNA was made in 1869 when it was first isolated by the Swiss scientist Miescher[2] (Fig. 14.1). He was born in the city of Basel. His father was professor of anatomy and Miescher began studying medicine as well. However, after a typhus infection, he went partially deaf and feared that this handicap would limit his ability to be a successful physician. As a

[1]Gregor Johann Mendel (1822–1884), an Austrian scientist and Augustinian monk who gained fame as the founder of genetics.

[2]Johannes Friedrich Miescher (1844–1895), a Swiss physician and biologist. He was the first researcher to isolate and identify nucleic acid. A prestigious research institute in Basel, the Friedrich Miescher Institute (FMI), is named after him.

J. Flammer et al., *Basic Sciences in Ophthalmology*,
DOI 10.1007/978-3-642-32261-7_14, © Springer-Verlag Berlin Heidelberg 2013

Fig. 14.1 Friedrich Miescher

Fig. 14.2 Salmon oocytes. Miescher first extracted DNA from these types of oocytes

result, he switched his major to chemistry, where he began analyzing the pus from wounds. This led to the discovery of DNA as Miescher found a previously unknown substance within the white blood cells of the pus. The new substance was unique since it contained phosphate. All other known organic molecules were made from carbon, oxygen, nitrogen, and hydrogen. Miescher named the newly found substance "nuclein" since it came from the cell's nucleus. A student of his, Altmann,[3] renamed nuclein "nucleic acid" since it was an acid molecule. Miescher later extracted DNA from many other cells, particularly from salmon sperm and oocytes (Fig. 14.2). At the time, however, it was not clear whether this "nucleic acid" was the molecule of heredity.

A breakthrough came about in 1928 when an army medical officer named Griffith[4] was trying to find a vaccine against *Streptococcus pneumoniae*. Griffith was experimenting with bacteria and mice. During his experiments, he made the interesting observation that characteristics of bacteria, even after the bacteria were killed, could still be transferred to other bacteria. This told him that certain molecules remained intact even after death and that they were capable of storing genetic information. However, it was not clear whether these molecules were proteins or other molecules.

Later, in 1944, Avery[5] was able to show that, indeed, DNA, not protein, was the hereditary molecule. Avery showed that he could transfer the information from cells after proteins had been digested but that, when DNA was digested, the hereditary transformation was blocked.

In 1950, Chargaff,[6] an Austrian biochemist who later became an American citizen, found that the amounts of adenine and thymine in DNA were roughly the same, as were the amounts of cytosine and guanine. Wilkins[7] and Franklin[8] applied the X-ray diffraction technique to DNA and concluded that DNA must be a long molecule with a regular, repetitive skeleton. The method of X-ray diffraction is based on the following principle. The molecule of interest is first purified and then crystallized. When the X-ray penetrates the crystalline structure, the rays are diffracted (see Sect. 2.10). When this refracted light is projected on a film, it results in an interference pattern. From this pattern, the three-dimensional structure can be calculated.

[3]Richard Altmann (1852–1900), a German pathologist and histologist who was credited with coining the term "nucleic acid."

[4]Frederick Griffith (1877–1941), a British bacteriologist whose focus was the epidemiology and pathology of bacterial pneumonia.

[5]Oswald Theodore Avery (1877–1955), a Canadian physician and medical researcher who showed that DNA transmits heredity.

[6]Erwin Chargaff (1905–2002), an Austrian biochemist who discovered two rules that helped lead to the discovery of the double helix structure of DNA.

[7]Maurice Hugh Frederick Wilkins (1916–2004), a molecular biologist whose research among others contributed to the scientific understanding of X-ray diffraction.

[8]Rosalind Elsie Franklin (1920–1958), a British biophysicist and X-ray crystallographer who contributed to the understanding of the fine molecular structures of various molecules, including DNA.

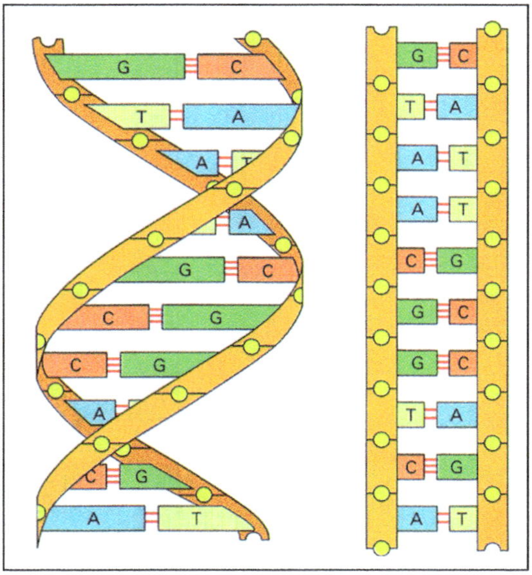

Fig. 14.3 Double helical structure of DNA. The structure of DNA is illustrated by a double helix (*left*) with about ten nucleotide pairs per helical turn. Each spiral strand, composed of a sugar phosphate backbone and attached bases, is connected to a complementary strand by hydrogen bonding (non-covalent) between paired bases, adenine (A) with thymine (T) and guanine (G) with cytosine (C) (*right*). (From Flammer J (2006) *Glaucoma*. Hogrefe&Huber, Bern. With permission)

Based on these types of X-ray diffraction images, in 1953, Watson[9] and Crick[10] were the first to propose a double helical structure for DNA (Fig. 14.3).

14.3 Structure and Function of DNA

The two helical strands of the DNA molecule form a backbone. The DNA backbone is a polymer with an alternating sugar-phosphate sequence. The deoxyribose sugars are joined at both the 3'-hydroxyl and 5'-hydroxyl groups to phosphate groups in ester links, also known as "phosphodiester" bonds (Fig. 14.4).

The human cell contains two types of DNA – nuclear and mitochondrial. The first constitutes the 23 pairs of chromosomes contained within the nuclei of human cells. Nuclear DNA is in the form of a double chain. The chains are complementarily arranged; i.e., there are fixed base pairings. Certain bases are always positioned across from each other. Thus, thymine is always across from adenine and cytosine is always across from guanine and vice versa, explaining the previously mentioned explanation of Chargaff. This permits a rapid doubling of the entire genetic information when a cell divides. When a double chain splits open, like a zipper, each chain is rapidly completed by a complementary chain. In humans, the entire genetic code is stored in the form of several billion base pairs that contain information for thousands of genes. Such an enormous amount of information is difficult to imagine. If one could lay out the DNA of a human being in a straight line and one base pair were placed every millimeter, then the entire length of the genetic information would stretch the distance between the East and West coasts of America. What is even more impressive is that every single cell in our body has the entire genetic information stored twice: once from the mother and once from the father. This is an amazing concentration of information, information that allows a human being, with all his/her physical, mental, and emotional qualities, to grow and develop from one fertilized egg.

This genetic information is stored in the chromosomes of the cell nucleus (Fig. 14.5). A chromosome is, thus, nothing more than a very long molecular chain of DNA. This long chain of DNA needs to be cleverly folded to fit into a short chromosome. This winding occurs around histone proteins, which package and order the DNA into structural units (called nucleosomes) (Fig. 14.6).

Mitochondrial DNA (mtDNA), which is about 0.3 % of the total DNA in a human cell, is found in the mitochondria. Mitochondria originated evolutionarily from bacteria. The reason that a bacterium became an organelle is believed to be linked to the increase in ambient oxygen tension in the

[9]James Dewey Watson, an American molecular biologist, geneticist, and zoologist, best known as a co-discoverer of the structure of DNA in 1953 with Francis Crick.

[10]Francis Harry Compton Crick (1916–2004), an English molecular biologist, biophysicist, and neuroscientist who was awarded the joint Nobel Prize for physiology or medicine in 1962 together with Watson and Wilkins.

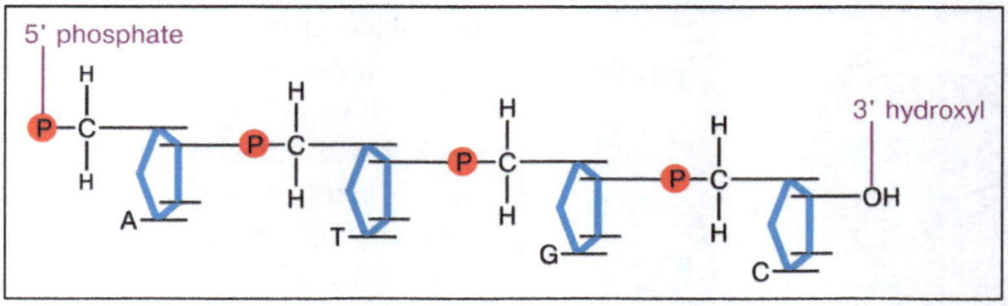

Fig. 14.4 DNA backbone. The basic structure of DNA can be divided into two parts: the external sugar-phosphate backbone and the internal bases. The backbone is constructed by alternating 2-deoxyribose sugar and phosphate molecules, which are highly polar. Because the backbone is polar, the molecule is hydrophilic

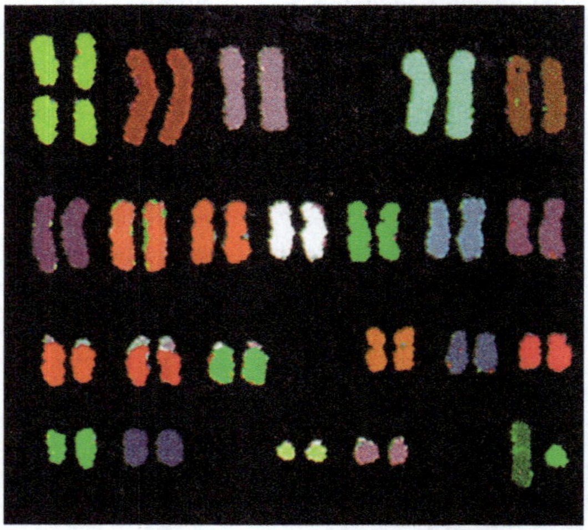

Fig. 14.5 Human chromosomes. Human cells have 23 pairs of large linear nuclear chromosomes (22 pairs of autosomes and one pair of sex chromosomes). (From Flammer J (2006) *Glaucoma*. Hogrefe&Huber, Bern. With permission)

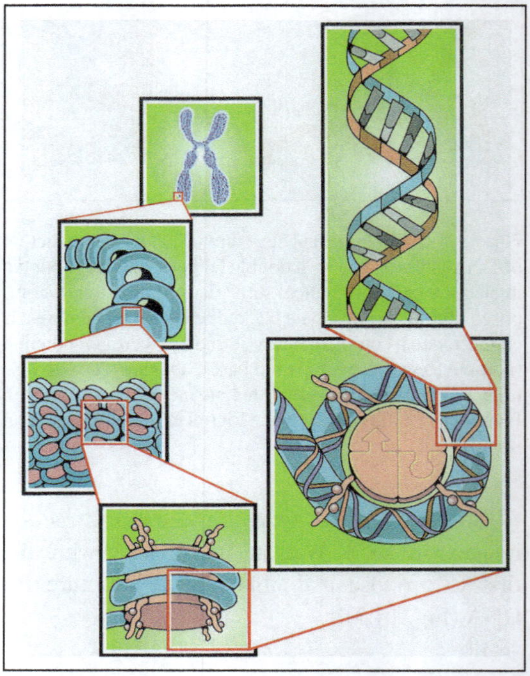

Fig. 14.6 Assembly of DNA into chromosomes. The long DNA molecule is wound around proteins called histones, allowing a large amount of DNA to fit into the chromosome. (From Flammer J (2006) *Glaucoma*. Hogrefe&Huber, Bern. With permission)

earth's atmosphere approximately two billion years ago. The primitive eukaryotes were unable to use oxygen for oxidative phosphorylation, in other words, to build ATP by burning molecules with oxygen. However, some bacteria were able to do so. Eukaryotes that carried such bacteria profited from their ATP production and, therefore, had an evolutionary advantage. That's how eukaryotes and bacteria survived symbiotically and, over time, bacteria evolved into organelles, namely mitochondria (Fig. 14.7). This explains why mitochondria contain their own circular DNA (Fig. 14.8).

Similar to bacteria, mitochondria have their own transcriptional and translational machinery. The mitochondria in a cell originate predominantly from the egg and only minimally from the sperm. Correspondingly, the mitochondrial DNA is inherited predominantly from the mother. Mitochondria have limited DNA repair capacity. Therefore, they use mechanisms of fission and fusion to prevent the accumulation of damaged DNA. These mechanisms, which are different from each other, are, therefore, required to maintain mitochondrial function. Whereas fusion protects mitochondrial function by, e.g., enabling mtDNA repair, fission acts to facilitate the equal

Fig. 14.7 Mitochondria originated evolutionarily from bacteria. (*1*) Primitive heterotrophic eukaryote with a bacterium nearby (*red ellipse*). (*2*) The eukaryote ingested but did not digest the bacterium. (*3*) The eukaryote retained the bacterium as a symbiont. (*4*) The symbiont lost many genes and became what is known today as a mitochondrion

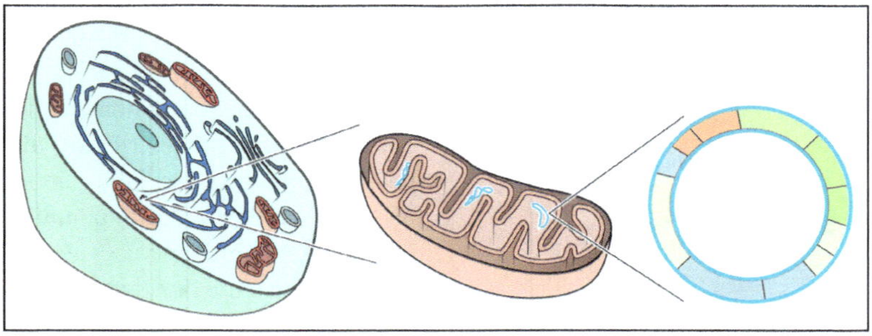

Fig. 14.8 Mitochondrial DNA (mtDNA) is circular. *Left*: cell with mitochondria. *Middle*: mitochondria containing loops of DNA. *Right*: circular DNA of mitochondria. In parallel, the cells developed mechanisms against the toxicity of oxygen (see Sect. 13.3)

segregation of mitochondria into daughter cells. When these protective mechanisms fail, mitochondrial fission can also promote apoptosis.

14.4 The Role of DNA Mutation

Mutations relevant to eye diseases can occur in either nuclear DNA or the mitochondrial DNA. We shall discuss one example from each.

An example of a nuclear DNA mutation is oculocutaneous albinism. This results from mutations in the genes involved in the biosynthesis of melanin pigment, such as the gene for the enzyme tyrosinase (Fig. 14.9).

Oculocutaneous albinism affects the eyes as well as the color of the skin and hair (Fig. 14.10).

Leber's hereditary optic neuropathy (LHON) is an example of a disease resulting from different types of mutations in the mitochondrial DNA, leading to degeneration of the retinal ganglion cells and their axons (Fig. 14.11). This disease predominantly affects young adult males and leads to an acute or subacute loss of central vision (Fig 14.12).

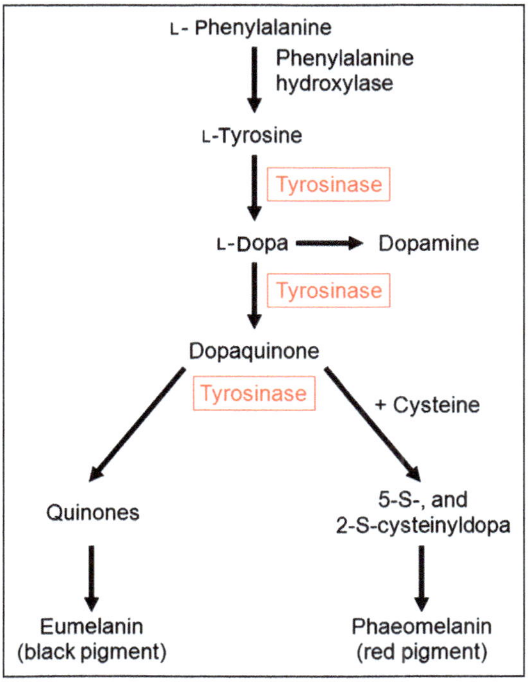

Fig. 14.9 Role of tyrosinase. A mutation of the enzyme tyrosinase leads to, among other changes, decreased production of melanin

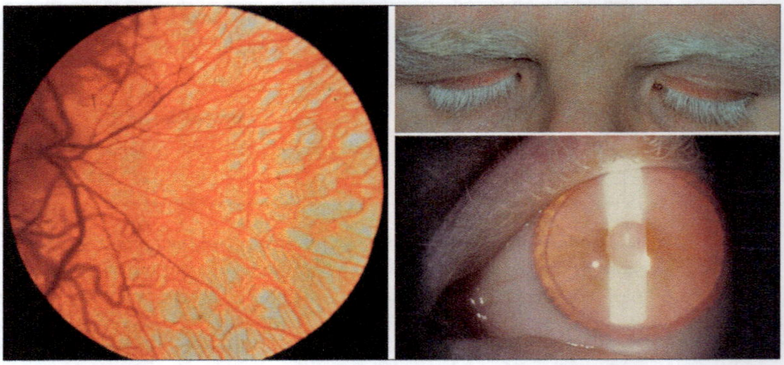

Fig. 14.10 Oculocutaneous albinism. *Left*: fundus with oculocutaneous albinism. The lack of melanin allows the visualization of choroidal vessels. *Right*: hair, eyebrows, eyelashes, and skin lack pigmentation. The iris is translucent

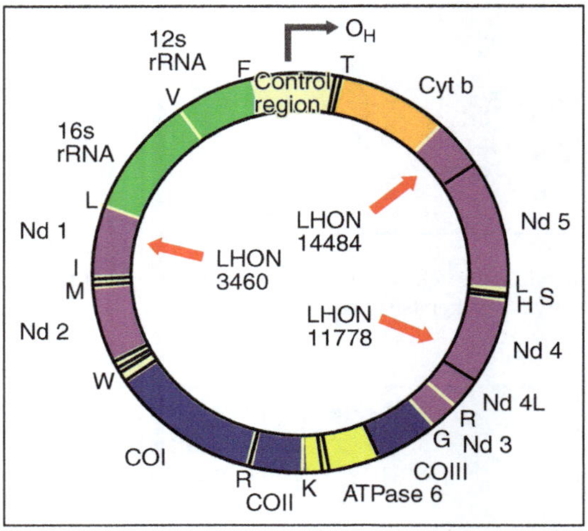

Fig. 14.11 Mitochondrial DNA (mDNA). Different mutations in mDNA can lead to different diseases, including Leber's hereditary optic neuropathy (LHON)

Although this mitochondrial mutation is present in the early stages of life, the disease manifests itself much later. This can only be explained by the fact that environmental factors such as smoking and alcohol use also play a role. Of interest is also the fact that the mutation, although present in all cells of the body, clinically manifests itself predominantly in the optic nerve. The manifestation in the RGCs is believed to be due, on the one hand, to the high energy demand of these cells compared to other cells and, on the other hand, to the unstable blood flow in the optic nerve head, leading to oxidative stress. In other cells, mitochondria still function enough to allow survival. The involvement of oxidative stress is also supported by the observation that patients with a genetic predisposition for LHON have a higher chance of a clinical manifestation if they suffer at the same time from a primary vascular dysregulation (PVD) syndrome.

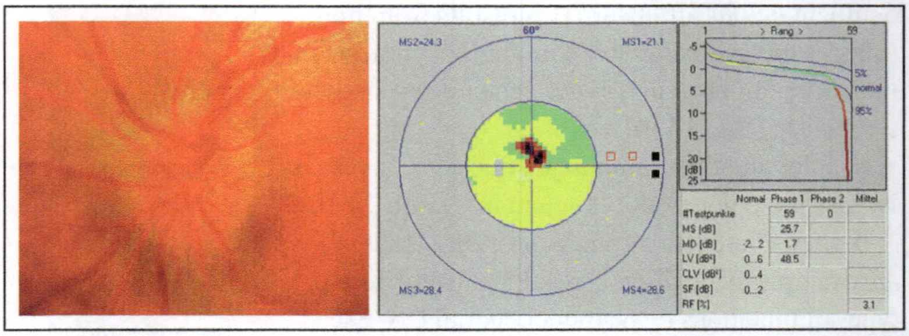

Fig. 14.12 Leber's hereditary optic neuropathy (LHON). *Left*: fundus showing optic nerve hyperemia and swollen nerve fiber layer with small peripapillar, teleangiectatic blood vessels. *Right*: central scotoma (Octopus program G1)

14.5 Acquired DNA Damage and Its Repair

Acquired alterations in DNA occur only in cells in which the mutation occurs and not in all cells of the body (Fig. 14.13). Such acquired DNA mutations can potentially cause diseases such as cancer.

Even under physiological conditions, we constantly acquire DNA breaks (Fig. 14.14). Fortunately, these can be repaired by various mechanisms. However, if the incidence of breaks is higher than the repair capacity, the number of DNA breaks will increase and this may finally contribute to the development of disease, including eye diseases. Decreased repair capacity is found, for example, in subjects with a mutation of the Xeroderma pigmentosum gene (XPG), belonging to the group of gene-repair genes (Fig. 14.15).

If, in such cases, the incidence of DNA breaks is relatively high (as, e.g., in the skin due to UV light) and the repair capacity low, alterations in the skin and, finally, even cancer result (Fig. 14.16).

To avoid the propagation of damage, nature has built in the so-called cell cycle control system, which prevents the propagation of damaged DNA from mother to daughter cells. If DNA is damaged, the p53 protein is upregulated. This, in turn, stops the cell cycle until DNA repair can be completed (Fig. 14.17).

By this mechanism, the transfer of DNA damage to daughter cells can be avoided. If the damage is extensive, p53 levels are increased even further, triggering apoptosis (Fig. 14.17). This control system is, therefore, capable of preventing cancer. The corresponding molecules involved, such as p53, are, therefore, called tumor suppressors and their genes are tumor suppressor genes. If these genes are mutated and thereby weakened in their function, the probability of cancer increases.

An increased incidence of DNA breaks occurs, among other things, as a consequence of oxidative stress (Chap. 13).

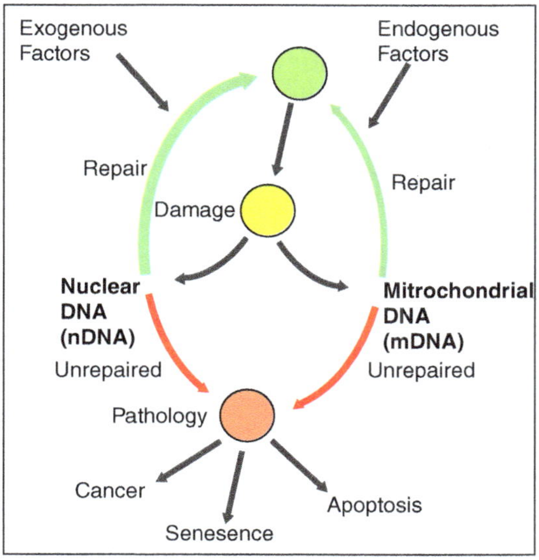

Fig. 14.13 Acquired DNA damage. Both exogenous as well as endogenous factors can damage nuclear DNA (nDNA) and mitochondrial DNA (mDNA). While nDNA can be repaired efficiently, mDNA is only repaired to a certain extent

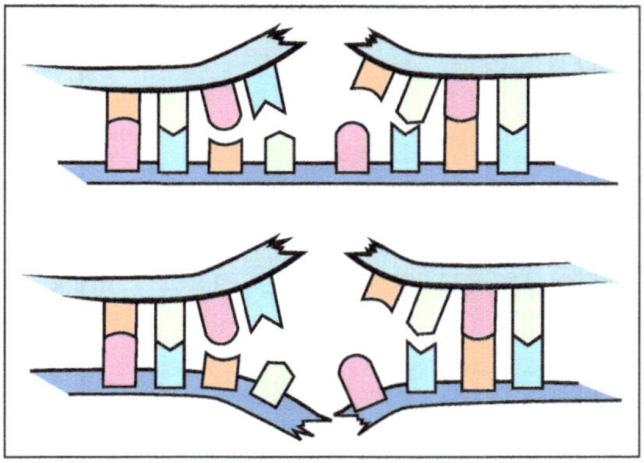

Fig. 14.14 DNA breaks. DNA breaks can occur either as single-stranded breaks (*top*) or less frequently as double-stranded breaks (*bottom*)

DNA breaks can be quantified by the method called the comet assay – also known as single-cell gel electrophoresis. The principle is simple and relies upon the fact that DNA molecules are negatively charged. In an electrical field

Fig. 14.15 DNA repair. If the XPG is mutated, DNA repair is reduced, leading to the clinical picture of Xeroderma pigmentosum. Normally, the damaged DNA lesion is recognized and excised. The missing DNA part is replaced with the help of the DNA polymerase and then inserted into the DNA with the help of a ligase

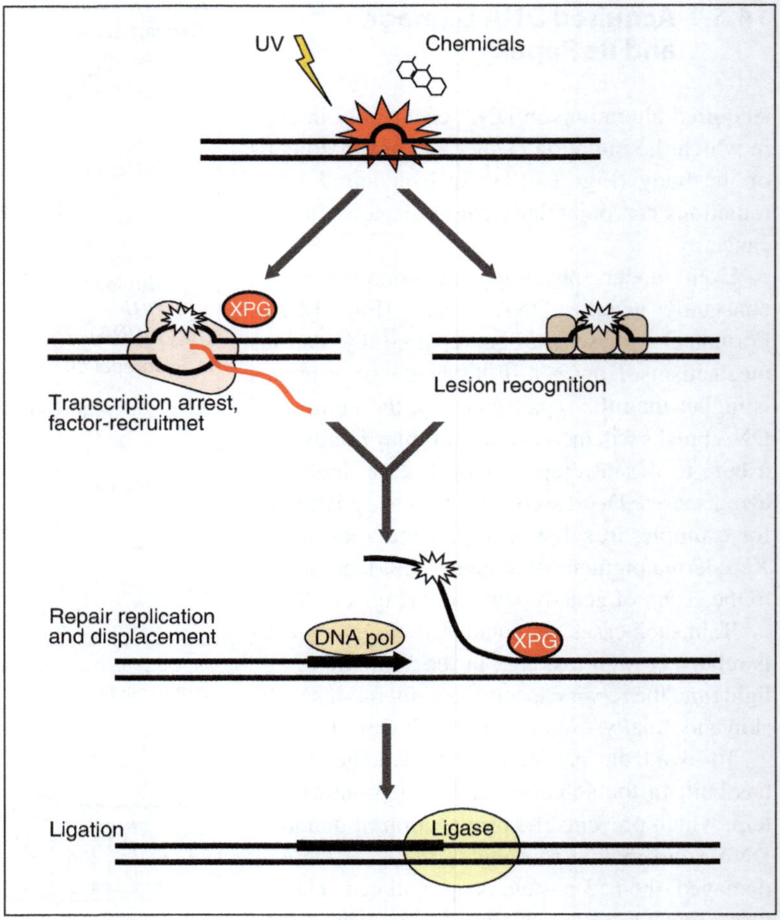

(electrophoresis), an intact DNA molecule has such a large size that it does not migrate in an electrophoresis gel. In contrast, smaller fragments that arise due to DNA breaks move in the electrical field toward the anode. The smallest fragments migrate the fastest. Since the size of the various fragments varies, the final result of the electrophoresis is not a distinct line but, rather, a continuum with the shape of a comet, as shown in Fig. 14.18. The comet assay, for example, allowed the demonstration that glaucoma patients with vascular dysregulation have higher average numbers of DNA breaks in their circulating leukocytes compared to healthy persons.

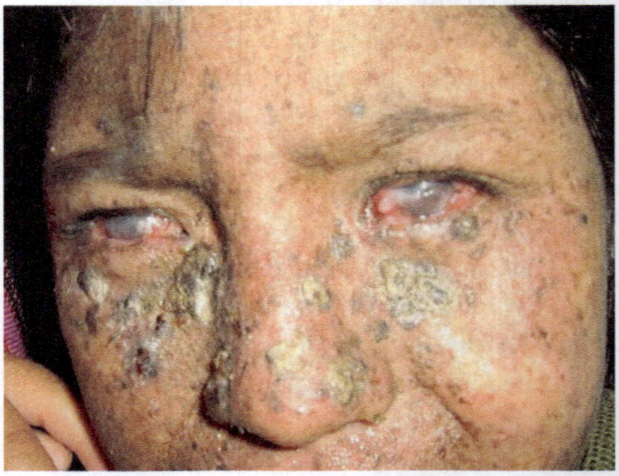

Fig. 14.16 Clinical picture of a patient with Xeroderma pigmentosum. (Copyright James Halpern, Bryan Hopping, and Joshua M Brostoff)

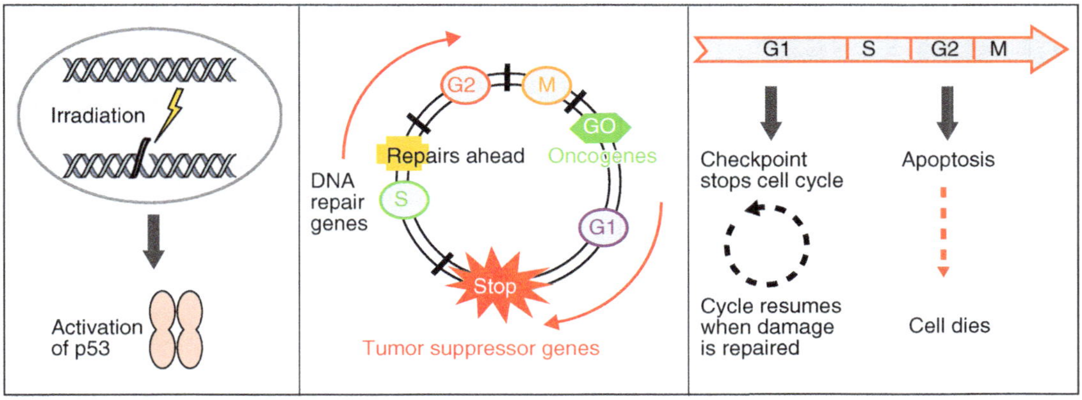

Fig. 14.17 Cell cycle control by p53 (tumor supressor gene). *Left*: DNA damage caused, for example, by radiation leads to the activation of p53. *Middle*: Activated p53 stops the cell cycle. *Right*: If p53 is markedly increased, the cells undergo apoptosis

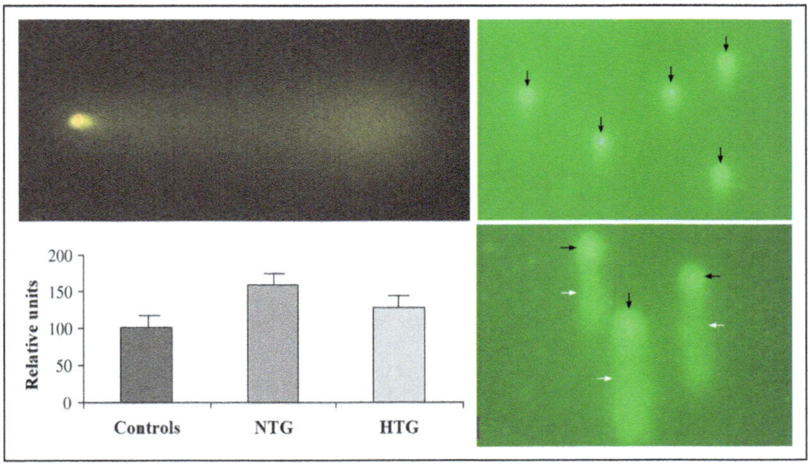

Fig. 14.18 Comet assay. *Left*: typical comet. *Right*: cells from a healthy person (*top*) and from a glaucoma patient (*bottom*). *Bottom left*: average DNA break values from healthy people and glaucoma patients. *NTG* normal-tension glaucoma, *HTG* high-tension glaucoma Modified after Mozaffarieh M, Flammer J (2009) Ocular blood flow and glaucomatous optic neuropathy. Springer Publ, Berlin

15 RNA

Ribonucleic acid (RNA), like DNA, is also a macromolecule made up of nucleotides, although it is shorter. However, in contrast to DNA, RNA has the base uracil instead of thymine and it contains ribose instead of deoxyribose sugar (Fig. 15.1).

The sequence of the nucleotides in RNA is determined by transcription from the DNA. The sequence in the RNA, in turn, determines the sequence of amino acids by translation. During translation, information in the language of

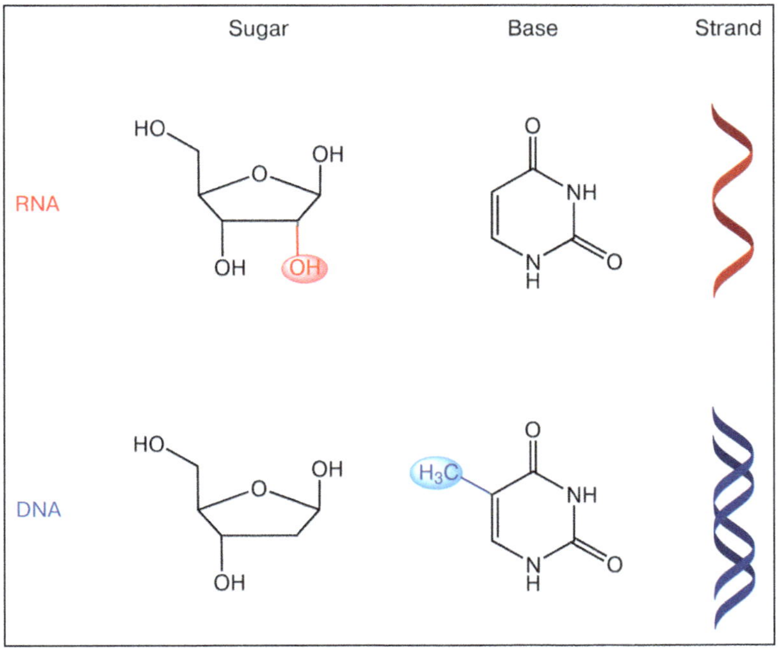

Fig. 15.1 Differences between DNA and RNA. *Left*: While the sugar present in a DNA molecule is deoxyribose, the sugar present in a molecule of RNA is ribose. *Middle*: DNA has the base thymine, whereas RNA has uracil as one of the bases. *Right*: DNA is double-stranded, whereas RNA is single-stranded

J. Flammer et al., *Basic Sciences in Ophthalmology*,
DOI 10.1007/978-3-642-32261-7_15, © Springer-Verlag Berlin Heidelberg 2013

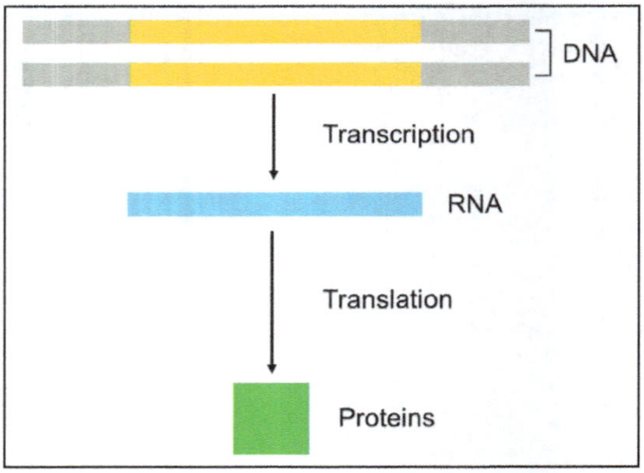

Fig. 15.2 Transcription and translation. The DNA becomes copied (transcription) onto an RNA chain; the RNA controls protein synthesis (translation)

nucleotides is "translated" into information in the language of amino acids (Fig. 15.2). As discussed later, three different kinds of RNA play a role in the synthesis of proteins.

15.1 Discovery of RNA

RNA was discovered in 1955 by a Spanish biochemist, Ochoa de Albornoz.[1] For this discovery, he received the Nobel Prize in 1955. The exact role of this newly detected molecule, RNA, was not known at that time. Nirenberg[2] and Matthaei[3] first to recognize its role in 1961.

[1]Severo Ochoa de Albornoz (1905–1993), a Spanish-American doctor and biochemist and joint winner of the 1959 Nobel Prize for physiology or medicine with Arthur Kornberg.

[2]Marshall Warren Nirenberg (1927–2010), an American biochemist and geneticist who shared the Nobel Prize for physiology or medicine in 1968 with Har Gobind Khorana and Robert W. Holley for "breaking the genetic code" and describing how it operates in protein synthesis.

[3]Heinrich Matthaei, a German biochemist, best known for his unique contribution to solving the genetic code.

15.2 Structure and Function of RNA

Three different kinds of RNA molecules are important. The role of the so-called messenger RNA (mRNA) is to carry the information from DNA to the ribosomes. The role of transfer RNA (tRNA) is to bring the amino acids to the ribosomes. Finally, the role of the third type of RNA, ribosomal RNA (rRNA), is to build a structural and functional component on which translation can take place. More details on these three kinds of RNA are now discussed.

15.2.1 Messenger RNA

Messenger RNA (mRNA) carries the information from the cell nucleus to the ribosomes in the cytoplasm, where protein synthesis takes place. Similar to DNA, genetic information in mRNA is encoded in the sequence of nucleotides that are arranged into codons. Each codon consists of three bases and codes for a specific amino acid.

15.2.2 Transfer RNA

Transfer RNA (tRNA), the smallest of the three major RNA molecules, brings the amino acids to the ribosomes. The tRNA carrying the amino acid makes up the anticodon. The anticodon binds to a specific amino acid and brings it to a ribosome. For every amino acid, there is at least one but often more tRNA molecules.

15.2.3 Ribosomal RNA

The ribosome itself interacts with both the mRNA and the tRNA, thereby catalyzing the formation of a peptide bond between the two amino acids, as illustrated in Fig. 15.3.

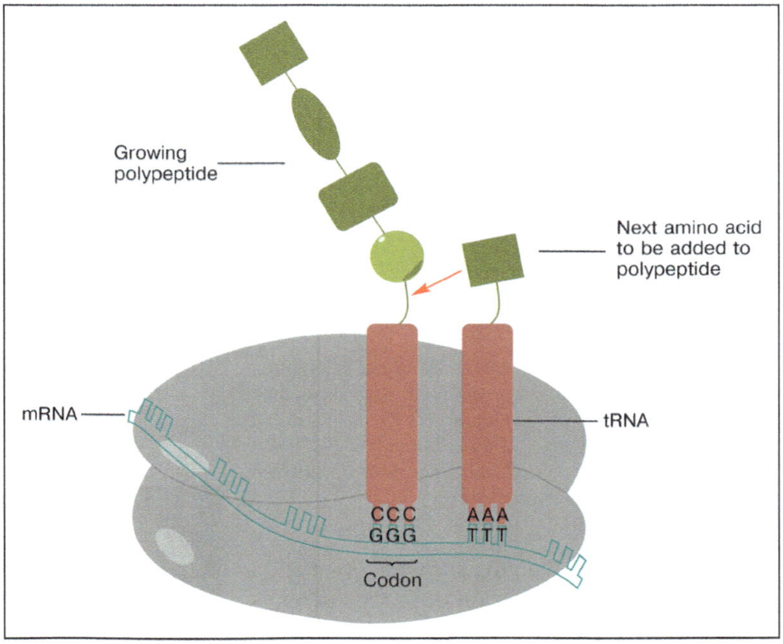

Fig. 15.3 Formation of a peptide. During translation, specific tRNAs pick up specific amino acids, transfer those amino acids to the ribosomes, and insert them in their proper place according to the mRNA code. This is done by the anticodon portion of the tRNA molecules complementarily base pairing with the codons along the mRNA

15.3 RNA and Cell Function

Gene expression is important for any living cell but has a particular impact during embryology where the different cells mature into different cell types. Think for a moment about how, for example, a cell, starting from just a single fertilized egg cell, becomes a living animal. Imagine that a fertilized egg cell has already divided and now consists of eight cells. How does each cell know in which direction it has to develop? Will this cell finally contribute to the growth of the head or the creation of a foot? Until recently, the answers to these questions were a great mystery to biologists. However, during the past several years, insights have been gained into the regulation of embryonic development. Today, we know that this is achieved by the establishment of a certain hierarchy among the genes involved, where the gene products of the first gene regulate the expression of subsequent genes.

Fig. 15.4 Christiane Nüsslein-Volhard

Much of our understanding in this area is owed to the work of the German biologist Nüsslein-Volhard (Fig. 15.4).[4]

[4]Christiane Nüsslein-Volhard, a German biologist who won the Nobel Prize in Physiology or Medicine in 1995, together with Eric Wieschaus and Edward B. Lewis, for their research on the genetic control of embryonic development.

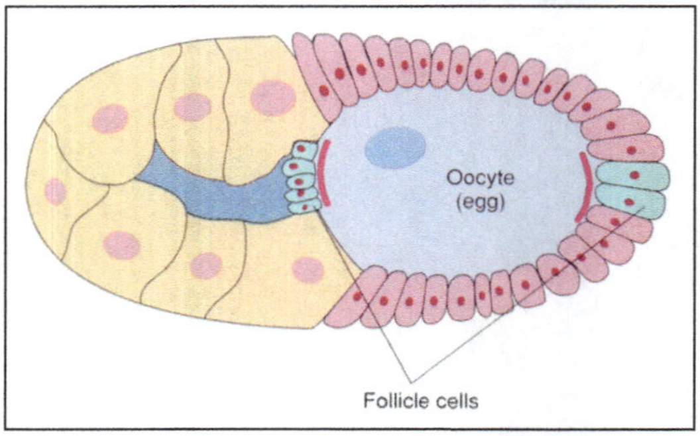

Fig. 15.5 Follicle of a *Drosophila*. Follicular cells (*blue*) at the tip of the oocyte secrete different substances that, after the first division of the egg, lead to the activation of a different set of genes. (From Tamarin RH (1996) *Principles of genetics, 5th edition*. Wm. C. Brown Publishers, Iowa. With permission)

Fig. 15.6 Walter Gehring

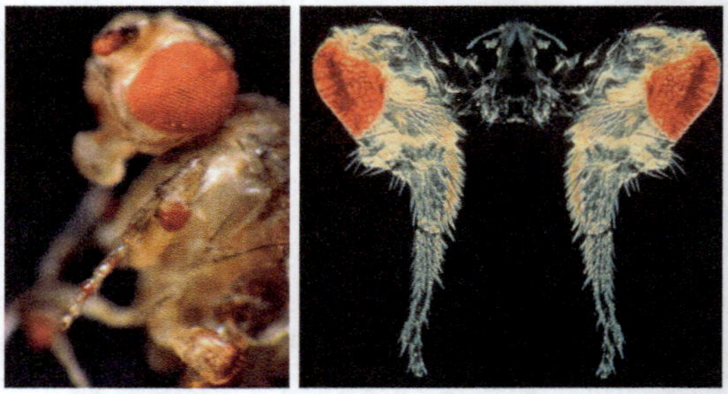

Fig. 15.7 Expression of the eyeless gene. Induction of additional eyes on the antennae (*left*) and legs (*right*) of the *Drosophila* (Courtesy of W. Gehring, Growth and Development, Biozentrum, University of Basel)

Nüsslein found that the primary orientation and control of the expression of further genes is determined by the gradients of certain proteins. Even the unfertilized egg cell inside the ovary contains concentration gradients that provide information as to how certain parts must develop should fertilization occur. The egg cell receives this information from the cells surrounding it. These different cells release different proteins, which then build up concentration gradients within the egg (Fig. 15.5).

The direction in which the daughter cells develop after division is, therefore, predetermined. After the first division, both cells contain different concentrations of these messenger substances and, therefore, "know" that they have different tasks to fulfill.

Let's consider later steps in the development of an embryo. What determines the development of an organ? Specific genes are responsible for initiating organ formation. When one of these genes is expressed, the development of an organ begins. These gene products lead to the expression of subsequent genes, finally leading to the development of a full organ. The different genes involved in the development of an organ are, therefore, expressed in a hierarchic order. This signifies that the expression of each gene controls not only certain aspects of metabolism but also the expression of the next gene downstream. Let us explain this with an example concerning the development of the eye.

A group of researchers at the University of Basel, led by Gehring[5] (mentor of Nüsslein), demonstrated that the PAX 6 homolog is responsible for the development of the visual system (Fig. 15.6). They proved their hypothesis by artificially switching on the so-called "eyeless" gene in other parts of the *Drosophila* fruit fly, such as the wings, legs, and antennae (Fig. 15.7)! Indeed, eyes developed in these places.

[5]Walter Jakob Gehring, professor at the University of Basel, Switzerland, was awarded the March of Dimes Prize in Developmental Biology in 1997, the Kyoto Prize for Basic Science in 2000, and the Balzan Prize for Developmental Biology in 2002.

Correspondingly, major mutations in the "eyeless" gene led to the absence of eyes in flies.

When Darwin described his theory of evolution, he found the eye to be the only example of an organ that did not fit into his theory. In the animal kingdom, the morphology of the different eyes varies so dramatically that they seem to have developed independently of each other. This would suggest that nature "invented" the eye more than once. The interesting observation now – that the same gene is responsible for the development of eyes in all animals – indicates that these different eyes have the same origin phylogenetically. This is indeed astonishing if we compare, for example, the faceted compound eye of a fly with a human eye. The differences are explained by differences in the genes that lie lower in the hierarchy.

Gene expression still plays a role in the mature organism. These genes expressed at a given moment (respectively, the mRNAs present in the cell) depend, on the one hand, on the function of the corresponding cell (e.g., rods produce the protein opsin) and, on the other hand, on the present condition of the cell (e.g., the production of VEGF increases during hypoxia). Therefore, the quantification of the mRNA gives insight into the type of cell and into its present cellular activity. This can be used for diagnostic purposes.

15.4 Diagnostics Based on RNA

Ponder, for a moment, how you would determine whether a particular gene is expressed in a certain tissue or certain cells of a tissue. This can be studied by making histological sections of the tissue of interest and checking to determine whether the corresponding RNA is present. This method, the so-called "in situ hybridization" (ISH), utilizes the fact that complementary DNA (cDNA) binds to RNA in the corresponding tissue. If the DNA is labeled (by either radioactivity or fluorescence), then it can be detected (Fig. 15.8).

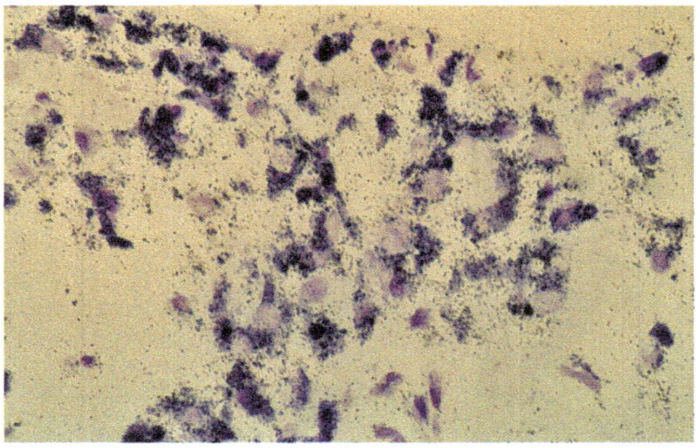

Fig. 15.8 In situ hybridization to visualize an mRNA of interest [in this example of connective tissue growth factor (CTGF) in a subretinal membrane]. Dark-stained purple areas indicate CTGF mRNA expression. (From Meyer P, et al. (2002) Ophthalmologica, 216. With permission)

To compare the gene expression in two groups of tissues or cells, we can use the method of "subtractive hybridization." From the first group, the mRNA is used to make cDNA, which is then labeled. The mRNA of the second group (called the target mRNA) is then hybridized to the cDNA of the first group (called the subtractor cDNA). The non-labeled mRNA molecules representing differentially expressed genes are then isolated. An example of the use of this method is in the comparison of gene expression in leukocytes from glaucoma patients with those from age- and sex-matched controls. The gene expression in the leukocytes of glaucoma patients was, indeed, shown to be different than that of controls.

If a difference exists between gene expressions in a particular tissue, the corresponding mRNA can be quantified with the real-time polymerase chain reaction (rtPCR) method. This technique is based on the "polymerase chain reaction" (PCR). The basic principle of PCR is the quantification of a specific DNA or RNA after its amplification. This amplification is achieved by primers (short DNA fragments) containing sequences complementary to the target region of interest, which are used along with a DNA polymerase (after which the method is named) to enable selective and repeated amplification. As PCR progresses, the DNA generated is itself used as a template for replication, setting in motion a chain reaction in which the DNA template is exponentially amplified (Fig. 15.9).

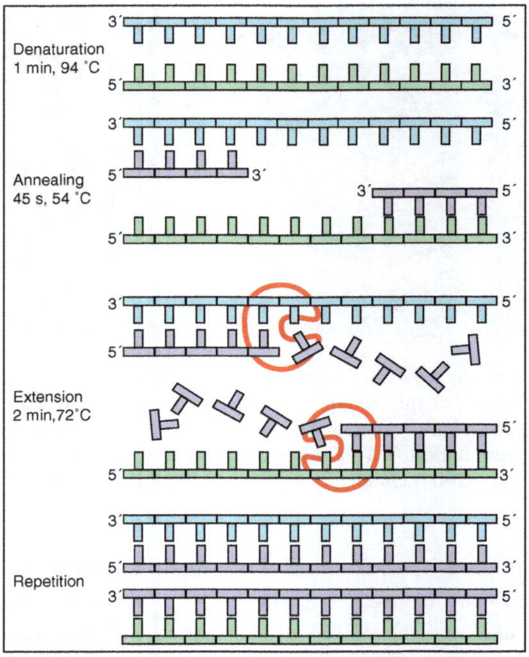

Fig. 15.9 Schematic drawing of the polymerase chain reaction (PCR) cycle. Heat (94-96 °C) is required to break hydrogen bonds (denaturation). The reaction temperature is lowered to 50-65 °C for about 40 s, allowing annealing of the primers to the single-stranded DNA template. With the help of the enzyme DNA polymerase (*red*), a new DNA strand complementary to the DNA template strand is synthesized (extension). In each cycle, the number of DNA strands of is doubled. The process is repeated for 30–40 cycles

The aforementioned rtPCR allows the quantification of the DNA or RNA of interest.

15.5 Therapies Based on RNA

If a specific gene product has an unfavorable effect or if a beneficial gene product is produced in excessive amounts, it is desirable to weaken the effect of that particular gene. In animals, the particular gene of interest can be "knocked out." The effect of the gene can also be reduced by binding the corresponding gene products (proteins) by antibodies (e.g., anti- VEGFs) by blocking the corresponding receptors (beta-blockers) or by weakening the effect of the gene at the level of mRNA so that the gene is translated less extensively. The application of siRNA to knock-down mRNA levels is one of the different methods of "gene silencing."

Gene silencing can take place at either the transcriptional or the post-transcriptional level. At the transcriptional level, the environment around a gene is modified (histone modification), making this gene less accessible to transcriptional machinery (RNA polymerase, transcription factors, etc.). Post-transcriptional gene silencing is the result of the mRNA of a particular gene being destroyed or blocked, thereby preventing translation.

Another method that holds promise for future therapy is interfering with alternative splicing. Alternative splicing is the process that leads to the formation of several different proteins derived from the same gene. The DNA and corresponding RNA are composed of several so-called "exons" and "introns." As the introns do not contribute to protein formation, only the exons are spliced and reconnected. Since the exons can be reconnected in multiple different ways, different mRNAs can result. Alternative splicing occurs as a normal phenomenon in eukaryotes. This explains why the number of different proteins is greater by far than the number of genes.

An example of a family of proteins generated by alternative splicing of the primary RNA transcript

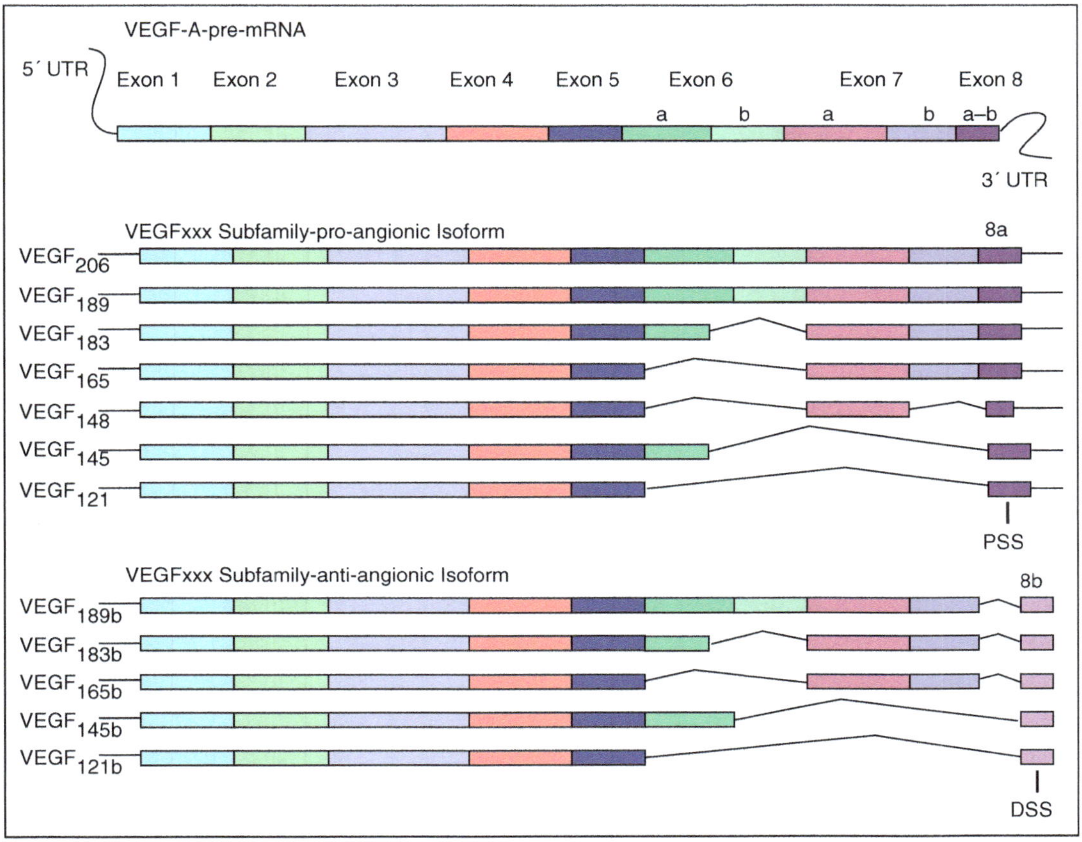

Fig. 15.10 The splice variants of human VEGF. The human VEGF gene, through alternative mRNA splicing, produces different isoforms with different biological activities. A change in the terminal amino acid (8a to 8b) changes the activity from pro- to anti-angiogenic. (Modified after VEGF-splicing. National Library of Medicine)

is the VEGF family. The resulting RNAs code for different proteins of different lengths of amino acids (Fig. 15.10). These VEGF isoforms differ in their biological properties with respect to the activation of various VEGF receptors. Knowledge of these isoforms potentially allows the development of highly specific anti-VEGF drugs, which bind to individual isoforms.

16 Proteins

The word "protein" is derived from the Greek word "protos," meaning "of prime importance." Proteins are involved in a variety of functions, all of which are essential to life. Proteins give strength and elasticity to skin and blood vessels. In the blood, they serve as protectors in the form of antibodies or as long-distance transporters of oxygen and lipids; in the nervous system, they form part of the communications network by acting as neurotransmitters; and throughout the body, they direct the work of repair, construction, and energy conversion.

The role of proteins in ophthalmology will be discussed in this section. First, however, we will briefly discuss how proteins were discovered in the first place.

16.1 Discovery of Proteins

Proteins were originally described in the early eighteenth century by the Dutch chemist Mulder[1] and the Swedish chemist Berzelius.[2] Mulder was the first person to ascertain the chemical composition of proteins and he observed that all proteins had the same empirical formula. At the time, his colleague Berzelius was the first to distinguish between organic and inorganic compounds. Berzelius proposed the name "protein" for Mulder's new findings and Mulder, in fact, used this terminology in his first publication.

Later, the German chemist and physiologist von Voit[3] stated that protein is the most important nutrient in the structure of the body. Voit was considered by many to be the father of modern dietetics. The enzymatic role of proteins was first discovered in the nineteenth century by Sumner,[4] who showed that the enzyme urease is a protein. Some years later, the first protein sequence, that of insulin, was presented by Sanger,[5] who won the Nobel Prize for this achievement in 1958. In the same year, the molecular structures of the proteins hemoglobin and myoglobin were discovered by Max Perutz[6] and Kendrew.[7]

[1]Gerardus Johannes Mulder (1802–1880), a Dutch organic and analytical chemist.

[2]Jöns Jacob Berzelius (1779–1848), a Swedish chemist who worked out the modern technique of chemical formula notation.

[3]Carl von Voit (1831–1908), a German physiologist and dietitian.

[4]James Batcheller Sumner (1887–1955), an American chemist who shared the Nobel Prize for chemistry in 1946 with John Howard Northrop and Wendell Meredith Stanley.

[5]Frederick Sanger, an English biochemist and a two-time Nobel laureate in chemistry.

[6]Max Ferdinand Perutz (1914–2002), an Austrian-born British molecular biologist who shared the 1962 Nobel Prize for chemistry with John Kendrew for their studies of the structures of hemoglobin and globular proteins.

[7]Sir John Cowdery Kendrew (1917–1997), an English biochemist and crystallographer who shared the 1962 Nobel Prize for chemistry with Max Perutz; their group at the Cavendish Laboratory investigated the structure of heme-containing proteins.

J. Flammer et al., *Basic Sciences in Ophthalmology*,
DOI 10.1007/978-3-642-32261-7_16, © Springer-Verlag Berlin Heidelberg 2013

Emil Fischer[8] (Fig. 16.1) proposed that enzymes had particular shapes into which substrates fit exactly. He first postulated in 1894 this specific action of an enzyme with a single substrate.

This model of exact fit is referred to as the "lock and key" model because an enzyme's binding to a substrate is analogous to the specific fit of a key into a lock (Fig. 16.2).

Fig. 16.1 Emil Fischer

Fig. 16.2 Lock and key model. Only the correctly sized key (substrate) fits into the keyhole (active site) of the lock (enzyme)

16.2 Structure of Proteins

The primary structure refers to the linear arrangement of amino acids (Fig. 16.3).

The sequence of amino acids determines the function of the protein. Variations in the amino acid sequence may occur physiologically, such as in polymorphism, which occurs when two or more clearly different phenotypes exist in the same population of a species – in other words, more than one form or "morph" occurs. Polymorphism allows for diversity within a population. A common example is the different allelic forms that give rise to different blood groups in humans.

However, variations in the amino acid sequence may also occur in the form of mutations. Mutations can result in several different types of changes in the amino acid sequence. These can have no effect or they may prevent the protein from functioning properly at all. Retinitis pigmentosa (Fig. 16.4) is one example of a genetic group of eye disorders that results from mutations. Multiple genes are known that, when mutated, can cause the retinitis pigmentosa phenotype. In 1989, a mutation was identified in the gene for rhodopsin (Sect. 16.5.4). Since then, more than 100 other mutations leading to RP have been found.

Corneal dystrophies are another example of eye disorders resulting from mutations. Some corneal dystrophies may be asymptomatic, whereas

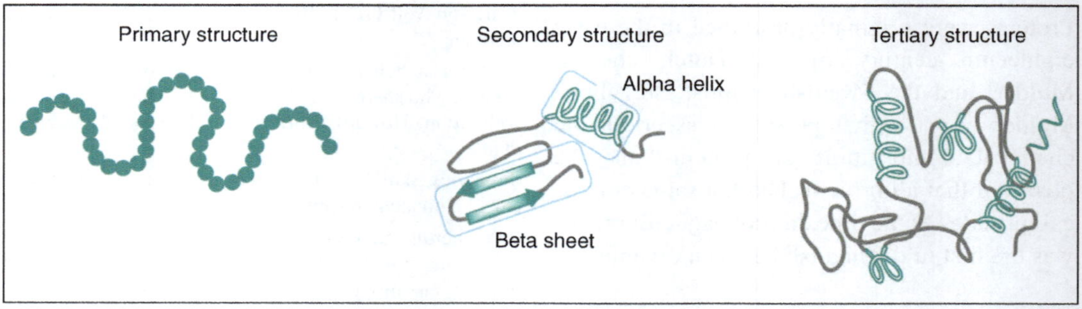

Fig. 16.3 Different structures of proteins

[8]Hermann Emil Fischer (1852–1919), a German chemist and 1902 recipient of the Nobel Prize for chemistry.

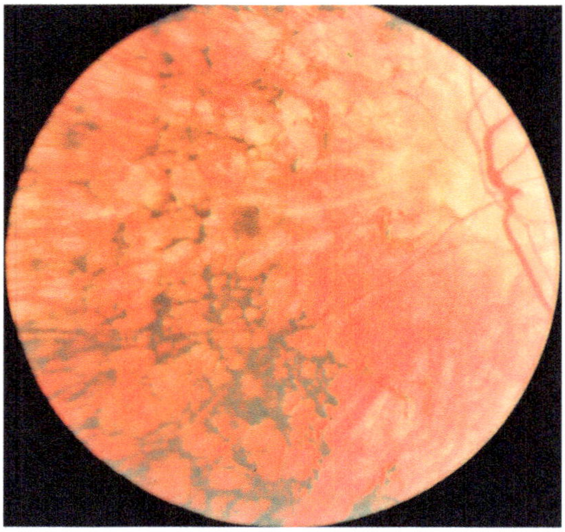

Fig. 16.4 Retinitis pigmentosa. Fundus of patient with retinitis pigmentosa. Bone spicule-shaped pigment deposits are present in the mid-periphery, together with retinal atrophy

others may cause significant vision impairment. An example is type 1 lattice dystrophy, one of the most common inherited corneal dystrophies, which is characterized by the accumulation of amyloid throughout the middle and anterior stroma. Type I lattice dystrophy is an autosomal dominant disorder that results from mutations in the TGFBI gene (5q31). The TGFB1 gene was first discovered in 1997 by Munier et al. in Lausanne, Switzerland. This gene encodes a member of the transforming growth factor beta (TGFB) family of cytokines, which are multifunctional peptides that regulate proliferation, differentiation, adhesion, migration, and other functions in many cell types. Many cells have TGFB receptors and the protein positively and negatively regulates many other growth factors. Mutations in the TGFB-1 gene can lead to various pathologies, including corneal dystrophies (Fig. 16.5).

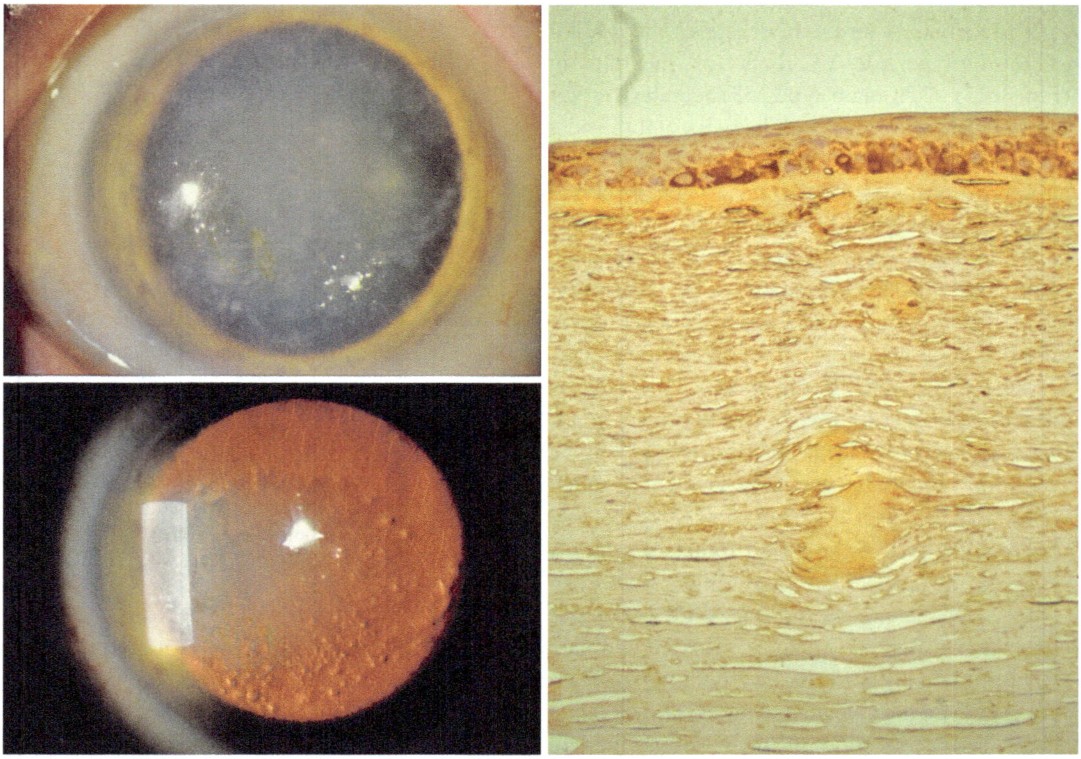

Fig. 16.5 *Left*: biomicroscopic views under diffuse light (*top*) and retroillumination (*bottom*) showing a lattice type 1 corneal dystrophy linked to a R124C heterozygote mutation in the TGFBI gene. *Right*: light microscopy from a lattice type 1 patient (TGFBI R124C) with kerato-epithelin immunostaining showing subepithelial and stromal amyloid deposits. (From Munier FL, et al. (1997) Nat Genet, 15. With permission)

Another example of a disorder resulting from mutations is galactosemia. Several types of galactosemias are recognized, each of which is caused by mutations in a particular gene and affect different enzymes involved in breaking down the sugar galactose. Infants with galactosemia who are not treated promptly with a low-galactose diet face complications. Affected children are also at increased risk of delayed development, speech difficulties, and intellectual disability, as well as the development of lens cataracts.

Lens dislocations that occur in the context of Marfan syndrome (Fig. 16.6) are often due to a mutation in the fibrillin-1 molecule. Fibrillin-1 is an extracellular matrix glycoprotein protein that is the main component of the microfibrils. Multiple autosomal dominant mutations in FBN1 have been described for a wide spectrum of disease, including complete and incomplete forms of Marfan syndrome.

The secondary structure describes the folding of the polypeptide backbone of the protein (Fig. 16.3). Two main types of secondary structures are recognized: the alpha helix and the beta sheet. The alpha helix takes the form of a spiral structure consisting of a tightly packed, coiled polypeptide backbone core with the side chains of the amino acids extending outward from the central axis. The keratins, for example, are a family of fibrous proteins whose structure is nearly entirely alpha-helical. The beta sheet is another secondary structure in which all of the peptide bond components are involved in hydrogen bonding. The surfaces of these beta sheets appear to be "pleated," which is why they are often called "beta-pleated sheets."

The third type of structure found in proteins is called the tertiary structure (Fig. 16.3). The tertiary structure is the final specific geometric shape that a protein assumes. This final shape is determined by a variety of bonding interactions between the side chains on the amino acids. These types of bonding interactions include hydrogen bonding, salt bridges, bisulfate bonds, and nonpolar hydrophobic interactions. Disulfide bonds, for example, occur between the sulfurs of two cysteine side chains. An example of a protein with many disulfide linkages between cysteines is found physiologically in the lens zonules. Therefore, a cysteine deficiency may affect normal zonular development (the zonules become brittle and can break), leading to lens dislocation.

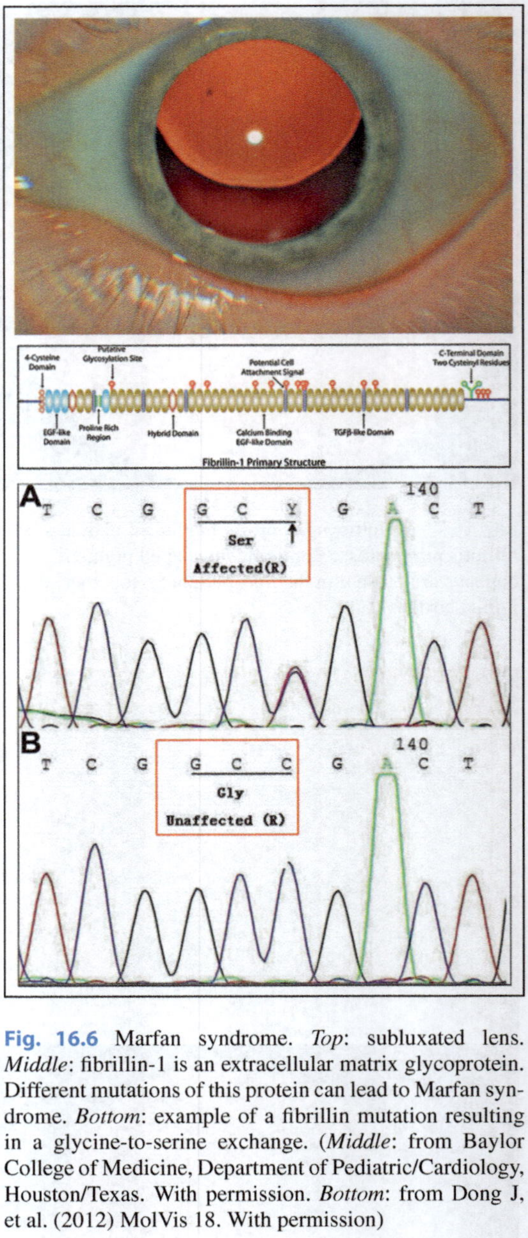

Fig. 16.6 Marfan syndrome. *Top*: subluxated lens. *Middle*: fibrillin-1 is an extracellular matrix glycoprotein. Different mutations of this protein can lead to Marfan syndrome. *Bottom*: example of a fibrillin mutation resulting in a glycine-to-serine exchange. (*Middle*: from Baylor College of Medicine, Department of Pediatric/Cardiology, Houston/Texas. With permission. *Bottom*: from Dong J, et al. (2012) MolVis 18. With permission)

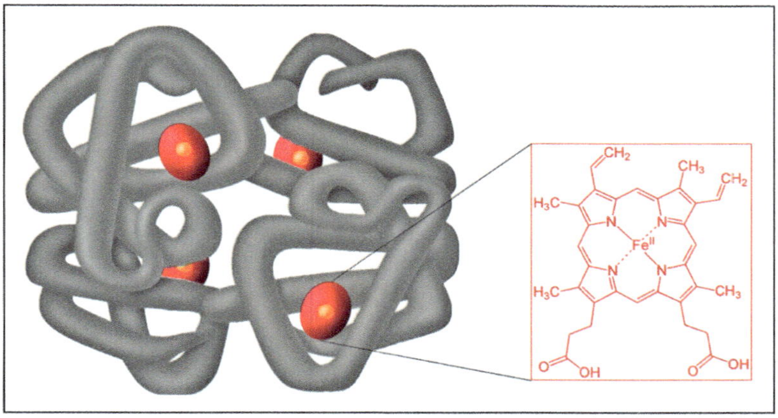

Fig. 16.7 Hemoglobin. The iron-containing heme groups in red are embedded in a protein that consists of two α and two β subunits

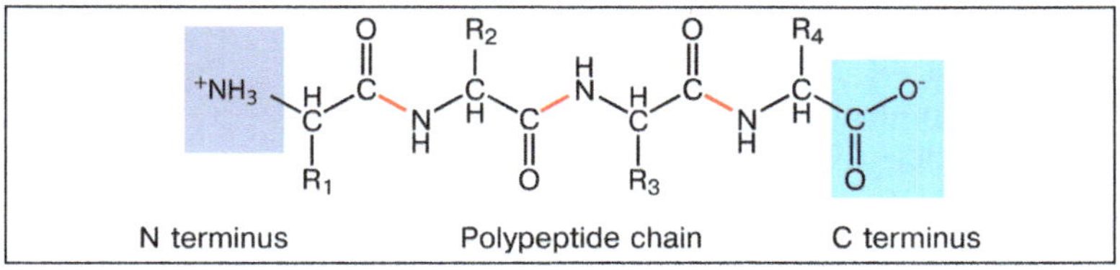

Fig. 16.8 Polypeptide chain. In a polypeptide chain, the amino acids are bound together by peptide bonds (*red*)

The quaternary structure of a protein is a larger assembly of several protein molecules or polypeptide chains. As in the tertiary structure, the quaternary structure is stabilized by noncovalent interactions (e.g., hydrogen bonds or ionic bonds). Complexes of two or more polypeptides are called multimers (e.g., a protein is a tetramer if it contains four subunits). These subunits may either function independently of each other or work cooperatively, as in the case of hemoglobin, where the binding of oxygen to one subunit of the tetramer increases the affinity of the other subunits for oxygen (Fig. 16.7).

16.3 Information Content of a Protein

All proteins are synthesized from only twenty naturally occurring amino acids, which belong to the group of α-amino acids (one carbon, the α-carbon, has all of the remaining groups attached to it). These groups include an amino group, a carboxyl (acidic) group, hydrogen, and one of twenty different R groups (side chains) that give each amino acid type its specific properties. Proteins are polypeptides; i.e., they are single linear chains of amino acids bonded together by peptide bonds (Fig. 16.8).

We shall first discuss the role of proteins in general and then in the eye.

16.4 Roles of Proteins

Proteins are the most common and manifold macromolecules in living cells. They are found in all cells. One single cell contains thousands of different proteins whose biological functions may differ greatly from each other. Proteins occur also outside the cells. The proteins in the blood, for example, serve as protectors (e.g., in the form of antibodies) or as long-distance haulers of molecules such as oxygen (e.g., hemoglobin). Some proteins function as structural

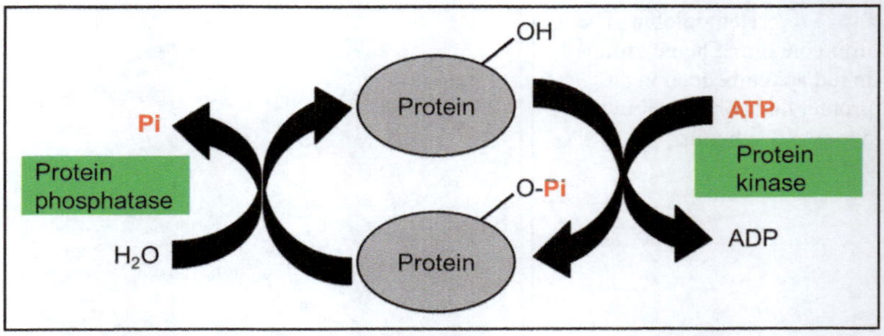

Fig. 16.9 Activation of proteins. Phosphorylation activates (or sometimes deactivates) proteins and dephosphorylation deactivates them

proteins that confer stiffness and rigidity to otherwise-fluid biological components. Other proteins, such as signaling proteins, form a part of the communications network of our nervous system. Certain proteins can also function as "chaperones" and assist in the folding, unfolding, and assembly of other proteins. The central role of all proteins, however – as we will show later – is their role as the most important end-products of information flow between and within cells. In a certain sense, proteins are the molecular instruments through which genetic information is expressed.

When a protein must be available at short notice or in large quantities, it is produced in its inactive form, called a pro-protein (e.g., proinsulin). This inactive form of the protein consists of a long chain of amino acids as well as one or more inhibitory peptides that can be activated when the inhibitory sequence is removed. This long chain of amino acids directs the transport of proteins to their specific location (e.g., insertion into membranes, as in the case of secretory proteins). A part of this long chain of amino acids, thus, acts as an "address." Once the protein has arrived at the specific location, these sequences of amino acids can be removed or processed. Further splicing can activate the protein.

The activity of other proteins and of enzymes in particular can be regulated by the addition or removal of phosphate groups. Protein phosphorylation (the addition of a phosphate group) is recognized as one of the primary ways in which cellular processes are regulated. Phosphorylation reactions are catalyzed by a family of enzymes called protein kinases that utilize ATP as a phosphate donor (Fig. 16.9).

Fig. 16.10 Edmond Fischer and Edwin G. Krebs

Phosphate groups are removed from phosphorylated enzymes by the action of phosphoprotein phosphatases. Depending on the specific enzyme, the phosphorylated form may be more or less active than the unphosphorylated enzyme. The discovery of reversible protein phosphorylation was made by the biochemists Fischer and Krebs[9] (Fig. 16.10).

Proteins can be grossly classified into those with functional and those with structural properties. We will not focus on the classification of proteins in any further detail but will, instead, concentrate on proteins that play a special role in the eye.

The overall role of proteins in the eye is principally the same as in all other tissues in the body. However, there are a few peculiarities.

[9]Edmond Fischer, a Swiss biochemist, and his collaborator Edwin G. Krebs (1918–2009), an American biochemist, shared the Nobel Prize for physiology or medicine in 1992 for describing how reversible phosphorylation works.

16.5 Roles of Proteins in the Eye

16.5.1 Proteins in the Cornea

It may be surprising that the transparent cornea consists mainly of macromolecules, in addition to water. The volume of the cornea comprises mainly of extracellular matrix, which, in turn, comprises macromolecules and water. The cornea provides a good example of the role of structural proteins. The transparency of the cornea is made possible by the spatial arrangement of its structural proteins. These proteins, therefore, maintain stability but, as a result of their alignment, they also allow light to pass through the cornea (Sect. 2.3). In contrast, the sclera scatters all wavelengths of light and, therefore, appears white. This is because the collagen fibers in the sclera are not aligned in a parallel array, so they scatter light.

Under physiological conditions, "cross-linking" occurs between the parallel-running collagen fibers of the cornea (Fig. 16.11).

Under physiological conditions, this cross-linking is achieved by the formation of covalent bonds (Fig. 16.12), which gives the cornea its characteristic stability and stiffness.

In certain pathological conditions (e.g., keratoconus) or after laser treatment (e.g., Lasik), the natural covalent bonds may not be strong enough to prevent the parallel-running collagen fibers from sliding against each other. This risk is reduced by introducing an iatrogenic cross-linking, which leads to the formation of additional bonds between the collagen molecules (Fig. 16.13).

This is achieved by induced oxidative stress: after abrasion of the corneal epithelium, eye drops containing riboflavin (a photosensitizer) are applied to the eye (Fig. 16.14). The eye is then illuminated by UV light (365 nm). This activates the riboflavin molecule, which can then transfer energy to a nearby oxygen molecule, thereby turning ground-state oxygen into the highly active singlet oxygen. Singlet oxygen then transfers this energy to form covalent bonds.

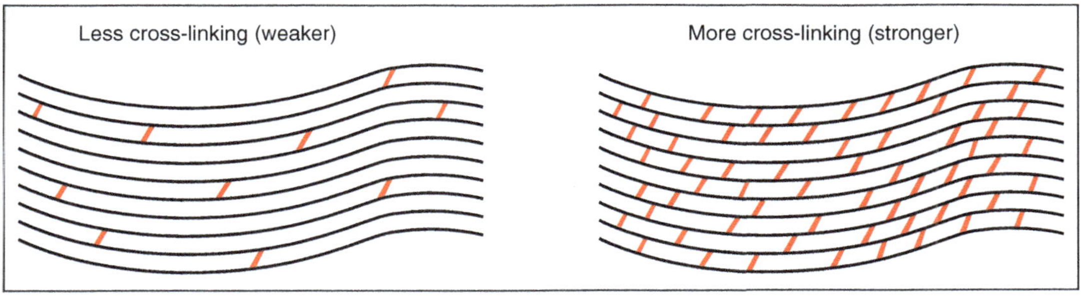

Fig. 16.11 Corneal stability. In the cornea, the fibers are linked together (cross-linking). These links can be weaker (*left*) or stronger (*right*)

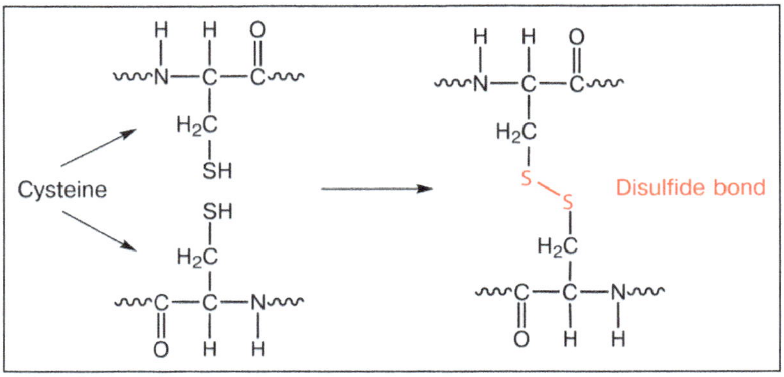

Fig. 16.12 Cross-linking. Cross-linking leads to the formation of covalent bonds (disulfide bonds) between the collagen molecules

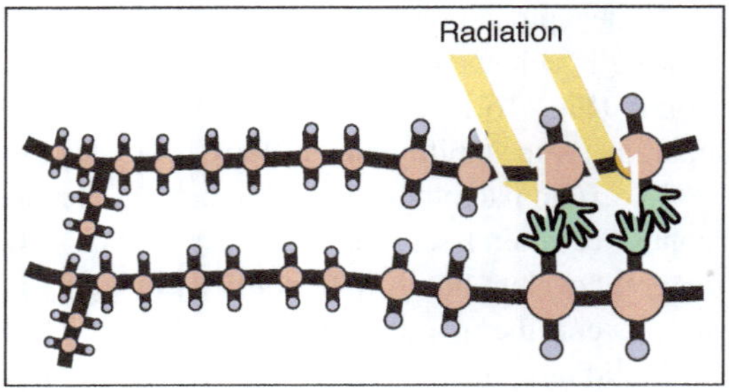

Fig. 16.13 Therapeutic cross-linking. Additional cross-linking can be achieved by the radiation of artificially added riboflavin, which acts as a photosensitizer

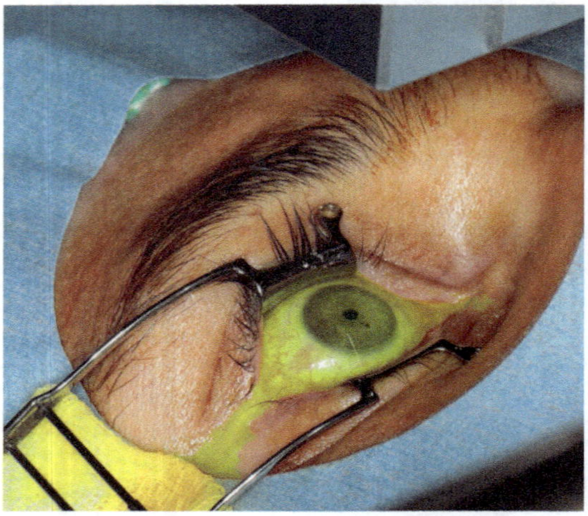

Fig. 16.14 Setting of a cross-linking procedure. The green color on the cornea is due to the riboflavin

However, the exact type of chemical bond that is induced is still not yet clear.

16.5.2 Proteins in the Lens

Our lenses grow throughout our lifespan. New fibers are continuously deposited appositionally, while the inner fibers become denser. The lens epithelial cells elongate and build intracellular structural macromolecules while doing so. The most important of these macromolecules are water-soluble proteins called crystallins. The epithelial cells eventually lose their nuclei, and in doing so, they lose the capability to build new proteins. The cells, however, remain alive and have minimal energy metabolism (in other words, they are not dead). The main component of the lens is, therefore, crystallins. When the lens crystallins become damaged or modified, a cataract results. Crystallins are the reason that the optical density of the lens and, therefore, its refractive index are higher than that of water (Sect. 2.4.1).

16.5.3 Proteins in the Vitreous

Although the vitreous is composed mainly of water, it is well structured by proteins. The internal structure of the vitreous explains a number of phenomena, such as the distribution of blood (in the case of a hemorrhage) or the distribution of cells (in the case of an inflammation). The vitreous body also acts as a relative diffusion barrier, which explains, for example, why the diffusion of oxygen increases after vitrectomy and why locally applied eye drops need a relatively long period (even days) to diffuse to the back of the eye. The vitreous also builds a compartment for drugs such as steroids or anti-VEGFs that are intravitreously applied.

The predominant structural macromolecule found in the vitreous is collagen. These are surrounded by other macromolecules, particularly glycosaminoglycans such as hyaluron. A particular property of these macromolecules is their capacity to bind water, which gives the vitreous its particular gel-like consistency. Proteins function optimally and retain the highest solubility when their secondary and tertiary structure is intact. If the secondary and/or tertiary structure of a protein is disrupted (by breaks in disulfide or hydrogen bonds), the protein loses its activity and becomes less water-soluble. In the case of the proteins in the vitreous, the disruption of secondary and tertiary structure reduces the proteins' water solubility and increases their aggregation among each other. If the structure of the protein is disrupted, it is said to be "denatured" (Fig. 16.15). Depending on the particular protein and the environment, denaturing can also lead to the loss of function (denatured antibodies,

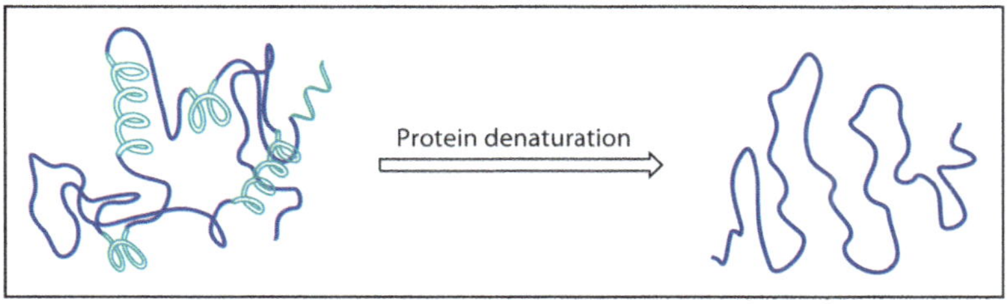

Fig. 16.15 Protein denaturation. The denaturation of proteins involves the disruption and possible destruction of both the secondary and tertiary structures

for example, cannot bind antigen). Denaturation can occur by different means, including physical factors (e.g., UV light or heat, as in cooking an egg), chemical factors (e.g., a strong acid or base), or oxidative stress, which is the most common pathway of the disruption of proteins in medicine.

Coming back to the eye, one important risk factor for the denaturation of proteins in the eye is simply age. As a consequence of age, the proteins in the vitreous aggregate or "clump" together. This leads to focal absorption and scattering of light, which, in turn, forms shadows on the retina. These shadows are perceived by the patient as "mouches volantes." The French word "volantes" means "moving." This particular movement of the aggregated proteins is due to the fact that the vitreous is relatively inert. Thus, when the eyes move in a certain direction, these proteins move more slowly than the rest of the eye.

Another consequence of the aggregation of the proteins is the separation between the fluid (liquid vitreous) and the protein phase (collagen) of the vitreous. As a result of this separation, lacunae are formed at a certain stage, which allows the liquid phase of the vitreous to pass through the prepapillary hole of the vitreous at a certain stage. The volume of the vitreous, thus, becomes smaller than that of the vitreous cavity and the vitreous collapses, causing the posterior vitreous to detach from the retina in all locations except the vitreous base, where it is anchored firmly. If the vitreous is pathologically adherent to the retina, movement of the eye leads to short tractions resulting from the inert vitreous. This is perceived by the patient as "lightning flashes." Normally, these flashes occur temporarily and the vitreous also detaches from these adherences. However, if the vitreous is adhered firmly to the retina, the traction can lead to the formation of horseshoe-shaped holes or tears (Fig. 16.16).

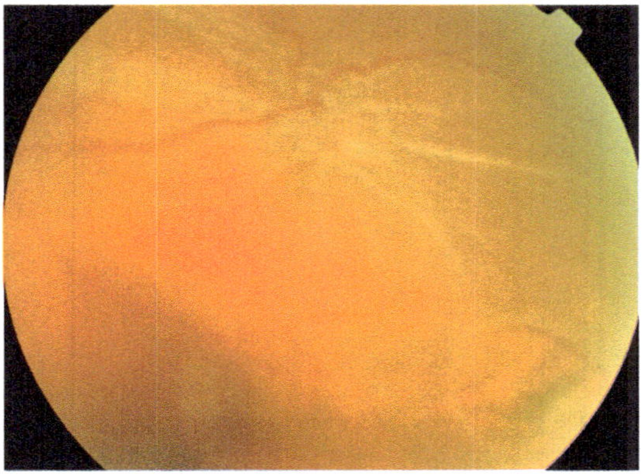

Fig. 16.16 Horseshoe tear. A retinal horseshoe tear in the periphery induced by vitreous detachment

16.5.4 Proteins in the Retina

The proteins in the retina, as in all other tissues, have a variety of functions. In this section, however, we shall focus on two examples: signaling and transport molecules.

The proteins involved in signal transduction serve to convert the information carried by light to information transported by neuro-electrical signals. The illustration shown in Fig. 16.17 is a schematic diagram of signal transduction in the retina. Why is such a cascade needed? This cascade enables a relatively small stimulus to be amplified, resulting in a large cellular

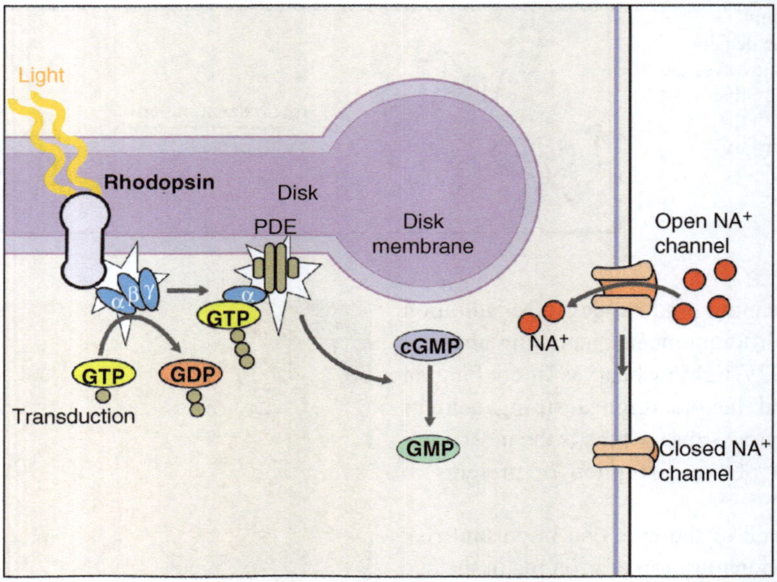

Fig. 16.17 Phototransduction in rod photoreceptors. Light stimulation of rhodopsin leads to the activation of a G-protein (transducin). The GTP-bound alpha subunit of transducin activates a phosphodiesterase (PDE), which then hydrolyzes cGMP into GMP, reducing the concentration of cGMP in the outer segment and leading to the closure of sodium channels

response that ends with the hyperpolarization of photoreceptors. This type of amplification of information is not specific for the photoreceptors, as similar amplification occurs, for example, when adrenalin binds to its receptor in a cell.

Nature has provided these basic mechanisms to allow the transfer of information to cells by a very low concentration of a stimulus such as light or a hormone. However, any type of activation makes biological sense only if it is subsequently deactivated promptly. This deactivation takes place, for example, with the aid of phosphatases that remove phosphate groups.

The cascade of phototransduction begins with the molecule 11-cis retinal, which is a derivative of vitamin A. The 11-cis retinal in photoreceptors is bound to an opsin protein to form the visual pigment rhodopsin (Fig. 16.18).

In general, all visual pigments that use 11-cis retinal as their chromophore are referred to as rhodopsins. Three different classes of rhodopsins in the cones react to different ranges of light frequency (blue, red, and green); this differentiation eventually allows the visual system to distinguish color (Fig. 16.19). The amino acid sequence of the protein in which retinal is embedded determines its absorption spectrum (Fig. 16.20).

Even in a healthy population, a slight variation occurs in the amino acid sequence of the opsin protein, so the absorption spectra of different individuals are not identical due to the fact that different amino acids influence the electrical field differently in and around the 11-cis retinal molecule. This electrical field, in turn, influences the distribution of the electron cloud and, thereby, the light absorption. However, a mutation in the opsin protein can also lead to color blindness. Mutations in any of the proteins involved in the visual cycle can lead to the clinical picture of retinitis pigmentosa (Sect. 16.2).

When the 11-cis retinal molecule is hit by a photon, it isomerizes to all-trans retinal (Fig. 16.17). However, for it to receive a subsequent

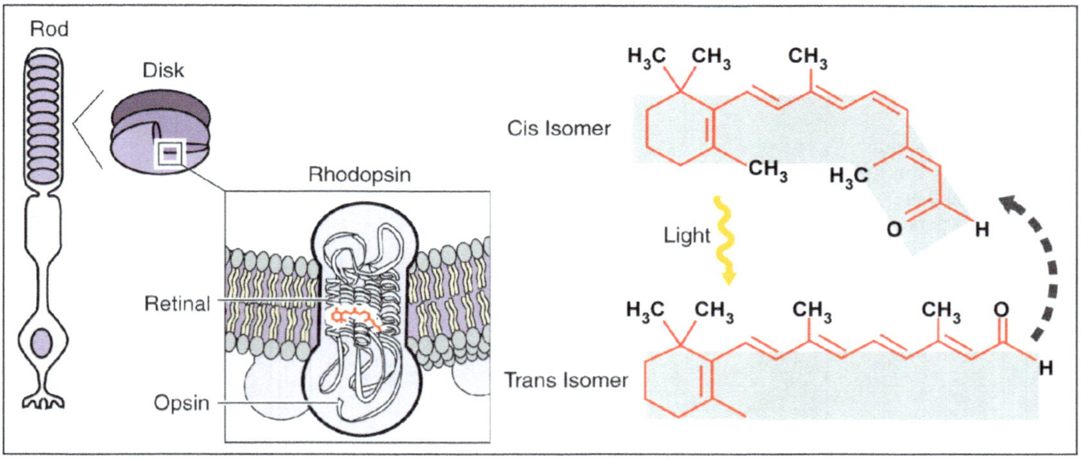

Fig. 16.18 Important elements of the visual cycle. The rods contain discs that contain the protein opsin, in which the retinal is embedded. Light converts 11-cis retinal into all-trans retinal. The all-trans retinal is recycled in many steps back into the cis form

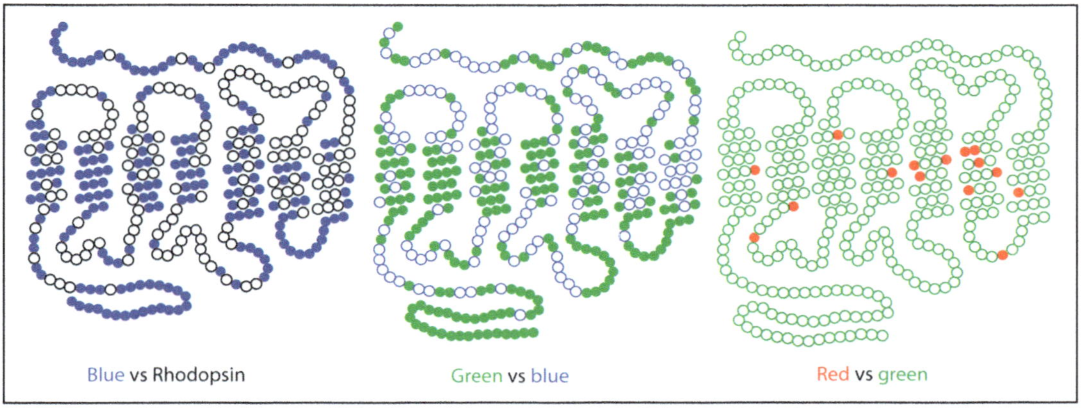

Fig. 16.19 The role of opsin. Retinal has different spectral sensitivities depending on the particular amino acid sequence of the opsin. The opsins for blue- (*left*), green- (*middle*), and red-sensitive (*right*) cones

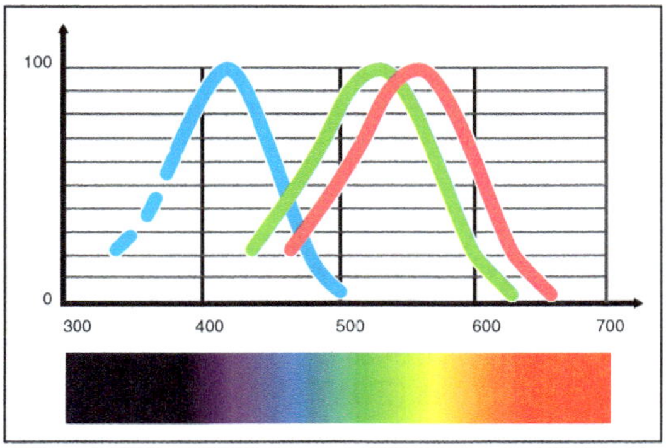

Fig. 16.20 Spectral sensitivity of the cones. The spectral sensitivity (color) resulting from the different opsins is shown in Fig. 16.19

photon, it has to be regenerated back to the 11-cis retinal form. This regeneration takes place in the retinal pigment epithelium cells. The all-trans retinal is first reduced to all-trans retinol and then transported by the cellular retinoid binding protein (CRBP) to the retinal pigment epithelium (RPE). At the RPE, it is converted to 11-cis retinal via several steps that include esterification, isomerization, and hydrolysis by different enzymes. The 11-cis retinal in the RPE is then transported back to the photoreceptor cells by the cellular retinaldyhyde binding protein (CRALBP). Mutations in this protein have been associated with severe rod-cone dystrophy,

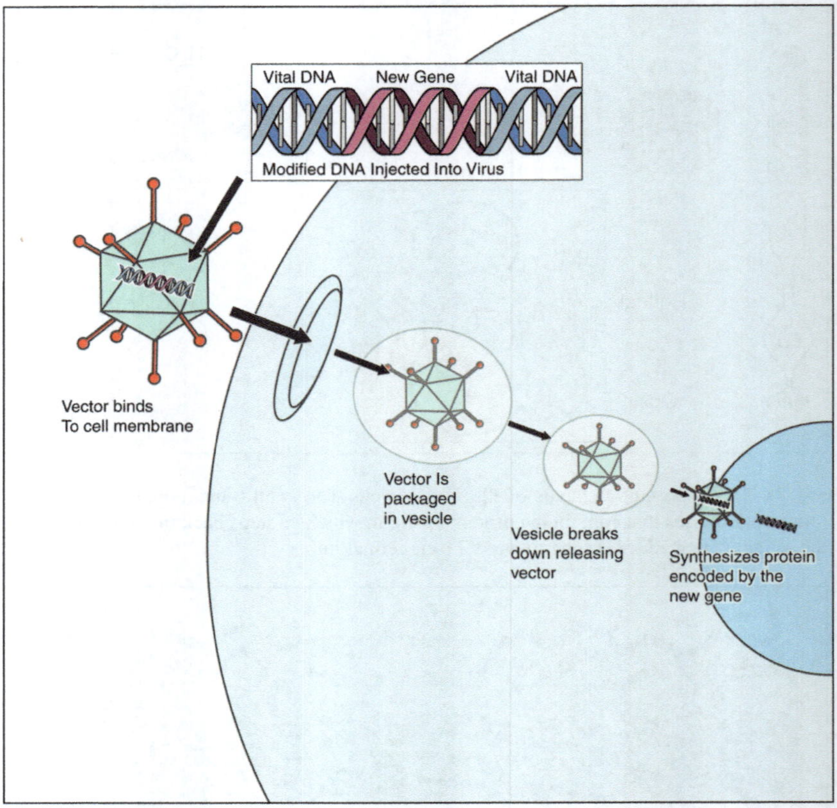

Fig. 16.21 Gene therapy using an adenovirus vector. A healthy gene is inserted into an adenovirus vector. The adenovirus carries this DNA into the nucleus and the DNA is incorporated into the DNA of the recipient

Bothnia dystrophy (autosomal recessive retinitis pigmentosa), and retinitis punctata albescens.

One of the proteins involved in the conversion of the all-trans-retinyl esters to 11-cis-retinal is RPE65 (retinal pigment epithelium-specific protein 65 kDa). The different types of mutations in this particular protein can lead to several diseases such as Leber congenital amaurosis. Whenever a mutation occurs, it is tempting to correct it with gene therapy. This is a complicated field in which a lot of research is being done at present. The first breakthrough was achieved in 2001 in an experimental study in dogs suffering from congenital stationary night blindness. RPE65 genes were cloned from dogs without the disease and injected, together with a viral vector, into the subretinal space of the eyes of the blind dogs. The results were promising: the dogs showed improved vision and were able to move around obstacles. In the meantime, gene therapy has been successfully used in clinical trials in patients. The most common form of gene therapy involves the use of DNA that encodes a functional, therapeutic gene to replace a mutated gene. As shown in the following illustration, in gene therapy, DNA that encodes a therapeutic protein is packaged within a "vector" (e.g., an adenovirus), which is used to get the DNA inside cells within the body. Once inside, the DNA becomes expressed by the cell machinery, resulting in the production of therapeutic protein, which, in turn, treats the patient's disease (Fig. 16.21).

We shall now focus on examples of specific proteins that play an important role in the vascular system.

16.6 Proteins in the Vascular System

16.6.1 Endothelial Derived Vasoactive Factors (EDVFs)

The vascular endothelium is a confluent monolayer of flattened cells that line the inner surface of the vasculature. This layer is not just a barrier but also contributes essentially to the regulation of vascular tone. Endothelial cells function like a relay station, receiving physical information (e.g., sheer stress) or chemical information (e.g., oxygen tension). All of this information is integrated, leading to the production and release of endothelial-derived vasoactive factors (EDVF) and either to vasodilatation or vasoconstriction (Fig. 12.14). These factors influence both smooth muscle cells and pericytes and regulate the size of arteries, veins, and capillaries. EDVFs work in conjunction with other systems, such as the autonomic nervous system, and consist of small molecules, lipids, and peptides. One example of a small molecule is nitric oxide (NO), an example of a lipid is prostacyclin (PGI_2), and an example of a peptide is endothelin (ET), which we shall discuss in more detail as follows.

16.6.2 Endothelin

The most important vasoconstrictive factor belonging to the group of EDVFs is the peptide endothelin 1 (ET-1). The endothelin (ET) peptide exists as three different isoforms: ET-1, ET-2, and ET-3. These are all members of a family of small 21-amino acid peptides with close structural analogies to bee and snake venom, suggesting an early phylogenetic origin of this molecule. ET-1 is essentially produced in vascular endothelial cells. It is formed from precursor peptides, as shown in Fig. 16.22.

ET has certain specific characteristics. First, the cells have no capacity for the storage of ET, so no pool exists that can be liberated in the short term. Therefore, an increase in ET concentration in the extracellular space can be achieved only by de novo synthesis. For this reason, changes in the concentration of ET always take time (approximately in the range of 10–20 min). Second, the half-life of ET in biological tissues is only about 10–15 min. Third, after ET binds to its receptor, the ligand/receptor complex is internalized into the cell and the receptor is then recycled. This explains the refractoriness to ET. If a tissue is stimulated with relatively high concentrations of ET, it can be re-stimulated only after the recycling of the ET receptor – in other words, after about 30–40 min.

Two different ET receptors are known: ETR-A and ETR-B. Their topography and function in vessels are shown in Fig. 16.23.

ET is secreted mostly abluminally but also, to some extent, intraluminally (Fig. 16.24),

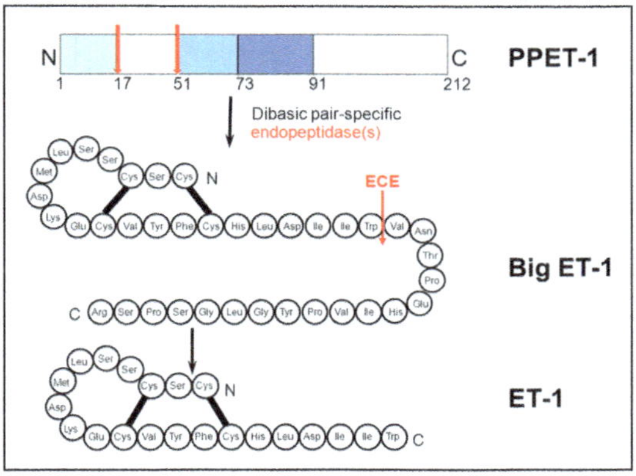

Fig. 16.22 Endothelin. Endopeptidases (enzymes) cleave the preproET-1 (PP ET-1) to form Big-ET-1, which is then cleaved by the endothelin converting enzyme (ECE) to ET-1

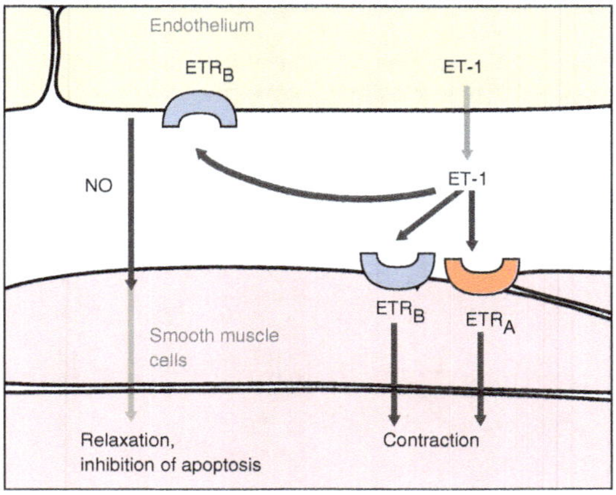

Fig. 16.23 Endothelin receptors. Two classes of endothelin receptors are recognized: endothelin A (ET-A) and endothelin B (ET-B) receptors, which play significantly different roles depending on their location. The binding of endothelin to ET-A or ET-B receptors located on smooth muscle cells causes vasoconstriction, whereas the binding of endothelin to ET-B receptors located on the vascular endothelium causes vasodilatation through the production of nitric oxide

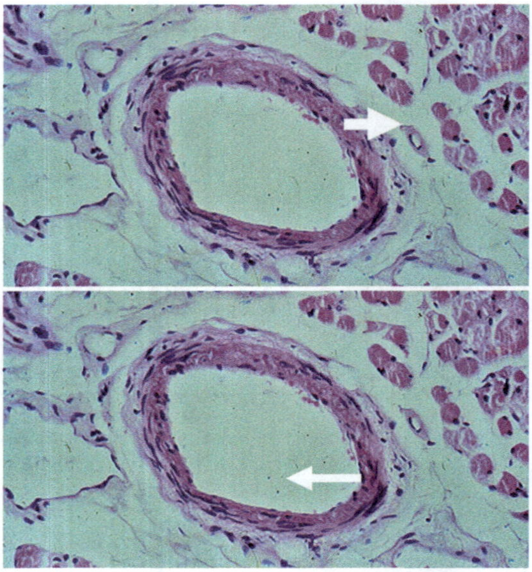

Fig. 16.24 Secretion of endothelin. In the case of blood vessels, ET is secreted mostly abluminally (*upper image*) but also, to some extent, intraluminally (*lower image*). (Courtesy of P. Meyer, University of Basel)

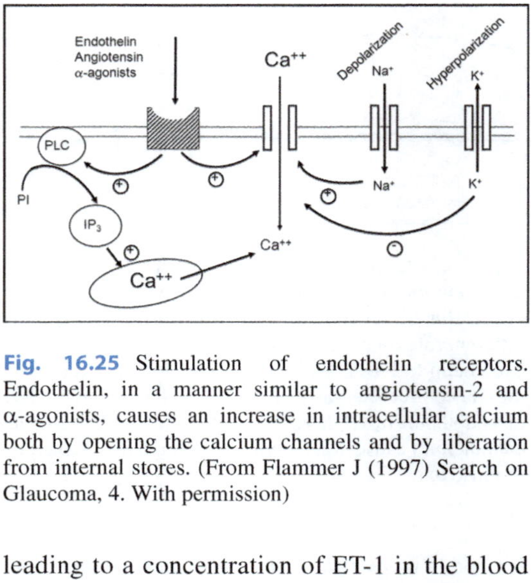

Fig. 16.25 Stimulation of endothelin receptors. Endothelin, in a manner similar to angiotensin-2 and α-agonists, causes an increase in intracellular calcium both by opening the calcium channels and by liberation from internal stores. (From Flammer J (1997) Search on Glaucoma, 4. With permission)

leading to a concentration of ET-1 in the blood of some pg/ml.

The activation of ET receptors on smooth muscle cells or pericytes increases cytoplasmic calcium: by influx into the cell as well as by liberation of calcium from the internal stores. This, in turn, results in constriction of the vessels (Fig. 16.25).

As previously mentioned, under physiological conditions, ET-1 is produced mainly by vascular endothelial cells. However, ET-1 can be produced by any cell if the cell is subject to stress (e.g., a hypoxic condition). The type of cell that will produce the additional amount of ET-1 depends on the underlying disease. In multiple sclerosis (MS), for example, the cells are the lymphocytes, while in HIV, they are macrophages, and in arthritis, they are synovial cells.

What is the clinical consequence of an increased level of ET in the circulating blood? It has no or little influence on the brain or retinal circulation as long as the blood–brain (blood–retinal) barrier is intact. Only the ETB receptors of the vascular endothelium are stimulated, which, in turn, have a more or less neutral effect (Fig. 16.23). If the blood–brain or blood–retinal barrier is disturbed, the same concentration of ET leads to vasoconstriction as it reaches the receptors (ETR-A, ETR-B) of smooth muscle cells. This can be depicted as shown in Fig. 16.26 for multiple sclerosis (MS).

Interestingly, the circulating ET (like other vasoactive peptides such as angiotensin II) has a major vasoconstrictive effect on the optic nerve head, even under physiological conditions (Fig. 16.27), because the capillaries of the choroid are fenestrated and, therefore, leak peptides. These leaked peptides can, in turn, diffuse into the optic nerve head and the adjacent retina, thereby bypassing the blood–brain barrier of the optic nerve head. This explains why diseases such as polyarthritis or MS quite often lead to a reduction in the blood flow to the optic nerve head.

In giant cell arteritis involving the eye, ET-1 circulating in the blood can be markedly increased. This, in turn, explains certain symptoms of these patients such as reduced feelings of thirst or the typical clinical precursor symptoms of amaurosis fugax or signs such as a reduction in the optic nerve head blood flow. The involvement of ET-1 in giant cell arteritis further explains why the optic nerve head and choroidal circulation are more involved compared to the retinal circulation. Interestingly, in giant cell arteritis, not only can ET be increased but the number of receptors is also upregulated, thereby increasing the sensitivity to ET.

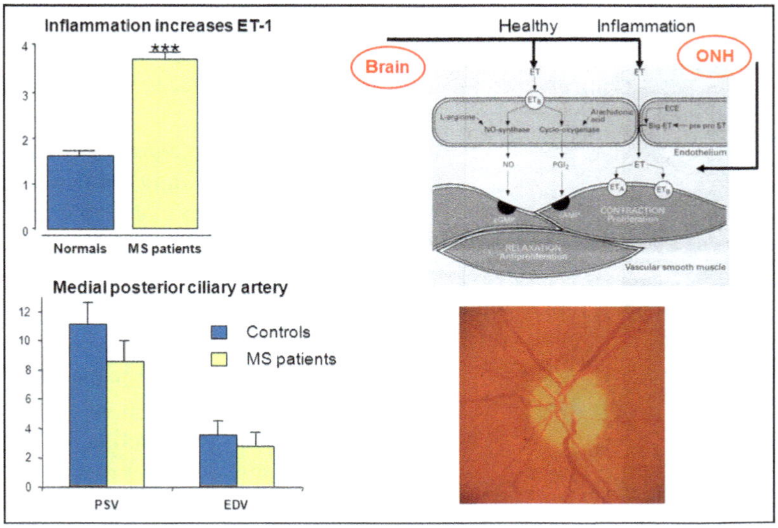

Fig. 16.26 The effect of endothelin-1 in circulating blood: Endothelin-1 (ET-1) is increased in many inflammatory diseases, as, for example, depicted in the upper left in MS patients. This has little impact on brain or retinal circulation as long as the blood–brain barrier is intact. However, it reduces blood flow in the choroid and ONH (*upper right*). This leads to a reduction of blood flow in the retroocular vessels as depicted for MS patients (*lower left*) and sometimes to a slight paleness of the optic nerve head (*lower right*). (From Flammer J, et al. (208) Can J Ophthalmol, 43. With permission)

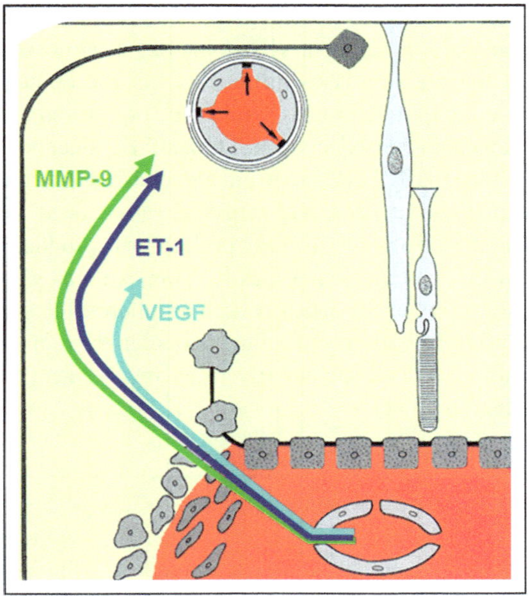

Fig. 16.27 Blood–brain barrier of the optic nerve head: Although the optic nerve head belongs to the central nervous system, the blood–brain barrier is insufficient, as molecules such as endothelin-1(ET-1) can diffuse from the choroid into the ONH. (Modified from Grieshaber et al. 2007)

Receptors can be stimulated not only by their corresponding hormones (ligands) but also by autoantibodies. This is particularly important in the pathogenesis of endocrine orbitopathy, where thyroid-stimulating hormone receptors (TSHRs) are stimulated. Another example of the stimulation of receptors by autoantibodies is the stimulation of ET receptors that occurs, for example, in pre-eclampsia.

ET, together with other molecules, also plays a major role in the pathogenesis of optic nerve splinter hemorrhages and retinal vein occlusions. Let's first discuss the pathogenesis of optic nerve splinter hemorrhages (Fig. 16.28).

In glaucoma, these hemorrhages typically occur at the border of the optic nerve head. The blood vessels of the retina form a blood–retinal barrier that is similar to the blood–brain barrier and this barrier is formed by tight junctions of the endothelial cells. When the tight junctions are partly opened, larger molecules such as fluorescein can leak out. This, for example, can be due to VEGF or to ET in the surrounding tissue of the vessel, produced locally or diffused from the choroid (Fig. 16.27). In this context, ET

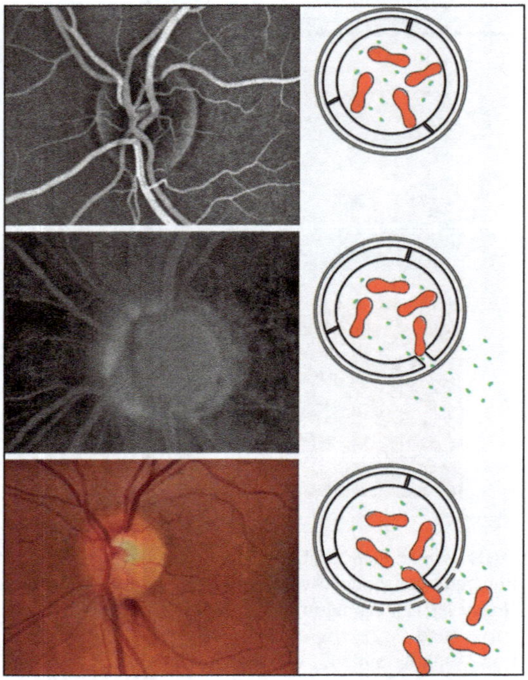

Fig. 16.28 Pathogenesis of optic nerve splinter hemorrhages. *Top*: under normal conditions, blood vessels build tight junctions between the endothelial cells. Middle: ET-1 or VEGF can open up the barrier at the level of the endothelial cells. *Bottom*: MMP-9 can digest the basal membrane, allowing erythrocytes to escape. (From Grieshaber MC, et al. (2007) Surv Ophthalmol, 52 Suppl 2. With permission)

acts via an increased concentration of prostaglandin-E2. Even though the blood–retinal barrier is partly disrupted, blood cells do not leak out since the basal membrane is intact (Fig. 16.28 middle). While white blood cells can actively pass through (e.g., during an inflammation), erythrocytes can only escape if both the blood–retinal barrier and the basal membrane are weakened (Fig. 16.28 bottom). This mainly occurs through the action of an increased concentration of MMP-9. Interestingly, the concentrations of these molecules (VEGF, MMP-9, ET-1) are increased in the circulating blood of patients with normal tension glaucoma (NTG), explaining why hemorrhages occur more often in NTG than in high tension glaucoma (HTG) patients. However, the same molecules can also be produced locally if hypoxia leads to an increased concentration of HIF-1 α. This explains why reducing IOP (and thereby increasing ocular blood flow) reduces the occurrence of splinter hemorrhages.

In a similar manner, ET also plays an important role in regulating retinal venous pressure and in the pathogenesis of retinal vein occlusions.

Although it is well established that retinal arteries are well regulated, the fact that the retinal veins are also regulated is less well known. The retinal veins even react to smaller concentrations of vasoactive molecules, particularly to ET. In the lamina cribrosa and in the "cross-over" sections of the vessels in the retina, the arteries and veins are anatomically very close and share a common adventitia (Fig. 16.29). Nevertheless, in normal conditions, they are regulated separately. If the concentration of ET around the vessels is increased, the vein locally constricts and, thereby, the retinal venous pressure upstream increases (Fig. 16.30).

Local constrictions of the veins lead to increased retinal venous pressure and, thereby, also often to dilatation of the retinal veins. Up to a certain degree of such increased retinal venous pressure, no clinical symptoms occur. However, if the pressure rises further, we get the clinical picture of a venostasis syndrome. The aforementioned increased concentration of ET occurs if either the ET diffuses from the choroid into the optic nerve head and adjacent retina or if an increased amount of these molecules is produced in the hypoxic retina locally. This explains why arterial types of diseases such as atherosclerosis, systemic hypertension, and vascular dysregulation, can lead secondarily to an involvement of the retinal veins.

16.7 Enzymes

Enzymes are proteins that catalyze different chemical reactions. They often contain an additional non-protein component or ion (e.g., Zn^{2+} and Mg^{2+}). Thousands of different enzymes exist; some of them function intracellularly (e.g., hexokinase), whereas others function extracellularly (e.g., matrix metalloproteinases, MMPs).

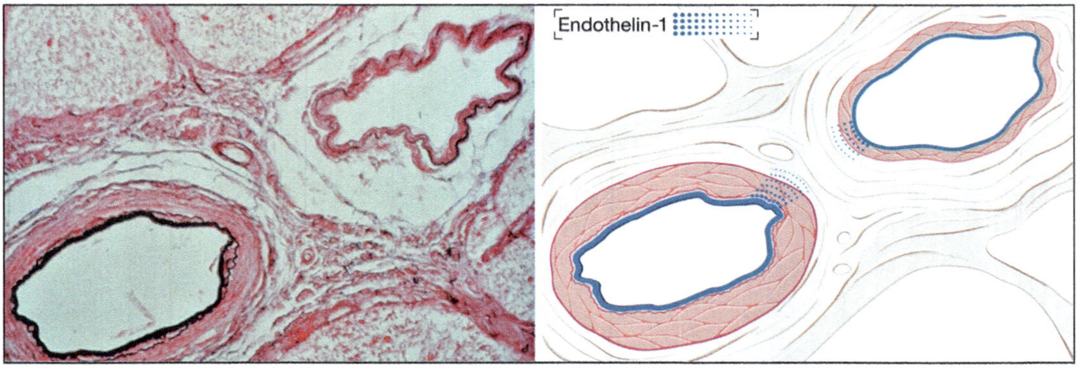

Fig. 16.29 Relationship between arteries and veins in the lamina cribrosa. The endothelial cells of both the arteries and veins secrete endothelin for local regulation. (From Fraenkl S, et al (2010) EPMA, 2. With permission)

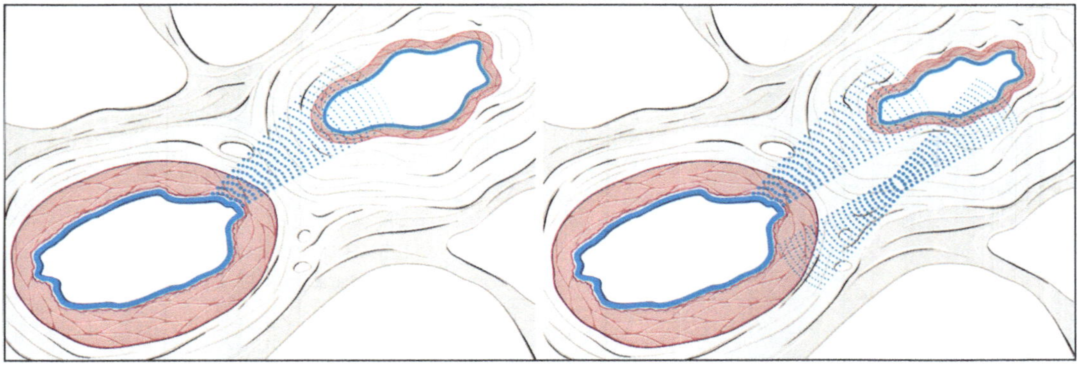

Fig. 16.30 Constriction of the retinal vein on the level of the optic nerve head. *Left*: a diseased artery can produce an amount of endothelin that influences even the neighboring vein. *Right*: endothelin can also be produced in the hypoxic tissue surrounding the vessels, also leading to local constriction of the vein. (From Fraenkl S, et al (2010) EPMA, 2. With permission)

Enzymes can also be grouped depending on their function, as shown in Fig. 16.31.

Through their activation and deactivation, enzymes play an important role in cellular metabolism. Enzymes themselves are normally synthesized in a precursor form and are then cleaved into their final enzymatic form. Most enzymes, in addition, need to be activated, which is often achieved by phosphorylation or dephosphorylation. In other words, the enzyme is either switched "on" or "off" by phosphorylation and dephosphorylation.

Other enzymes are activated when they are brought into their final form. This can be shown by the example of angiotensin (Fig. 16.32). The renin–angiotensin system is a hormone system that regulates blood pressure and water balance.

The precursor protein in this system, angiotensinogen, is made by the liver and converted by plasma renin (mainly produced in the kidney) into a second precursor protein, angiotensin I. The circulating angiotensin I is then locally converted by the membrane-bound angiotensin-converting enzyme (ACE) into the biologically active angiotensin II. The inhibition of the ACE is used clinically mainly to lower blood pressure but also in the treatment of diabetic retinopathy.

Enzymes also play an important role in phototransduction, a process in which the information of light is converted into information in the form of electrical signals in the retina. Catching this light and converting its minute amount of energy into a neural response distinguishes the

Class	Function	Example
Hydrolases	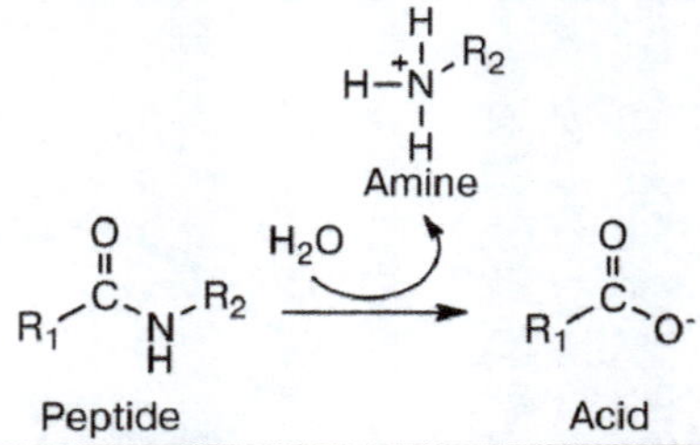	Peptidase
Isomerases	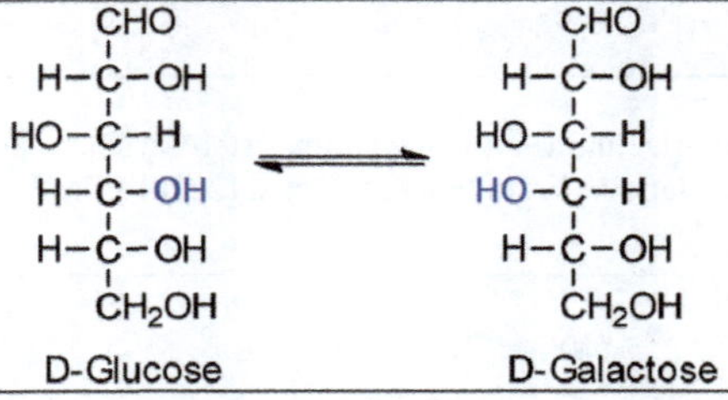	Epimerase
Ligases	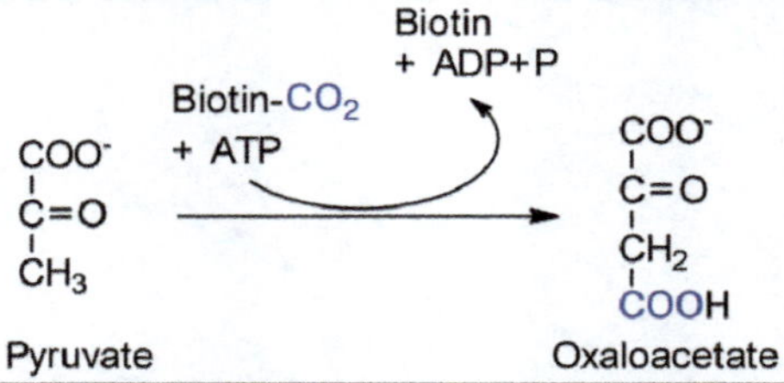	Pyruvate carboxylase
Lyases	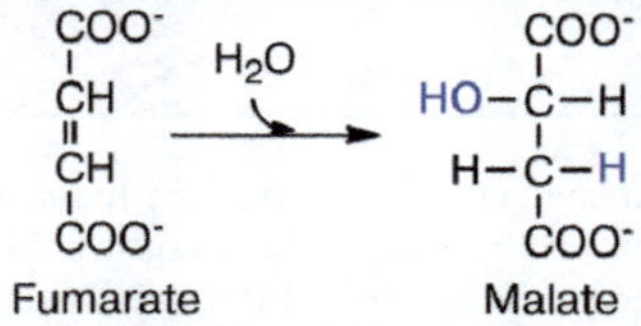	Fumarate hydratase
Oxidoreductases	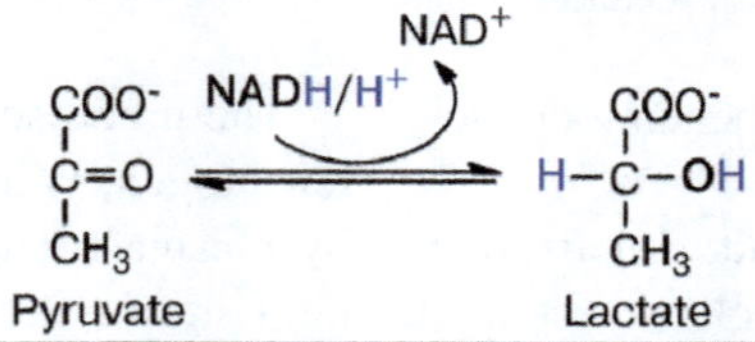	Lactate dehydrogenase
Transferases	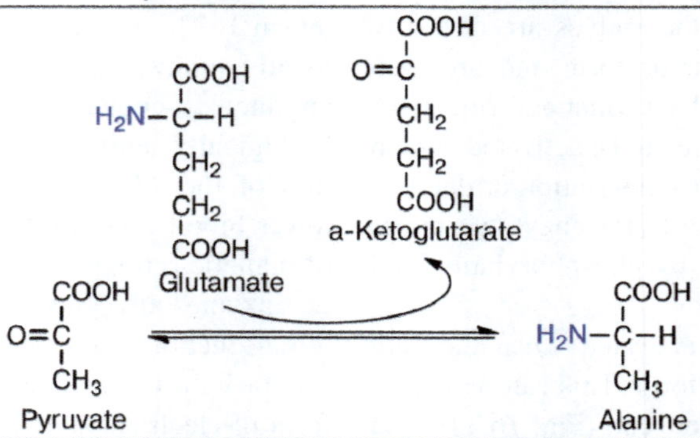	Aminotransferase

Fig. 16.31 Different classes of enzymes and their functions

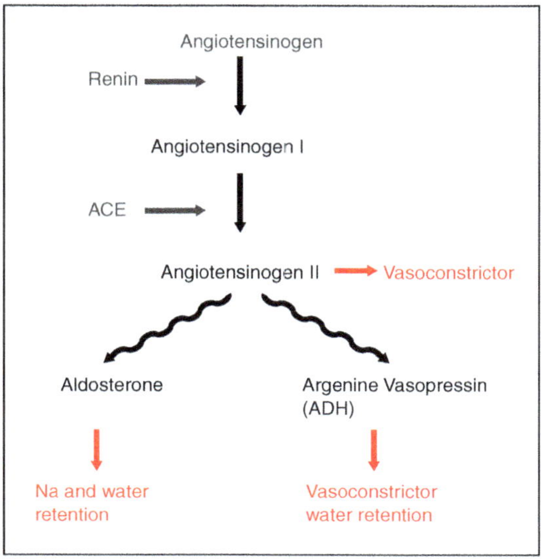

Fig. 16.32 Renin–angiotensin–aldosterone system. Plasma renin converts angiotensinogen to angiotensin I. Angiotensin I is subsequently converted to angiotensin II by angiotensin-converting enzyme (ACE). Angiotensin II causes vasoconstriction but also stimulates the secretion of the hormone aldosterone from the adrenal cortex

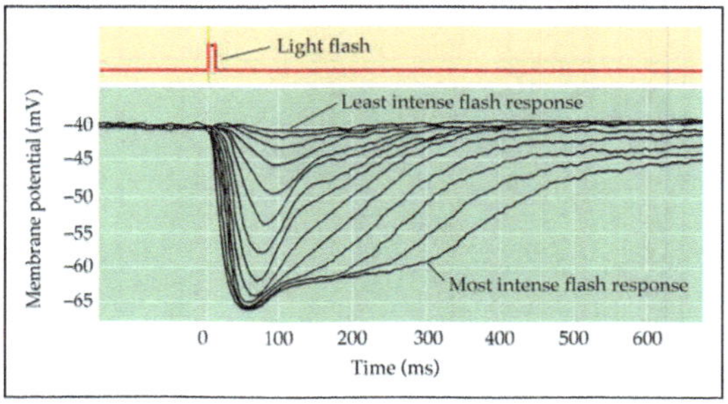

Fig. 16.33 Stimulus-induced hyperpolarization of photoreceptors. In the dark, the cells are only slightly polarized (–40 mV). Stimulation with light, however, leads to a hyperpolarization of –70 mV because the enzyme phosphodiesterase hydrolyzes cGMP into GMP, thereby closing the channels. (From Purves D, et al. (Eds) (2001) *Neuroscience. 2nd edition*. Sunderland (MA): Sinauer Associates Inc. With permission)

retina from all other neural structures. It is interesting that the absorption of a single photon finally leads to a broad cascade of reactions that amplify the information. Yet how does this process actually occur?

Two main types of photoreceptors are distributed unevenly across the retina: the rods and the cones. The rods have the unique capacity to adapt themselves within a huge range of brightness, ranging from daylight to dim light. In contrast, cones cannot adapt as well to dim light; however, in daylight, they contribute more to color vision and spatial resolution. Both rods and cones contain opsin. The light-sensitive molecule is the 11-cis-retinal (an aldehyde of vitamin A), which is embedded in the opsin (Fig. 16.18). The opsin plays a key role in determining the absorption spectrum of retinal. Each photoreceptor expresses only one type of opsin. Rhodopsin is present in rods, while photopsins are present in cones.

Normally, in a sensory organ, the corresponding stimulation leads to a depolarization that induces a cascade of events. In the case of photoreceptors, however, the opposite occurs. In the dark, the cells are only slightly polarized (–40 mV) and light stimulation leads to a hyperpolarization of –70 mV (Fig. 16.33).

This stimulus-induced hyperpolarization is a distinctive characteristic of photoreceptors. A key molecule responsible for maintaining a low membrane potential in photoreceptors is the nucleotide cyclic guanosine 3'-5' monophospate (cGMP). High cGMP levels keep cGMP-gated ion channels in the open state, which allows an inward Na^+ and Ca^{++} ion flow in the dark. In the light, the levels of cGMP decrease, leading to closure of the cGMP ion channels and thereby to an increase in membrane potential (membrane hyperpolarization). Yet how does light stimulation lead to a decrease in cGMP?

When light hits an 11-cis-retinal molecule, the molecule undergoes an isomerization to all-trans retinal. For this to occur, the following three requirements must be met: (1) since the precursor for 11-cis-retinal is all-trans-retinol (vitamin A), the diet must be rich in vitamin A; (2) the correct absorption of vitamin A in the gastrointestinal tract must occur, and (3) normal transport of vitamin A must occur from the blood via the pigment epithelial cells into the rods and cones. If one of these three requirements is not met, adverse visual symptoms will result.

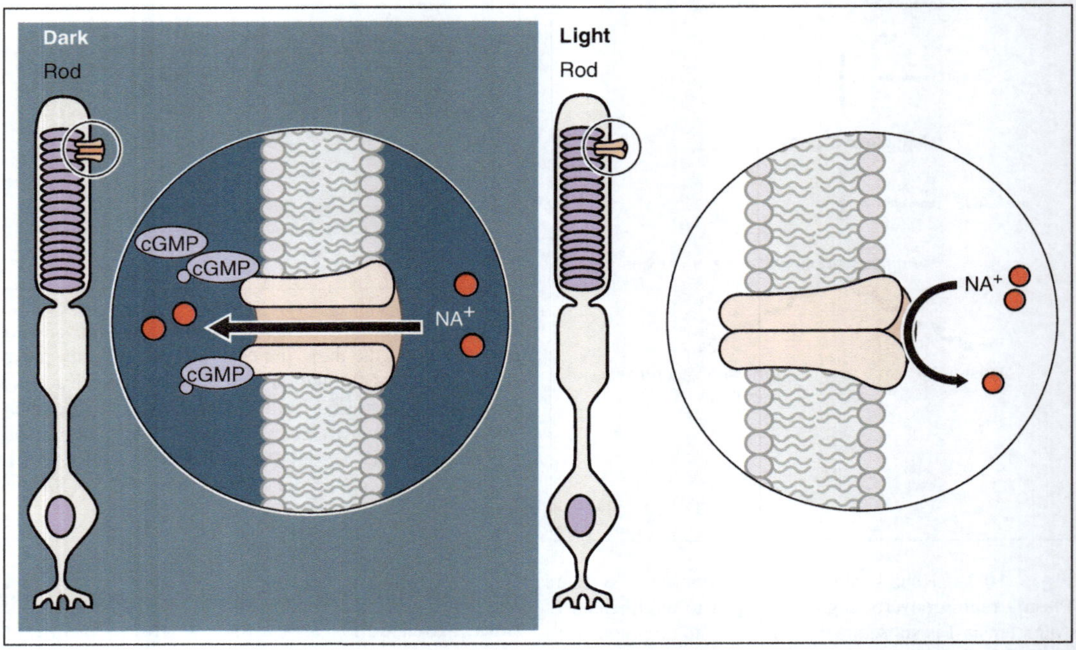

Fig. 16.34 cGMP-gated channels. A key second messenger molecule responsible for maintaining a relatively low polarized rest state in photoreceptors is cGMP. High cGMP levels keep the ion channels in the open state. Decreased levels of cGMP cause the closure of ion channels, which leads to membrane hyperpolarization

When the photoreceptor is exposed to light, metarhodopsin II stimulates the activation of a G-protein (transducin), which, in turn, activates the phosphodiesterase (PDE) enzyme. Interestingly, one rhodopsin molecule can actually activate up to a hundred transducin molecules, amplifying this reaction. The activated PDE enzyme then hydrolyzes cGMP, reducing its concentration in the outer segment and leading to the closure of sodium channels (hyperpolarization) in the outer segment membrane (Fig. 16.34).

The enzyme phosphodiesterase (PDE), thus, plays a crucial role in phototransduction. PDEs belong to a superfamily of enzymes that break phosphodiester bonds (bonds between two sugar groups and a phosphate group). The PDE superfamily of enzymes is classified into 11 families: PDE1-PDE11. The PDE5 inhibitors, such as sildenafil, are used to treat erectile dysfunction. PDE5 inhibitors are also expressed in the retina, explaining why these types of drugs can potentially provoke visual symptoms.

16.8 Antibodies

Antibodies, also known as immunoglobulins (Fig. 16.35), are large Y-shaped proteins produced by B-cells as a part of the humoral immune system (the cellular immune system will not be discussed here).

Their main but not sole task is to identify and neutralize foreign objects such as bacteria and viruses. The antibody recognizes a unique part of the antigen. The antibody contains a paratope (a structure analogous to a lock) that is specific to one particular epitope (similarly analogous to a key) on an antigen, allowing these two structures to bind together with precision.

When antigens are recognized by antibodies, a chain of reactions occurs that leads to inflammation. However, certain sites in the body can tolerate the introduction of antigens without eliciting an inflammation response. These sites are said to be "immune-privileged." Immune privilege is thought to be an evolutionary adaptation to protect vital structures from potentially damaging inflammation.

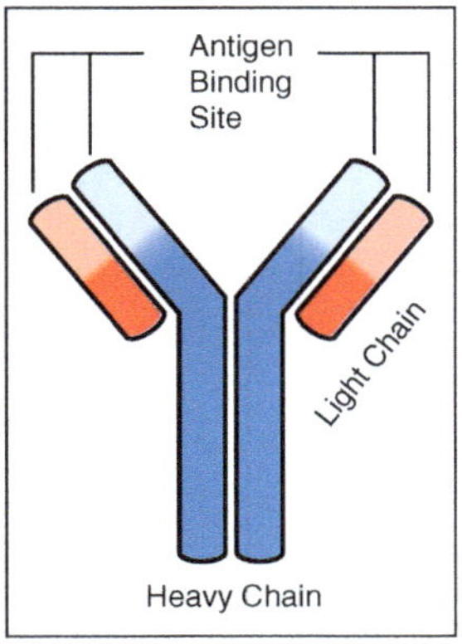

Fig. 16.35 Immunoglobulin (Ig). Example of Ig G consisting of two light and two heavy chains

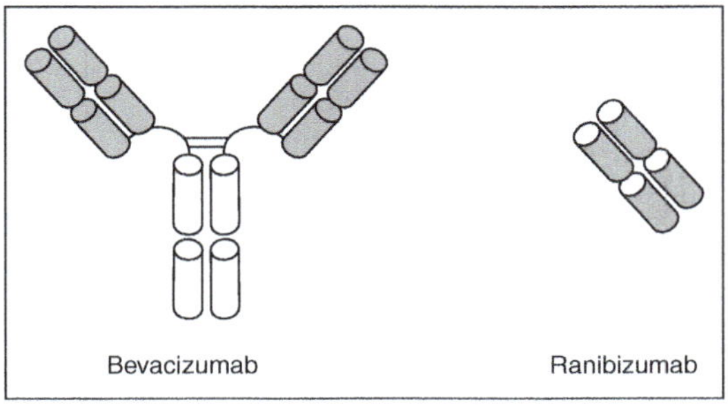

Fig. 16.36 Bevacizumab (complete antibody) and Ranibizumab (fragment of an antibody)

The testes, the placenta, and the central nervous system are immune-privileged sites, as is the eye. From a therapeutic point of view, corneal transplantation is currently the only procedure that takes advantage of this immune privilege.

Obviously, immune privilege has advantages as well as disadvantages. Under normal circumstances, T cells encounter self-antigens during their development, when they move to the tissue draining lymph nodes. This leads to so-called immune tolerance and prevents an autoimmune response in the future. However, as a result of immune privilege and the lack of antigen presentation to T cells during their development, T cells fail to produce sufficient amounts of energy to the self-antigens. If, during a pathological process, the self-antigens are exposed to the immune system, no tolerance exists and an autoimmune response is provoked. An example of this type of autoimmune response is sympathetic ophthalmia. This is a rare disease in which trauma to one eye causes an autoimmune attack on both the damaged, traumatized eye and the non-traumatized or "sympathizing" eye. Usually, trauma to one eye induces the release of eye antigens that are recognized and picked up by local antigen presenting cells (APCs) such as macrophages and dendritic cells. These APCs carry the antigen to local lymph nodes to be sampled by T cells and B cells. Entering the systemic immune system, these antigens are recognized as foreign and an immune response is mounted against them.

However, autoimmunity can also develop when certain similarities exist between different antigens. A number of antibodies against bacterial antigens can also attack human antigens. Another mechanism of autoimmunity can be explained by the example of endocrine ophthalmopathy. Through pathologies in the thyroid gland, antibodies against thyroid tissue can also attack fibroblasts in the orbit.

Antibodies can also be used for diagnostic purposes. Normally, the presence of an antibody is a surrogate for the presence of the corresponding antigen. This is particularly used to diagnose present or past viral infections. Examples include cytomegalovirus (CMV), herpes simplex virus (HSV), and Epstein-Barr virus (EBV).

Antibodies can also be used to detect certain molecules in histological sections, a method called immunohistochemistry (IHC). IHC provides additional information to the classical histology; for example, it is applied to tumors to detect and quantify certain molecules.

Finally, antibodies can be used for therapeutic purposes. The two drugs now commonly used in therapy for exudative age-related macular degeneration are Bevacizumab and Ranibizumab. The drug Bevacizumab is the whole antibody – two Fab parts together with an Fc part – whereas the drug Ranibizumab is just a Fab part of the antibody (Fig. 16.36).

Experimentally, even much smaller parts of the antibody, specifically the epitope-recognizing tip of the Fab fragment, are used. The challenge in the use of these fragments lies in stabilizing them to keep the form of the paratope specific to the epitope.

17 Lipids

Lipids are a heterogeneous group of hydrophobic or lipophilic molecules, respectively. Some of them are amphiphilic in nature, which allows them to form structures such as vesicles, liposomes, or membranes in an aqueous environment. Biochemically, lipids are made up of two subunits (or "building blocks"): ketoacyl and isoprene groups. Lipids may be classified into eight categories: fatty acids, glycerolipids, glycerophospholipids, sphingolipids, saccharolipids, polyketides, sterol lipids, and prenol lipids (Fig. 17.1). The main biological functions of lipids include energy storage, maintenance of the structure of cell membranes, and signaling. In the eye, lipids have numerous different functions. One example is the vitamin A-derived retinal, which is involved in the visual cycle (Chap. 16). Another example is the prostaglandins, which are often involved in both physiological and pathophysiological processes. In this chapter, we will focus on the role of lipids in the tear film and in the retina.

17.1 Tear Film

In humans, the tear film of the eye, known as the precorneal film, has three distinct layers (Fig. 17.2).

The lipid layer is the outermost surface and provides a smooth tear surface; the aqueous layer, consisting mainly of water, promotes spreading of the tear film, whereas the innermost mucus layer increases the wettability of the corneal surface. The tear film creates a smooth surface for light to pass through the eye, nourishes the front of the eye, and provides protection from infection.

The lipid layer of the tear film contains nonpolar lipids as well as lipids with a polar head. The polar heads of the phospholipids form the interface between the aqueous–lipid layers (Fig. 17.2). The nonpolar lipid components allow other lipids, such as cholesterol, to dissolve in the tear film. Tear lipids are not susceptible to lipid peroxidation because they contain extremely low levels of polyunsaturated fatty acids (Sect. 13.3). The lipids in the tear film are secreted primarily by the meibomian glands. There are approximately 30–40 meibomian glands in the upper lid and 20–30 smaller glands in the lower eyelid. As shown in the illustration (Fig. 17.3), each gland opens onto the skin of the eyelid margin, between the tarsal gray line and the mucocutaneous junction. A second group of glands also secrete lipids: the sebaceous glands of Zeis. These glands are located at the lid margin in relation to the lash roots and the secreted lipids are also partly incorporated into the tear film. The lipids in the tear film have several functions. They provide a smooth surface, reduce the evaporation of water, prevent tear overflow onto the lids, and provide a watertight seal during lid closure when sleeping.

Various different diseases such as meibomitis (Fig. 17.4) and blepharitis (Fig. 17.5) can lead to lipid alteration.

Other dysfunction of the meibomian glands also leads to reduced lipid secretion, which may

J. Flammer et al., *Basic Sciences in Ophthalmology*,
DOI 10.1007/978-3-642-32261-7_17,

Class	Structure	Example
Fatty acyls	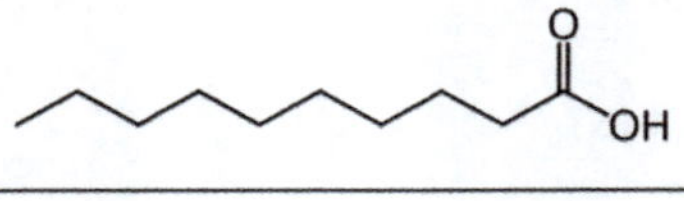	Decanoic acid
Glycerolipids	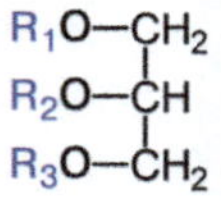	R_1, R_2, R_3: Fatty acids
Glycerophospholipids	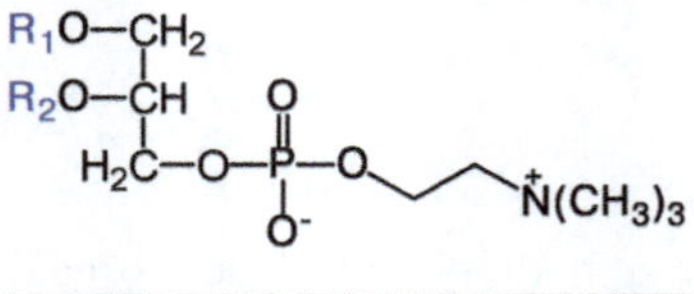	Phosphatidylcholines R_1, R_2: Fatty acids
Sphingolipids	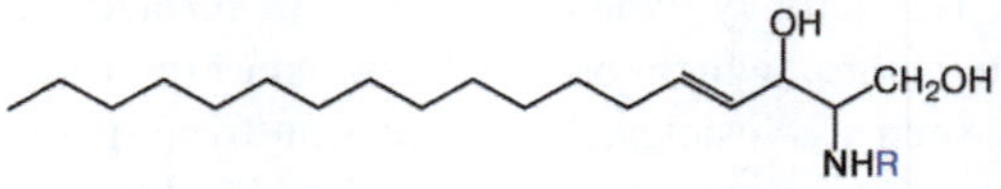	Ceramide R: Fatty acid
Sterol lipids	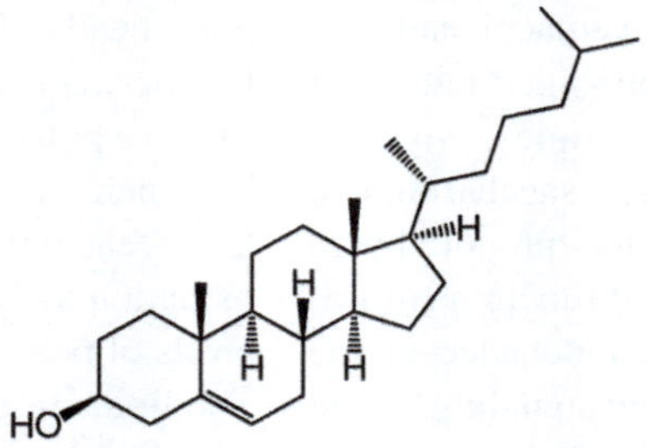	Cholest-5-en-3β-ol
Prenol lipids	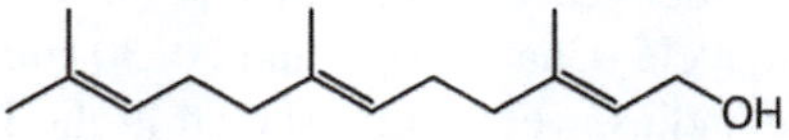	2E,6E-farnesol
Saccharolipids	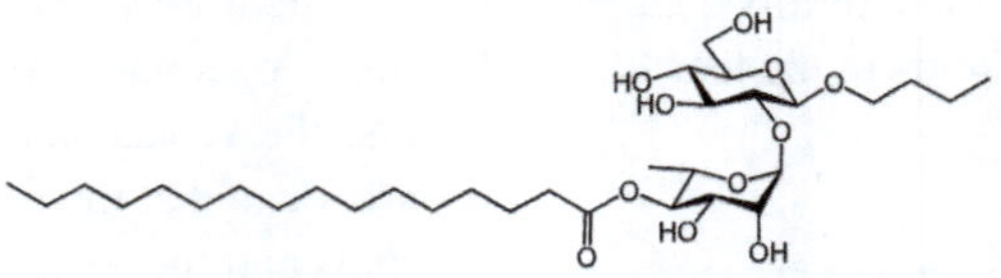	Butyl 4'-O-hexadecanoyl-neohesperidoside
Polyketides	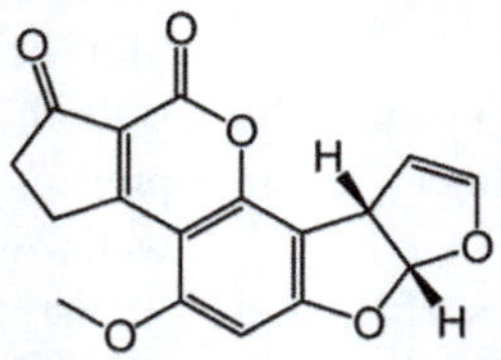	Aflatoxin B1

Fig. 17.1 The eight categories of lipids

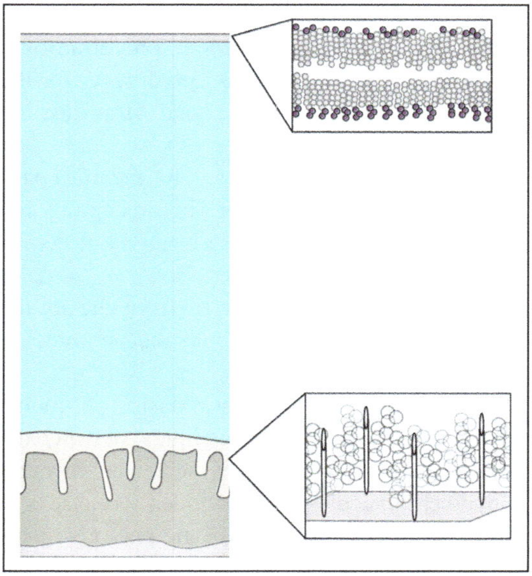

Fig. 17.2 Tear film. Schematic drawing of the tear film showing the outermost lipid layer, with the polar and non-polar heads, middle aqueous layer, and inner mucus layer containing transmembrane glycoproteins and mucins

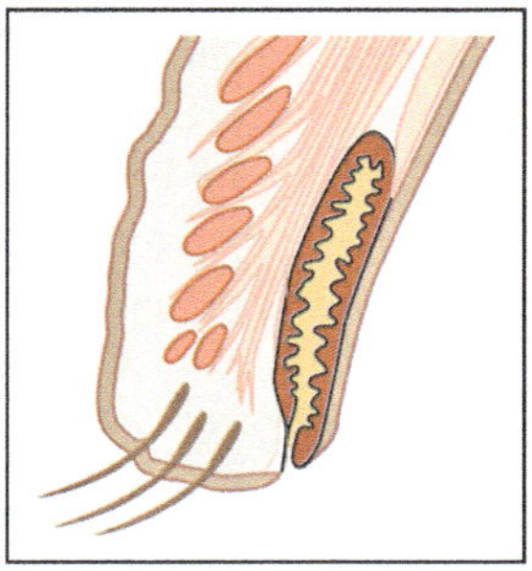

Fig. 17.3 Schematic drawing of the meibomian glands of the lid. The orifices of the ducts for the glands are located along the posterior margin of the eyelid

result in dry-eye syndrome. A partial obstruction of the gland ducts facilitates the degradation of the lipids by bacterial lipases, resulting in the formation of free fatty acids, which irritate the eyes and sometimes cause punctate keratopathy (Fig. 17.6).

Therapeutically, we can open up the ducts of the glands by cleaning the lid margin and by applying warm, wet compresses to the eyelid. The latter also increases the fluidity of the tear film, which can be further improved by the intake of omega-3 fatty acids. Tetracyclines also have a beneficial effect both for blepharitis as well for

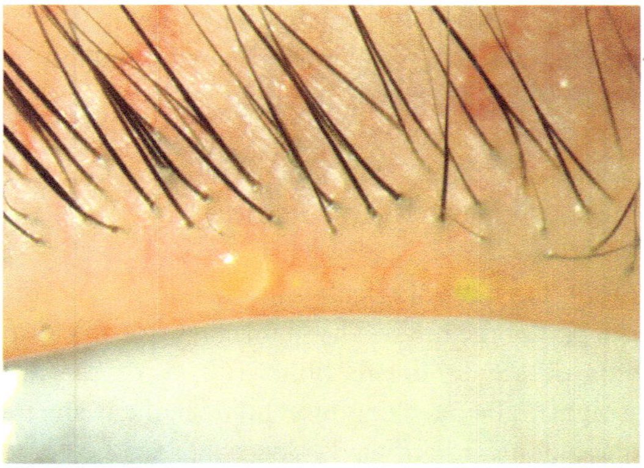

Fig. 17.4 Lid margin in a patient with meibomitis

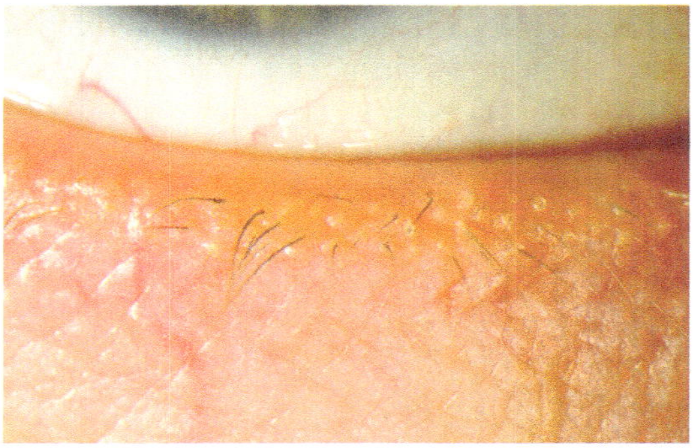

Fig. 17.5 Lid margin in a patient with blepharitis

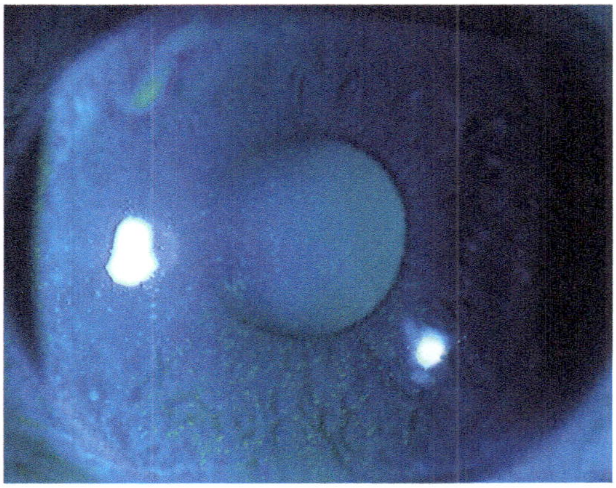

Fig. 17.6 Punctate keratopathy. Epithelial erosions in a pinpoint pattern revealed by staining with fluorescein

the tear film. They act by inhibiting bacterial growth, reducing inflammation, and inhibiting bacterial enzymes.

17.2 Lipids in the Retina

From an embryological point of view, the retina is a part of the brain. The main chemical constituents of the brain are lipids (60 % of the dry weight). The most common lipids in the brain are arachidonic acid and docosahexaenoic acid (Fig. 17.7).

The highest concentration of DHA in the body is in the retina, particularly in the retinal pigment epithelial cells and in the retinal photoreceptor outer segment disc membranes. DHA has a protective role, preserving mitochondrial activity, increasing RPE acid lipase activity and inducing antioxidative, antiproliferative, and antiapoptotic effects. DHA is a long-chain polyunsaturated fatty acid with a 22-carbon chain, a carboxylic acid, and six cis double bonds; the last double bond is located at the third carbon from the omega end (Fig. 17.6). Therefore, it belongs to the omega-3-fatty acid group. Besides DHA, other lipids occur in the retina, including arachidonic acid, phospholipids, and cholesterol. These lipids together account for good membrane fluidity. The lipid composition of the photoreceptor membranes plays an important role in phototransduction since these lipids act as "platforms" for the molecules involved in signal transduction (Fig. 17.8).

These lipids are synthesized, for the most part, in the inner segment of the photoreceptors and are then delivered to the base of the photoreceptor outer segments in vesicles. However, some of the lipids are also transported from the retinal pigment epithelium (RPE) into the photoreceptors (Fig. 17.9).

The outer segment discs consist of double folded membranes (containing lipids) in which the light-sensitive visual pigment molecules (together with other molecules) are embedded. The stacks of discs in the outer segments of the photoreceptors are constantly renewed (their synthesis occurs mainly at night) and new discs are added at the base of the outer segment at the cilium. There is a burst of rod disc shedding in the morning and of cone disc shedding in the evening. These lipids carry a number of other molecules, including the protein opsin. Therefore, this shedding leads to a turnover of a myriad of opsin molecules every day. These outer membrane discs are then phagocytosed by the RPE cells and broken down by lysis (Fig. 17.10). The RPE, therefore, plays a crucial role in the degradation of these lipids.

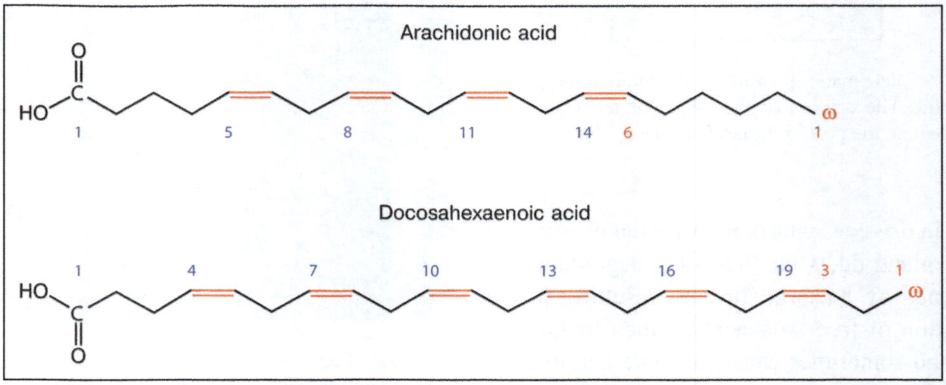

Fig. 17.7 Arachidonic acid (AA) and docosahexaenoic acid (DHA). AA is a carboxylic acid with a 20-carbon chain and four cis-double bonds, whereas DHA is a carboxylic acid with a 22-carbon chain and six cis double bonds. DHA is an omega-3 fatty acid; the first double bond is located at the third carbon counted from the methyl end, which is also known as the omega (ω) end. AA is an omega-6 fatty acid; the first double bond is located at the sixth carbon counted from the omega (ω) end

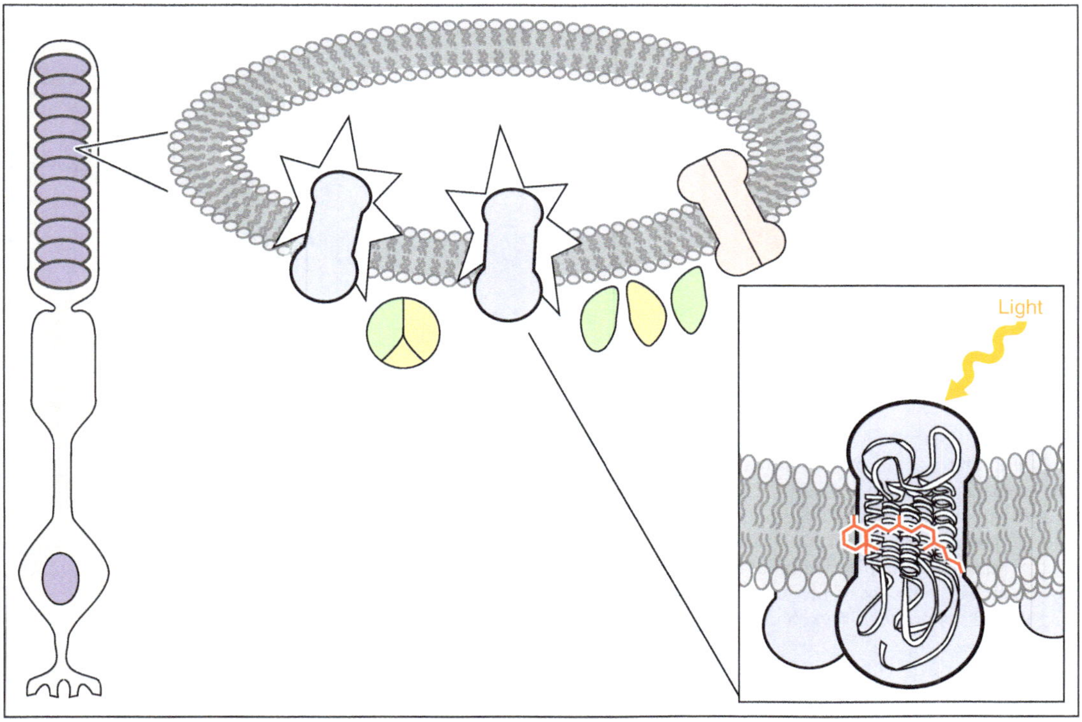

Fig. 17.8 Composition of photoreceptor membranes. *Left*: rod cell. *Middle*: the lipids in the photoreceptor membrane act as "platforms" for molecules involved in signal transduction. *Right*: the opsin molecule containing retinal (*red*) is embedded within the membranes of the rod cell disks

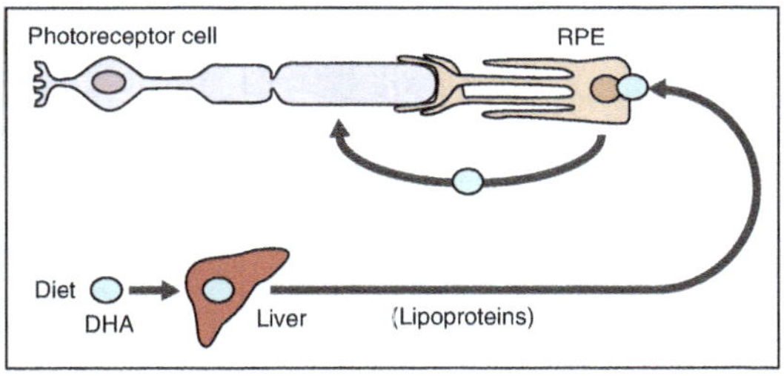

Fig. 17.9 Docosahexanoic acid (DHA) transport. DHA from nutrition is absorbed and delivered to the liver and then transported with lipoproteins into tissues, particularly into the retinal pigment epithelial cells (RPE). From there, it is introduced into the photoreceptor cells. This continuous renewal maintains a high concentration of DHA in the photoreceptor

A problem arises if not all the lipids sequestered from the photoreceptors are degraded. The incompletely digested residues of photoreceptor outer segments can become chemically modified and converted into the so-called "lipofuscin" (Fig. 17.11).

Chemically, lipofuscin is the product of oxidation of lipids and proteins. Lipofuscin also occurs in ageing skin, which led to its name: "age-pigment." Both the numbers of lipofuscin granules and their sizes increase with age. These granules are bounded by a single membrane and have diameters in the range of 1–5 μm. Many different types of lipids are present in the lipofuscin granules, including triglycerides, phospholipids, and cholesterol as well as an equally wide variety of proteins. Lipofuscin also contains a high concentration of metal ions such as zinc, copper, and especially iron. In addition, lipofuscin shows autofluorescence characteristics, as it is composed of different fluorophores. However, as depicted in Fig. 17.12, the fundus autofluorescence is also a physiological characteristic.

An accumulation of waste products also occurs in the extracellular space between the

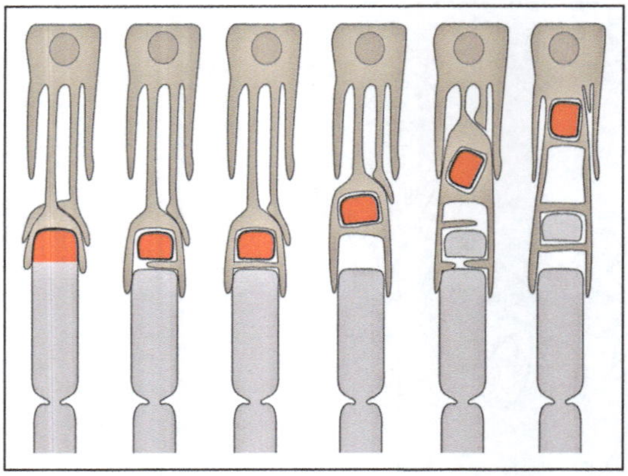

Fig. 17.10 Disc shedding in the retinal pigment epithelial cell. The stacks of discs containing visual pigment molecules in the outer segments of the photoreceptors are constantly shed. New discs are added at the base of the outer segment at the cilium. Old discs (*red*) are pushed out of the outer segment, pinched off at the tips, and engulfed by the apical processes of the pigment epithelium. These engulfed discs are then broken down by lysis within the phagosomes

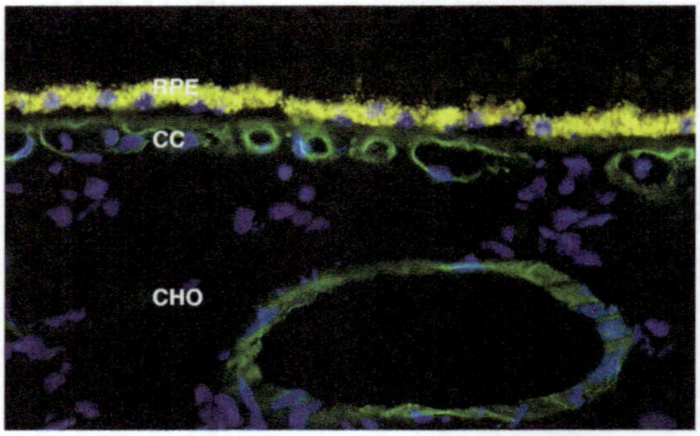

Fig. 17.11 Immunofluorescent image. *RPE* retinal pigment epithelium, *CC* choriocapillaris, *CHO* deeper choroid. Yellow staining of the lipofuscin in the RPE. (From Mullins R, University of Iowa. With permission)

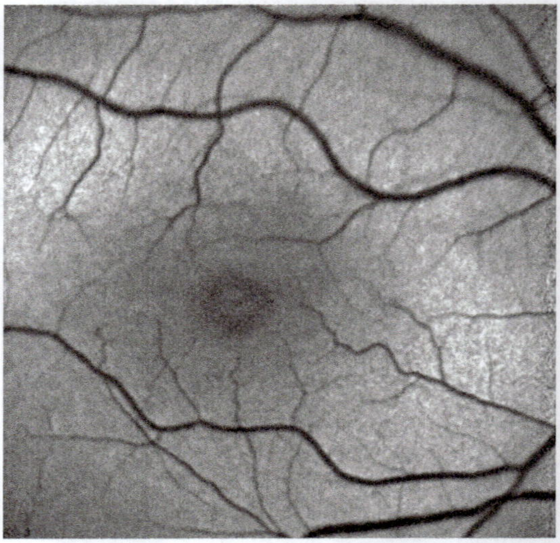

Fig. 17.12 Fundus autofluorescence. Shown here is the slight autofluorescence of the fundus of a healthy person. (Courtesy of S. Wolf, University of Bern)

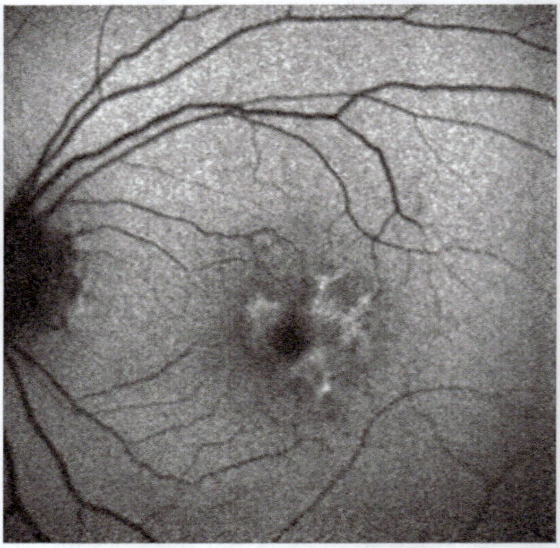

Fig. 17.13 Fundus autofluorescence. Shown here is the marked autofluorescence of drusen in an elderly woman. (Courtesy of S. Wolf, University of Bern)

RPE and the photoreceptors. These fatty accumulations are called drusen (Fig. 17.13) and they also have autofluorescent properties.

The abnormal accumulation of intracellular lipids and extracellular drusen material is associated with the development of degenerative diseases. The exact causal relationship needs to be further clarified. These abnormal accumulations of intra- and extracellular material not only make cellular metabolism more difficult and impede the transport of molecules from the RPE to the photoreceptors but also further stimulate oxidation by absorbing light. This leads to a vicious cycle where, on the one hand, these accumulations are products of oxidation themselves and, on the other hand, they themselves further stimulate the oxidation of other molecules. Oxidative stress plays an important role in age-related macular degeneration (AMD).

This oxidative stress is brought about mainly by blue light. It is also worth mentioning that the lipophilic pigment carotenoids, lutein, and zeaxanthin that are found in the circulating blood, skin, and brain have a particular dense distribution in the macula, leading to the name "macula lutea" (Fig. 17.14).

These carotenoids serve as scavengers of free radicals, absorb blue light, and improve visual acuity. The concentration of these carotenoids can be increased in the macula by an increased intake either through nutrition or by supplementation. Whether this has a protective effect against AMD is still under investigation.

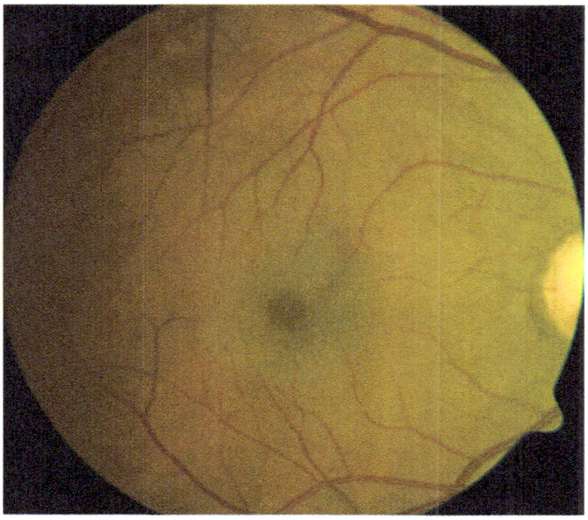

Fig. 17.14 Macula lutea. Fundus photograph showing the macula of the retina. The macula lutea (derived from the Latin macula, "spot," and lutea, "yellow") is an oval-shaped, pigmented yellow spot around the fovea of the retina

18 Matter: Using Water as an Example

In many chapters, we have had "light" as our central theme – as a phenomenon of nature, as our connection to the outer world, as an instrument for ophthalmological examinations, and as a tool for therapeutic interventions. In this chapter, we would like to discuss some concepts regarding the physics of matter that can be encountered frequently in practice. Since we do not wish to present the broad systematic physics of matter, we limit considerations to the fluid phase and explain some concepts, taking water as our example.[1] However, water itself is also a topic. The more one deals with its properties, the more one gets the impression that this material was created to enable life and its development billions of years later.

Just like light, water is taken for granted – but where does the earth's water as a compound of hydrogen and oxygen come from? As we know, according to the Big Bang model, hydrogen nuclei (protons), helium nuclei, and electrons were created in the very first minutes. When, after a million years, the cooling had progressed far enough, the nuclei and the electrons combined to form neutral hydrogen and helium atoms. Heavier elements, among them oxygen, arose only in fusion processes in massive stars and were hurled out into outer space by supernova explosions millions of years later. The materials of the solar system, including the elements of life on earth, originate from earlier star generations. Water arose in outer space – in the Orion cloud, visible to the naked eye, unimaginably large amounts of water have been discovered and its creation continues.

Ice, water, and steam consist of identical water molecules. The electrical forces between the charge distributions of adjacent molecules determine all the phenomena that we observe daily, such as crystalline bonding to form solid ice, the surface tension of water, the specific heat capacity of water,[2] the energy necessary to melt ice or to evaporate water, melting and evaporation temperatures, the viscosity of water (internal friction), the expansion of water as a function of temperature, the solubility of other substances, and numerous additional subtle phenomena.

18.1 The Isolated Water Molecule

The structure of the water molecule is very well known. The distance between the two hydrogen nuclei (protons) from the oxygen nucleus amounts to 0.074 nm and the angle between the bonds of the oxygen nucleus to the protons is 104.5°. In principle, the properties of the ground state as well as of the excited states are predicted

[1]See Chap. 10 for the chemical aspects.

[2]Energy input per mass and per increase in temperature (4180 J/kg·K).

J. Flammer et al., *Basic Sciences in Ophthalmology*,
DOI 10.1007/978-3-642-32261-7_18, © Springer-Verlag Berlin Heidelberg 2013

theoretically with absolute precision by quantum theory.[3] The bonding energy manifests itself in oxyhydrogen (a mixture of O_2 and H_2) explosions; it must be introduced again to split the hydrogen from the oxygen.

One of the most important characteristics of the water molecule is its strong polarity due to the fact that the negative charges (electrons) are shifted with respect to the positive charges (nuclei). The side with the two H atoms has a net positive charge and the side beyond the O atom has a negative one. A very simplified but easily remembered model is sketched in Fig. 18.1. It shows the main features of the polar charge distribution.

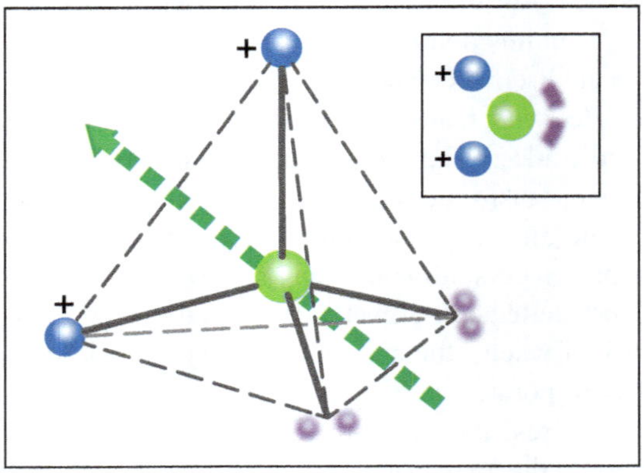

Fig. 18.1 A model of the polarity of the water molecule. The broken lines represent a tetrahedron (pyramid with six edges with the same length). The oxygen atom is located at its center of gravity and the two protons (nuclei of the H atoms – positive) in two of the corners. The two other corners of the tetrahedron are negatively charged and – in water or ice – they form bonds with the protons of other water molecules. The *arrow* indicates the electrical dipole moment. The four electrons of the covalent bonds of hydrogen to the oxygen as well as the two innermost oxygen electrons are not shown here. *Insert*: Simplified scheme representing the non-uniform charge distribution. *Blue* protons, *purple* electrons, *green* oxygen

18.2 The H-Bond in Ice and Water

The interaction between molecules in ice and water is dominated by H-bonds. The prototype can be studied in the water dimer (Fig. 18.2). The H-bond results mainly from the electrostatic attraction between a positive charge of the one molecule (proton) and a non-bonding electron pair of the other molecule. The strength of an H-bond amounts to approximately 0.2 eV, which is about 8 times the energy of thermal agitation.[4] It is the energy needed to separate a pair of water molecules (mainly in evaporation).

The structure of ice reflects the tetrahedral symmetry: each molecule has four nearest neighbors that form a tetrahedron (Fig. 18.3). In liquid water, each molecule is connected to its neighbors by, typically, three to four H-bonds. In a "snapshot," only a minority of the molecules would be singles or dimers. Most are bound into smaller and larger networks with structures similar to that of ice. These clusters break apart permanently and then reform in other combinations. The motor of this dynamism is the thermal translation and rotation of the molecules. Thermal agitation sets the lifetimes of these

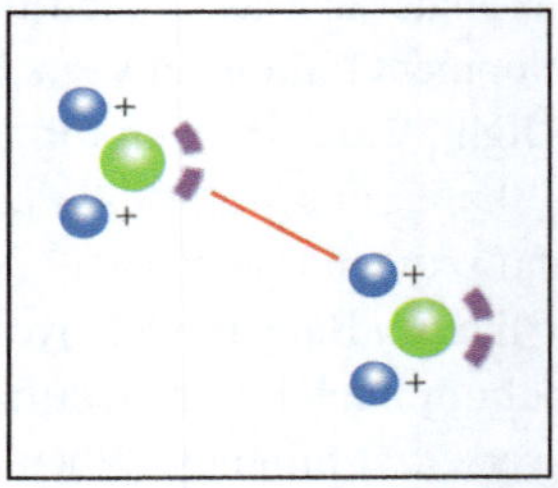

Fig. 18.2 The H-bond in the water dimer (symbolized by the *red line*) results from the electrostatic attraction between a proton of one molecule and electrons of the other. The distance between the oxygen nuclei amounts to about 0.3 nm

[3] Quantum theory – in the formulation that is still valid today – was created in the 1920s. Energy levels and electron states are determined by the Schrödinger equation and Pauli's exclusion principle. The main ingredients are the electromagnetic forces between charged particles (nuclei and electrons).

[4] The order of magnitude of the energy of thermal agitation is $kT = 0.026$ eV at the ambient temperature ($T = 293$ K). The Boltzmann constant k is given in the Appendix. The energy of thermal agitation includes the kinetic energies of molecular translation, rotation, and vibration.

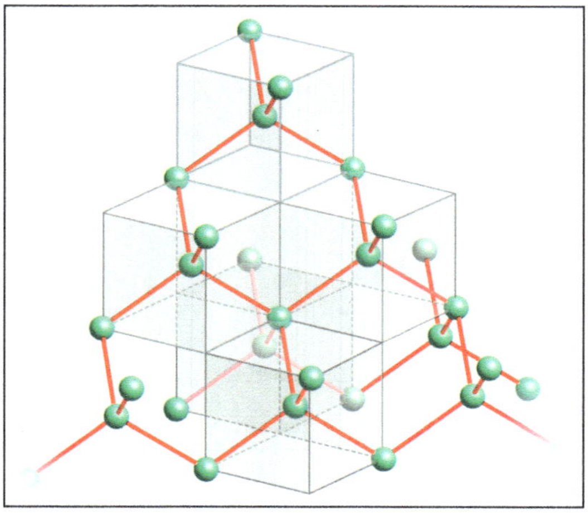

Fig. 18.3 Structure of ice. The *green spheres* represent the positions of the water molecules and the *red links* symbolize the H bonds. Each oxygen atom has four nearest neighbors; these form the corners of a tetrahedron. One obtains the occupied positions in a cubic raster: The center of every other cube is occupied. Four of the eight corners of each cube are occupied

bonds to an order of magnitude of 10^{-12} s. A measure of the mobility is so-called self-diffusion[5]: in water that is absolutely free of any currents, the random walk of a given molecule leads to a mean displacement of about 0.1 mm in a second, or 1 mm in a minute.

18.3 Heat and Temperature

An alteration of the body temperature by 2 K (3.6 °F, 2 °C) has a strong and direct or indirect influence on life processes.[6] For lifeless matter, this type of a temperature difference often means very little; for example, the pressure of a gas in a given volume changes by less than one percent. On the other hand, only a tiny temperature change across the freezing temperature suffices to change ice into liquid water and vice versa. What, then, is temperature – this variable that we seem to be so dependent on? What do the molecules of a stone in water have in common with water molecules when the stone takes on the water's temperature? What do steam molecules and molecules of the boiling water have in common when both have the same temperature? Temperature is undoubtedly one of the more difficult concepts in physics.

It is only with noble gases that the question can easily be answered: the mean kinetic energy E_{kin} of the atoms can be understood as the measure of the temperature. Not only are these two variables proportional to each other but the mean energy per atom for a given temperature is the same for all noble gases. This is expressed in the equation $E_{kin} = 3kT/2$, where k is Boltzmann's constant and T is the absolute temperature as expressed in Kelvin.[7] The factor 3 corresponds to the three degrees of freedom and the three spatial dimensions. With molecular gases, comparable shares of rotation and, with higher temperatures, molecular vibrations are added in. Steam also obeys this rule: the mean thermal energy per molecule amounts to roughly $3kT$. It is stored in the movements of the centers of gravity of the water molecules and their rotations, while additional degrees of freedom (intramolecular vibrations) are still little stimulated at the boiling temperature of water.

However, the simple proportionality between temperature and energy is true only for ideal gases. The relationship between heat energy in water and its temperature is shown in Fig. 18.4. It is not at all linear. The latent heat in the ice ↔ water transition stabilizes the climate and weather at temperatures around freezing. The latent heat of evaporation has its origin in the break-up of bonds between neighboring water molecules. This is the main mechanism for the cooling of our bodies through the evaporation of sweat. Water, for example, shows that we cannot simply identify temperature with energy per molecule, although it is true that the thermal energy of every system increases with rising temperatures. No doubt the reader will have noticed that we have

[5] The self-diffusion coefficient amounts to $2.3 \cdot 10^{-9}$ m²/s.

[6] The various temperature scales are summarized in the Appendix.

[7] At room temperature ($T = 293$ K), $3kT/2 = 0.04$ eV.

left open the question as to what temperature is in terms of molecular physics.[8]

18.4 Solubility of Gases: Partial Pressure

Water can contain dissolved gases. If we open a bottle of mineral water, CO_2 is released. Blood serum contains dissolved N_2, O_2, and CO_2. However, in comparison with the oxygen that is transported by the erythrocytes, the quantity of dissolved O_2 in the blood is quite small. The basic situation is shown in Fig. 18.5: water in contact with a gas mixture (e.g., air). Here, the gases are present with their associated partial pressures, such as nitrogen in air with a partial pressure of 0.8 bar and oxygen with a partial pressure of 0.2 bar. If none of these is dissolved in water at the beginning, they gradually diffuse into it. With the growing concentration of the gases in water, a flow back into air builds up until equilibrium is attained between the flows of molecules in both directions through the water's surface.

The dissolved quantity of a given gas is proportional to its partial pressure in the gas mixture that the water is in contact with. If the partial pressure is doubled, twice as much per unit time flows in through the water surface and the equilibrium is reached only when, due to the doubled concentration of the dissolved gases in the water, twice as much also flows out through the surface. Some concentrations in equilibrium are presented in Table 18.1, corresponding to a partial pressure of 1 bar.

Compared to O_2, the great solubility of CO_2 in water is conspicuous. In a one-liter bottle of mineral water, 6–9 g are dissolved, corresponding to a partial pressure of about 5 bar. At any given partial pressure, approximately the same amount of CO_2 is dissolved in water as in the same volume of a gas mixture. In contrast, roughly 60 times less N_2 is dissolved in water as in the same volume of a gas mixture (with any amount of N_2).

The term "partial pressure" is not only applied to the gases that are in equilibrium with a fluid, but also on those that are dissolved in the fluid. We explain this extension of the concept using an example. A glass of water stands on a table in

[8]Temperature can be understood within the framework of statistical thermodynamics. *The same temperature* means that the given total energy of two bodies is distributed in the most probable way among and within them.

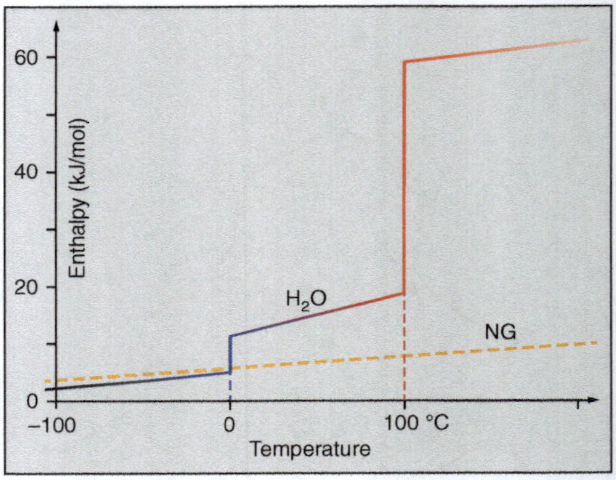

Fig. 18.4 Enthalpy (energy content at constant pressure) of water as well as that of a noble gas (NG) as a function of the temperature for 1 Mol at pressure of 1 bar. The two phase transitions (ice ↔ water, water ↔ steam) stabilize the temperature in that these also remain constant when the system takes in or gives off energy. In its liquid phase, water also stabilizes the temperature because the specific heat capacity is relatively large

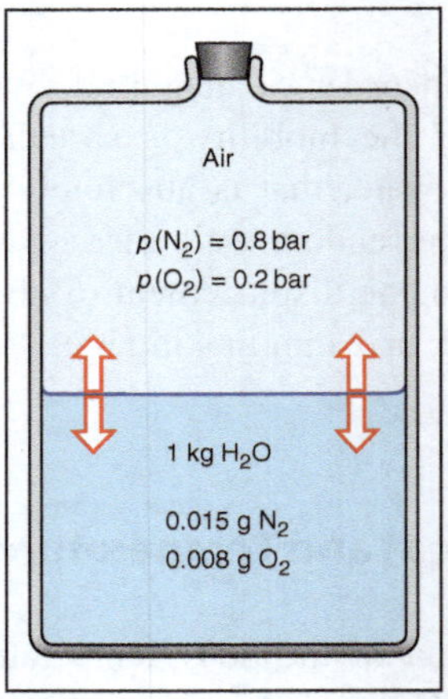

Fig. 18.5 Dissolved gases in water in equilibrium with a gas mixture. The dissolved quantities per kg of water can be derived from Table 18.1 by multiplying the solubility by the associated partial pressure, for example for N_2 at 20 °C: 0.8·0.019 = 0.015 g/kg

contact with air for a long time. The partial pressure of N_2 in the air amounts to 0.8 bar. An equilibrium forms between the N_2 in water and the N_2 in air. In this situation, the partial pressure of N_2 *in water* amounts to 0.8 bar. Using this terminology, it is simply stated that the N_2 in water is at equilibrium with the gaseous N_2 of 0.8 bar. It is helpful to imagine the partial pressure of a gas in water more in the sense of this equilibrium and less as mechanically effective pressure (as a force per unit area).

The aforementioned principles play a role when intraocular gases are used in retinal detachment surgery. In the beginning, a bubble is injected consisting of a mixture of perfluoropropane (C_3F_8) and air (Fig. 18.6). It is subject to atmospheric pressure plus intraocular pressure (e.g., 760 + 15 mmHg). Immediately following the injection, the partial pressure of N_2 inside the bubble is less than in the surrounding fluid, where it is practically the same as the partial pressure of N_2 in blood and in air (0.8 bar). In an attempt to equalize the partial pressure, N_2 flows from the surrounding fluid into the bubble, while C_3F_8 diffuses much more slowly into the surrounding fluid. Since the bubble is under constant pressure, its volume increases. Later, equilibrium is established and, finally, a resorption of all the gases occurs (Fig. 18.6). If the injected gas initially consisted of pure C_3F_8, its volume would increase by roughly a factor of 4. To prevent this, a mixture that includes air is used. We do not wish to enter into a discussion of the various gases in detail at this point. It suffices to say that they differ with respect to their expansion factors and in the rapidity of their resorption. Intraocular air is resorbed within a few days and perfluoropropane within a few weeks.

How does an intraocular gas behave in response to a relatively rapid decrease in the external pressure, e.g., in an airplane after it takes off? We must now distinguish between the absolute external pressure p_{atm}, the IOP (as it is understood in ophthalmology as the difference between absolute interior and exterior pressures), and the absolute interior pressure $p_A = p_{atm} + IOP$ in the eye. To give an example, these values are $p_{atm} = 760$ and $p_A = 775$ mmHg (760 + 15) at take-off, respectively. When the airplane cabin pressure decreases to 700 mmHg, the IOP inside an eye without gas remains the same due to normal regulation, implying that the absolute interior pressure p_A decreases to 715 mmHg. In an eye with intraocular gas, however, the pressure changes are different. When the absolute interior pressure is reduced, the gas bubble has a tendency to expand. This becomes possible only to the necessary extent with the

Table 18.1 Some solubilities (in g of gases per 1 kg water) at a partial pressure of 1 bar. The dissolved quantities decrease with rising temperature. It is conspicuous that far more CO_2 is dissolved than O_2 or N_2. The solubility of N_2O (laughing gas) is close to that of CO_2

Temperature		Solubility (g/kg H_2O/1 bar)		
°C	°F	O_2	N_2	CO_2
20	68	0.042	0.019	1.7
36	96.8	0.032	0.015	1.0

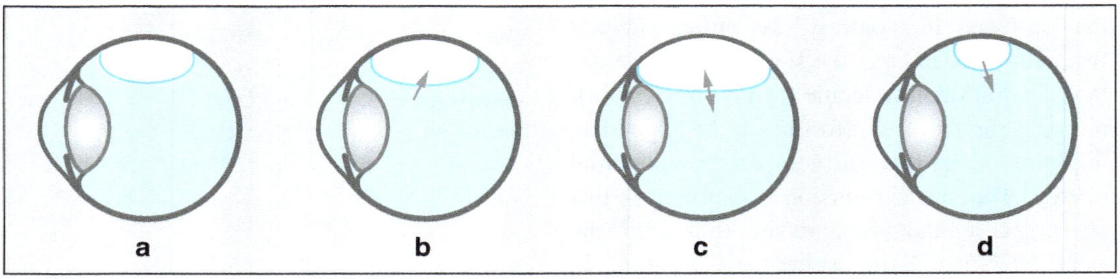

Fig. 18.6 Intraocular gas bubble. (**a**) Immediately after the injection. (**b**) Expansion due to the inflow of N_2, O_2, and CO_2, as these gases in the bubble have lower partial pressure than the surroundings. (**c**) Equilibrium. (**d**) Resorption. The form of the bubble is determined by the surface tension of the intraocular fluid and the buoyancy (see the comments in Sect. 18.6). The arrows indicate the diffusion and resorption of the gases

outflow of the aqueous humor. Therefore, the pressure of the intraocular gas initially causes the absolute interior pressure p_A to decrease less rapidly than the external pressure, which means that IOP increases beyond the initial 15 mmHg. During landing, the reverse process occurs. The possible problems with flying are fewer if the bubble volume is smaller. Aside from this purely physical phenomenon, potential pathophysiological mechanisms may simultaneously occur. One potential effect could be a forward dislocation of the iris diaphragm by the gas bubble, which may lead to an occlusion of the anterior chamber angle and – in extreme cases – cause an acute glaucoma attack.

18.5 Surface Tension

At the interface between water and the surrounding air, the water's surface tension becomes evident in numerous phenomena that can be observed with the naked eye. Water strider insects can flit back and forth on water (Fig. 18.7). The spherical form of drops of water also derives from surface tension. A dangling water drop, for instance, looks as if its surface is encompassed by an invisible, thin skin. We shall also discuss how surface tensions (and related interfacial tensions) influence the form of intraocular gas bubbles.

The molecular origin of the surface tension lies in the attraction between neighboring molecules. A molecule in the interior of a liquid is pulled in all directions by its neighbors, so these forces cancel one another out. A molecule on the surface, in contrast, is pulled inward (Fig. 18.8). To enlarge the surface area by moving an additional molecule from within the liquid onto the surface, effort has to be expended, implying that the surface carries potential energy. The surface tension is defined as this energy per area and its units are, thus, J/m^2 (the same as N/m). Some values are provided in Table 18.2. Aside from mercury, water has the largest surface tension among all liquids – a result of the strong forces between the polar water molecules. For this reason, the effects of the surface tension are more marked with water than with oil, for example. To achieve the smallest possible surface energy, the surface of a fluid assumes as small an area as possible. This gives

Fig. 18.7 Water strider on a water surface

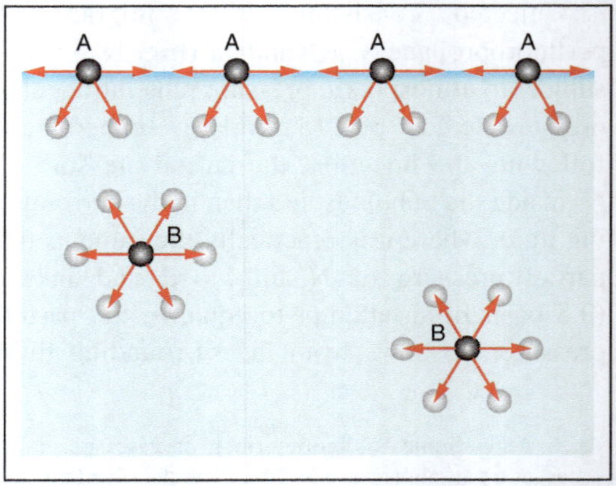

Fig. 18.8 The molecular cause of the surface tension is the attraction between neighboring molecules. Molecules at the surface of a material are pulled inward. This is true for both fluids and solids that border a vacuum or a gas. To create more surface, energy has to be expended

Table 18.2 Surface tension σ of some liquids and of glass at 20 °C

	σ (J/m^2)
Water	0.073
Soap water	0.025
Mercury (pure)	0.480
Silicone oil	0.021
Olive oil	0.033
Glass	0.5–1

rise to the spherical shape of a small water droplet or soap bubble because a sphere has the smallest possible surface for a given volume. However, this may be modified by other forces, such as air friction on falling rain drops, the weight of dangling drops, or the buoyancy of an intraocular gas bubble.

Surface tension is one of the determining factors in the size of drops falling from small openings. For small drops that are released from a vertical cannula with very thin walls, the rule of thumb is that the drop volume is proportional to the cannula diameter. Most drugs are delivered to the eye in the form of drops that then admix with the precorneal tear film. As the capacity of this precorneal area is limited, a constant and relatively small drop of about 20 μL is desirable, corresponding to a diameter of 3.5 mm. Most drops, however, are rather larger and reproducibility is limited. Factors that influence the size of the drops include surface tension, the design and material of the dropper tip, and the dispersing angle. The surface tension, in turn, is influenced by the dissolved substances and this includes not only the drug itself but also other molecules such as the antimicrobial preservatives (Fig. 18.9).

Closely related with surface tension is the more comprehensive concept of interfacial tension.[9] Here, the energy involved is ascribed to the boundary layer between two different materials (e.g., water and silicone oil). It is derived from the forces that adjacent molecular layers of the two materials exert on one another as well as the forces with which they are pulled inward, as found in surface tension. The interfacial tension is also measured in energy per area. The interfacial tensions of some material pairs are found in Table 18.3.

Intraocular gas bubbles of varying sizes were already depicted in Fig. 18.6. The strong buoyancy pushes the bubble upwards. In addition, the three interfacial tensions (gas-retina, gas-water, and water-retina) operate on the relevant contact areas. These forces determine the bubble's form, especially the boundary angle, which is the same for all sizes of the bubble and amounts to approximately 40°. Small bubbles are almost spherical because their buoyancy is negligible compared with the surface tension. With increasing size, the bubble departs more and more from the spherical shape, and the boundary area to the intraocular fluid becomes practically flat. A volume of 1 cm^3 yields a contact angle between the bubble and retina of roughly 90° (roughly 70° with 0.5 cm^3). The pressure that the bubble exerts on the retina is slight, on the order of 1 cm water column (≈1 mmHg). The effect consists in closing a break in order to separate the water in front of and behind the retina.

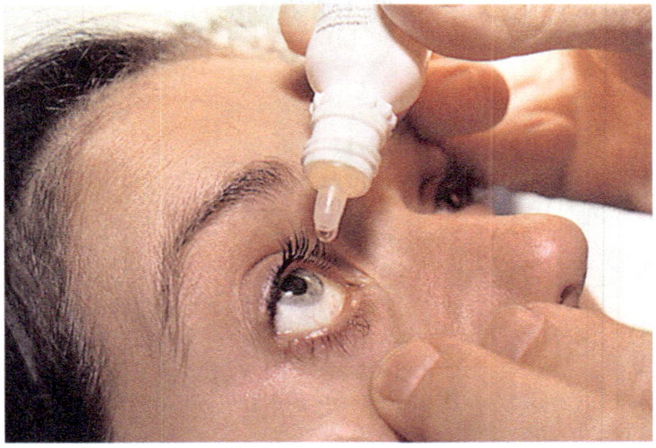

Fig. 18.9 The size of a drop depends on its surface tension and on the diameter of the dropper tip

Table 18.3 Interfacial tensions σ with water

	σ (J/m^2)
Water to silicone oil (25 °C)	0.035
Water to glass	0.208
Water to olive oil	0.018
Water to mercury	0.375

18.6 Silicone Oil–Water Interface

The silicone oils that are most frequently utilized for eyes have densities just slightly below that of water (0.97 vs. 1.0). The buoyancy forces are

[9] In the literature, "surface tension" is often used instead of "interfacial tension." Strictly speaking, the term "surface tension" is reserved for the interfacial tension between a material and air.

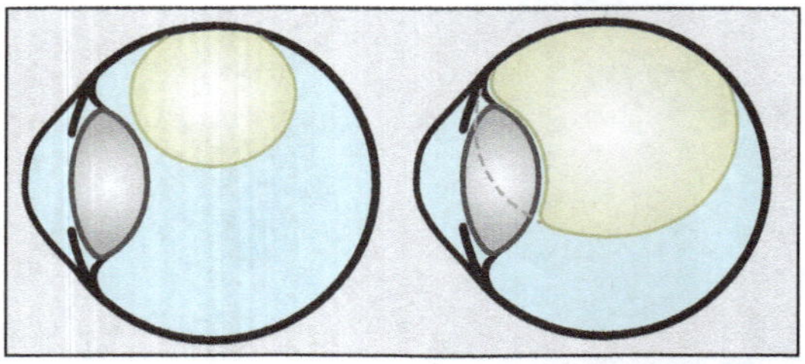

Fig 18.10 Silicone oil with a density less than that of water. The surface tension tries to form a sphere and this is only slightly distorted by the buoyancy forces. The angle to the retina amounts to about 20°

thus much less than with an intraocular gas.[10] Although the interfacial tension between silicone oil and water is much smaller than water's surface tension, the ratio of surface forces to the buoyancy is much larger than with gas bubbles so that a silicone oil bubble has approximately a spherical shape (Fig. 18.10). However, the boundary angle is flatter than with intraocular gas: only roughly 20°. This means that the same volume of silicone oil covers a smaller area of the retina than does a gas.

Silicone oil with a specific weight slightly higher than water sinks downward, driven by a lower force and shows similar shapes to rising silicone oil.[10] A heavy oil, the specific weight of which is twice as large as that of water, sinks downward with the same net forces as an air bubble rises upward.

One problem with silicone oil is its tendency to emulsify with time. In individual cases, this process sometimes takes place rather rapidly. With this very undesirable side effect, oil drops can get into the anterior chamber and into the trabecular meshwork. The emulsification consists of the fragmentation of the oil volume into smaller drops. Because the total surface area increases, energy proportional to the interfacial tension between silicone oil and water must be expended. This process will not happen spontaneously at rest. It occurs more easily when the interfacial tension is lower. The fragmentation can be triggered by mechanical forces, e.g., by saccadic movements and by a flow along rough boundary areas. Blood components such as proteins, lipids, and phospholipids have been proposed as substances that might reduce the interfacial tension between silicone oil and intraocular fluid, facilitating emulsification.

[10]Archimedes' principle states that the resulting force is equal to the difference between the body's weight and the weight of the displaced liquid.

18.7 Viscosity

Water is also subject to internal friction. This is evident, for example, in the small vibrations seen on the surface of a glass of tea, which can be observed very sensitively by their reflection and come to rest after a short time. The strength of the internal friction of a fluid is expressed in a material constant, the viscosity (Table 18.4). Thick fluids such as oil or honey have high viscosity; in comparison, that of water is low. Viscosity decreases with rising temperature. In certain fluids (e.g. water), the internal friction can be described with a single constant (the viscosity). These types of fluids are called Newtonian fluids (blood is not a Newtonian fluid; see below).

An important consequence of the internal friction is the pressure difference between the ends of a pipe to pump water through. The pressure required to pump through olive oil at a given speed is 100 times larger than that required to pump through water at the same speed. Thus, the viscosity of olive oil is 100 times greater than that of water (Fig. 18.11).

The friction inside a pipe derives from the fact that the velocity of flow is not the same at every point of the pipe cross-section. The velocity is largest along the axis and vanishes at the wall of the pipe. Neighboring lines of flow, thus, do not have the same speed. The slower ones decelerate the faster ones. In general, instead of a pipe, any obstruction of the flow can be

Table 18.4 Viscosity of some fluids[a] at 20 °C

	Dynamic viscosity	
	kg m^{-1} s^{-1}	mPa·s
Water (20 °C)	0.0010	1
Water (37 °C)	0.0007	0.7
Blood plasma	≈0.0015	≈1.5
Blood (37 °C), vessel diameter 10 μm	≈0.002	≈2.
Blood (37 °C), vessel diameter >0.2 mm	≈0.006	≈6
Silicone oil	1–5	1,000–5,000

[a]Blood has more complex flow properties than water and cannot be described by a single viscosity alone. The viscosity depends on the capillary diameter (Fåhraeus–Lindqvist effect; see text). The relation between various units of viscosity is explained in the Appendix.

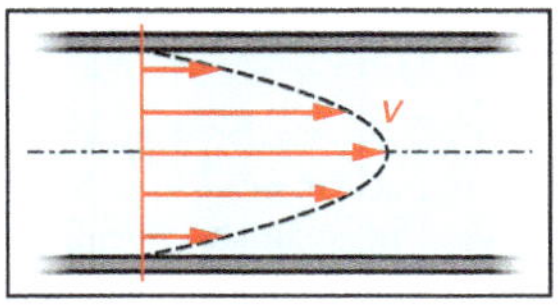

Fig. 18.11 Parabolic velocity profile in a pipe (Newtonian fluid). The velocity is largest on the axis and vanishes at the wall of the pipe. Friction is due to unequal velocities of adjacent lines of flow

involved. The trabecular meshwork in the chamber angle represents such an obstacle for the outflow of the aqueous humor. The flow F through such an obstacle (pipe, trabecular meshwork, blood vessel, etc.), described as the throughput volume per unit time, is proportional to the pressure difference p. In medicine, this pressure difference is called perfusion pressure. The proportionality can be expressed by the equation $F=p/R$. Here, the constant R characterizes the resistance of the associated obstacle. For the same pressure difference, larger values of the resistance R yield a smaller flow.[11] Various applications of this simple law are illustrated in Fig. 18.12. For a given geometry of the obstacle, the resistance R is proportional to the viscosity of the liquid.

As an example, let us consider the flow of aqueous humor. The production amounts to $F=4{\cdot}10^{-11}$ m^{3}s^{-1}. From the pressure difference $p\approx 8$ mmHg$\approx$0.01 bar$=10^3$ Pa, a resistance of $R=p/F=0.25{\cdot}10^{14}$ kg/(s·m^4) results. A doubling of the resistance requires twice the pressure difference (an additional 8 mmHg) to achieve the same outflow. Interestingly, within a large range of IOP, the production of aqueous humor remains constant. The situation corresponds to Fig. 18.12c. Under this condition, the IOP is proportional to the outflow resistance.

The flow characteristics of blood are complicated and cannot be described with a single viscosity constant. Blood is, thus, not a Newtonian fluid. For given conditions (vessel diameter and flow velocity) an effective viscosity can be defined.[12] In larger vessels (>0.2 mm diameter) and a higher velocity, blood behaves like a fluid with an effective viscosity that is roughly six times larger than that of water. With decreasing vessel diameter, the effective viscosity decreases. In the case of a vessel

[11]For a pipe with a circular cross-section (radius r), length L, and fluid dynamic viscosity η, the Hagen-Poiseuille formula $R=8\eta L/(\pi r^4)$ holds. The fourth power means that the resistance R increases strongly with decreasing pipe diameter.

[12]In any given situation, an effective viscosity η is derived by means of the Hagen-Poiseuille formula $F=(\pi/8\eta)\cdot(p/L)\cdot r^4$ from the pressure difference p over the length L, the half-diameter r of the vessel, and the observed flow F (volume per unit time).

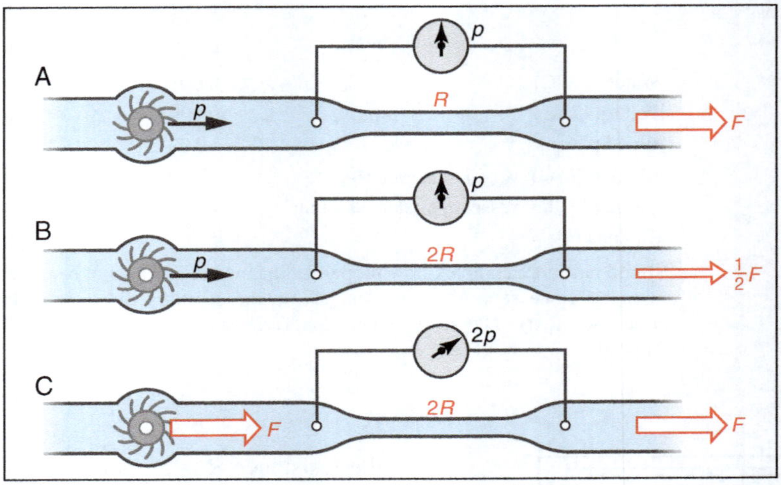

Fig. 18.12 (**A**) Reference situation. For a given pressure difference (or given pressure p of the pump), the flow F through the resistance R is determined by $F = p/R$. The resistance depends on the geometry of the obstacle and on the viscosity of the fluid. (**B**) For the same pump pressure as in (**A**) but with more resistance, the flow is decreased. A reduction in the pipe diameter of 15 % doubles the resistance ($R \rightarrow 2R$). (**C**) When the pump enforces the same flow as in (**A**), it must deliver greater pressure; i.e., greater pressure is created across the obstacle. The manometer indicates the pressure difference along the obstacle

diameter of 10 μm, the lowest viscosity value is reached: only slightly above the viscosity of plasma (approx. 50 % above that of water). This is the Fåhraeus–Lindqvist effect,[13] which was discovered in 1931 and contributed much to the understanding of hemodynamics. It is based on the fact that the erythrocytes change their arrangement into "rolls" aligned with the axis of flow so that the viscosity is essentially determined by the plasma layer between this roll and the vessel wall. This also explains why the diameter of retinal vessels is smaller in a photo (in reality not showing the vessel but, rather, the column of erythrocytes) than in fluorescence angiography, where the dye is diluted in the plasma.

The ocular blood flow (e.g., the retinal blood flow) depends on the perfusion pressure, local resistance, and blood viscosity.

Concerning viscosity, increased blood viscosity is one of many known risk factors for retinal vein occlusion. Isovolemic hemodilution improves blood flow in such cases. The whole blood viscosity is often estimated based on the hematocrit. While this is relevant for larger vessels, the resistance in the capillaries, however, is influenced more by the plasma viscosity.

Concerning local resistance, the resistance in a given vascular bed (like the retina) is often incorrectly considered to be a constant for a given person. It has to be kept in mind that ocular vessels are not passive tubes. Both arteries and veins are highly regulated and the resistance is an intra-individual variable depending on factors such as blood pressure, blood sugar level, emotional status, environmental temperature, and other factors. Besides the physiological regulation of these vessels, ocular blood vessels can be involved in primary vascular dysregulation (PVD) syndrome, which in the past has often been termed vasospastic syndrome.

Perfusion pressure (PP) is the difference between the pressure in the central retinal artery and the central retinal vein. The pressure in the central retinal vein is often assumed to

[13] Robin Fåhraeus, Swedish pathologist and hematologist (1888–1968). Torsten Lindqvist, Swedish physician.

be equal to the intraocular pressure (IOP). While this is often (but not always) correct in healthy subjects, retinal venous pressure is above the IOP in many diseases, particularly in conditions where the level of Endothelin in the circulating blood is increased. In addition, this retinal venous pressure can be influenced by drugs.

19 If You Are Interested in More ...

19.1 Ray Optics or Wave Optics?

Ray optics (or geometrical optics) describes light propagation in terms of rectilinear paths in homogeneous media. The simple concepts of ray optics are very powerful in explaining image formation by mirrors, lenses, and the eye and in a multitude of optical instruments. The design of photographic objectives, which may consist of a dozen lenses to optimize image quality, is done within the framework of ray optics. However, we know that ray optics does not account for all phenomena in the propagation of light. Diffraction and interference cannot be understood without taking into account the wave nature of light.

How are these two models, light as rays and light as waves, related? Physics offers a very simple basic explanation: first, as long as the quantum theory is of secondary importance, all optical phenomena can be completely explained by the electromagnetic wave nature of light (Maxwell's theory). Second, the concept of light rays represents a surprisingly good approximation to the wave theory. As usual, approximations are simpler to handle, but they have their limitations. In this chapter, we illustrate the relationship between the two models in terms of the image formation of a source infinitely far away by a collecting lens (Fig. 19.1).

It terms of ray optics, the position of the focal point F is determined by the law of refraction as the ray leaves the curved back surface of a lens (Fig. 19.1, left). The focal length, thus, follows from the refractive index of the lens and from the curvature of its surfaces. Nevertheless, focusing can also be explained using wave optics: the focal point is exactly the location where the light waves – as they travel down all the ray paths – have experienced exactly the same number of periods. In other words, all contributions arrive at the focal point in phase and produce constructive interference (Fig. 19.1, right): off axis, the light travels a shorter distance across the lens and, thus, for fewer periods than those that travel along the axis within the lens. This difference is exactly made up for by the longer distance from the point of exit from the lens to the focal point. For one point behind the lens – the focal point – this condition is fulfilled for all the rays. In locations that lie closer to or further away from the lens, the contributions from the various lens zones arrive out of phase and thereby almost completely cancel each other out.

While both diagrams are suitable for understanding how focusing occurs, the wave version offers a lot more additional information: if one shifts point F a little bit laterally away from the axis, the waves from the various lens zones still arrive almost in phase – the light intensity can, then, not be limited to the geometric point F. An analysis shows that only with a lateral offset on the order of $\lambda \cdot f/(n \cdot D)$ or higher do the various contributions become substantially out of phase (λ: wavelength, f: focal length, D: diameter of the lens, n: refractive index of the medium between the lens and focus). This is the basis for understanding the Airy disk[1]: a parallel beam of light, passing through a lens, is not imaged as a point

[1] George B. Airy (1801–1892), English astronomer.

J. Flammer et al., *Basic Sciences in Ophthalmology*,
DOI 10.1007/978-3-642-32261-7_19, © Springer-Verlag Berlin Heidelberg 2013

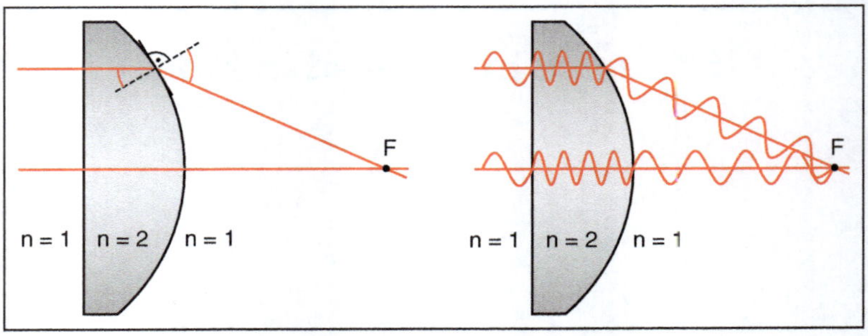

Fig. 19.1 Ray optic and wave optic explanations of the focus for parallel rays passing through an ideal collecting lens. *Left*: According to ray optics, the focusing results from the law of refraction when the rays leave the lens. *Right*: According to wave optics, the focus is that place where all contributions arrive with the same phase and thereby interfere constructively (the same number of periods along all the paths). Lens with a refractive index $n=2$, in vacuum

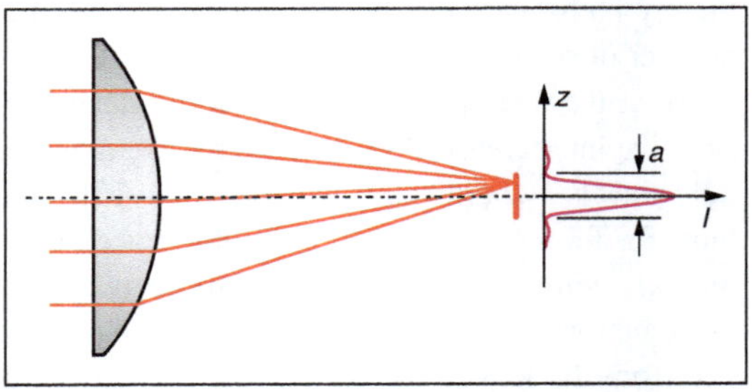

Fig. 19.2 Image of a parallel beam passing through an ideal lens in the focal plane, as explained with wave optics. Essentially, a disk (Airy disk) is illuminated in which constructive interference occurs. Outside the Airy disk, the various contributions cancel each other almost completely out. On the right, the intensity profile is indicated. The diameter between the zero locations is given by the equation $a=1.22\lambda \cdot f/(n \cdot D)$, where f is the focal length, D is the lens diameter, and n is the refractive index of the medium between the lens and the focal point. In typical lenses, the diameter of the Airy disk amounts to a few wavelengths

Table 19.1 Diameter of the airy disk in the eye for various pupil diameters

Pupil diameter D	2 mm	4 mm
Airy disk diameter a	12 µm	6 µm
a, in minutes of arc	~2′	~1′

but, rather, as a small disk. Figure 19.2 shows the intensity profile. This blurring of the image cannot be avoided, even in the case of an ideally shaped lens.

For the eye ($n=1.33$, $f=23$ mm, $\lambda=0.55$ µm), the diameter a of the Airy disk for various pupil sizes D is listed in Table 19.1 – the smaller the pupil is, the larger the Airy disk will be. In addition to the wave optical limitations regarding the sharpness of the image that forms at the retina, imperfections also exist in the imaging media and interfaces. These increase with the pupil diameter. For a pupil size of 2 mm, in eyes with good visual acuity, both effects have roughly the same strength; in smaller pupils, the wave effect outweighs the other. However, for an ametropic eye, the view through an aperture of 1–2 mm in diameter can be beneficial – e.g., when trying to read a menu without glasses.

Intuitively, this phenomenon can also be understood as a consequence of diffraction at the pupillary edge. The smaller the pupil is, the larger the ratio of the pupil circumference to the pupil area will be since the circumference increases linearly while the area increases quadratically as a function of its diameter. Therefore, in very small pupils, the relative portion of the diffracted light increases and the image becomes more blurred.

19.2 Simple Lenses and Achromats

Often, the effects caused by the wave nature of light are of minor importance, so the approximation of ray optics can be applied. Examples include eyeglasses, optical instruments, and camera objective lenses. In principle, one follows the light rays on their pathways through the system,

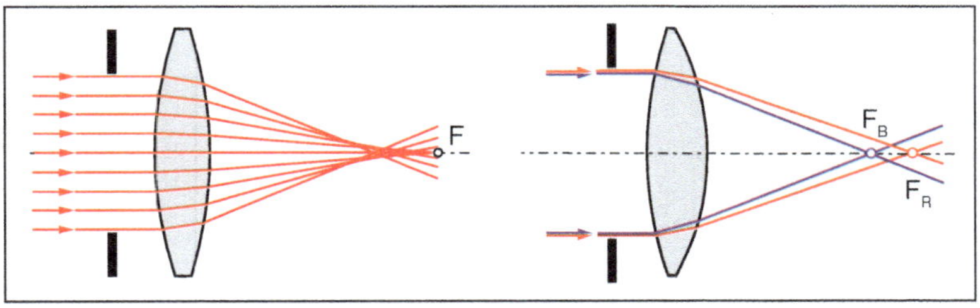

Fig. 19.3 Simple lens. *Left*: spherical aberration. With a larger *f* number (ratio of the focal length to the diameter of the lens), the spherical aberration is less important. For this reason, we see more sharply when there is good illumination (making our pupils smaller). *Right*: chromatic aberration. Due to the dependence of the refractive index on the wavelength, the focal length is smaller for shorter wavelengths

whereby the law of refraction determines the change in direction at each transition into a new medium. Camera objective lenses are optimized in this way even though a compromise between various residual errors must be accepted. Camera objective lenses owe their high degree of perfection to the large number of refractive surfaces, which permit a large amount of freedom in the optimization process. It should not be forgotten that the creation of objective lenses with practically no devastating residual stray light has been possible only with the invention of antireflective coatings.

A simple lens with spherical surfaces exhibits various aberrations. The most obvious one that occurs on the optical axis is spherical aberration (Fig. 19.3). Rays from the lens edges pass through the axis in closer proximity to the lens so that a parallel incident ray results in a spot at every image plane rather than a point. This error stems from the spherical shape of the lens surfaces. Non-spherical surfaces, which are more difficult to produce, are employed only for special purposes. They are also used for eyeglasses for greater degrees of ametropia. For a given focal length, the spherical aberration is less disturbing in lenses with smaller diameters. Conversely, this imaging error is at its largest with a ball-shaped lens. Nevertheless, fishes with ball-shaped lenses have sharp imaging because the refractive index gradually diminishes with increasing distance from the optical axis. The rays at the edges are less strongly deviated than in a lens with a homogeneous refractive index. In this way, sharp imaging is possible for every direction.

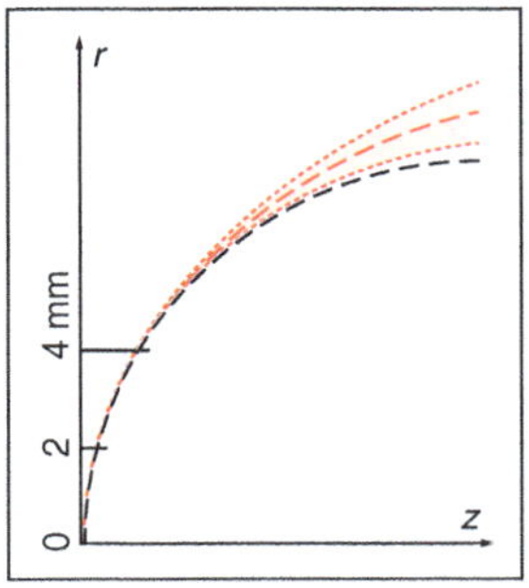

Fig. 19.4 Models of the anterior surface of the cornea. The deviation of the real elliptical profile (*red*) from a circle with the same radius of curvature in the center (*black*). *Shaded area*: Mean ± 1 standard deviation of the inter-individual variation of the (extrapolated) elliptical profile of the cornea. For a distance of 2 mm from the axis, the axial difference between the ellipse and the circle amounts to about 1 μm at a distance of 4 mm from the axis to 23 ± 15 μm (horizontal distances between the *black* and *red lines*). The spherical and elliptical profiles are extrapolated beyond the real pupil size. *z* and *r* tangential and axial coordinates

In human eyes, various factors help to attenuate the spherical aberration. The cornea becomes flatter toward the periphery. Its profile is more elliptical than spherical (Fig. 19.4). In addition, the index of refraction of the crystalline lens decreases toward the outside, but less than in fish eyes. The spherical aberration of the human eye for light that enters 2 mm off axis, which, if uncorrected, would have a value of 2 D, is thereby reduced by half. In addition, peripheral rays are perceived with lower sensitivity than axial ones are because of the angle at which the light reaches the photoreceptors (Stiles-Crawford effect).

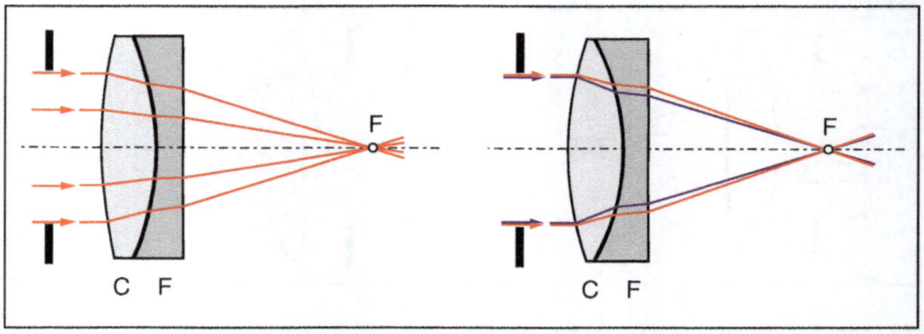

Fig. 19.5 The principle of achromatic objectives. Usually, a positive lens made of glass with low dispersion is combined with a negative lens of glass with high dispersion in such a way that the dispersions (a positive and a negative) are offset, but the refractive power of the collecting lens prevails. The chromatic aberration is almost completely eliminated (*right*). In addition, spherical aberrations are also substantially reduced (*left*)

The second axial error is the longitudinal chromatic aberration. Blue light is more strongly refracted than red (dispersion). For this reason, the focus of blue light lies closer to the lens (Figs. 19.3 and 2.20). In human eyes, over the whole visual range, the difference amounts to roughly 2 D (see Sect. 2.4.2). Normally, we do not perceive this aberration. One reason is the fact that blue-sensitive rods are missing in the innermost part of the fovea. A number of explanations have been put forward to rationalize how it is that we are, nevertheless, able to recognize a tiny blue object as such.

In Newton's opinion, it was not possible to correct the chromatic error of lens telescopes. With the invention of achromats in London in the 18th century, it was actually possible to manufacture a 2-lens system, which eliminated chromatic aberrations (Fig. 19.5). Even more important, in many applications, is the fact that the achromat eliminates longitudinal spherical aberrations. The three radii of curvature of an achromat are determined by three requirements: the given focal length, first-order elimination of longitudinal chromatic aberration, and elimination of the longitudinal spherical aberration. One can experience the quality of an achromat by taking apart an inexpensive pair of binoculars and then using the objective as a high-grade magnifying glass.

Strictly speaking, an achromat compensates for the chromatic aberration for only two wavelengths (usually chosen in red and in green-blue). The residual error for other wavelengths is very small. More elaborate constructions that compensate for even three wavelengths are termed apochromats. Binoculars have achromats as objective lenses. Their residual dispersion is seen only in special circumstances, e.g., as colored edges in the peripheral field.

19.3 Adaptive Optics

With the Hubble telescope, stars were much more sharply imaged than with terrestrial telescopes where the turbulence of the atmosphere – perceived with the eye as the twinkling of the stars – strongly impairs the sharpness of the images. For this reason, astronomical telescopes are stationed at very high locations.[2] In observational astronomy, the application of adaptive optics has had sensational success in that the atmospheric turbulence can be largely eliminated in the imaging of terrestrial telescopes. Basically, the same technique can measure the individual aberrations of an eye and simultaneously correct it up to the highest possible visual acuity with external optical corrections – at least experimentally – limited only by diffraction

[2] The two Keck telescopes, the mirrors of which are each 10 m in diameter, are installed on the peak of the 4,200-m dormant volcano Mauna Kea on the island of Hawaii.

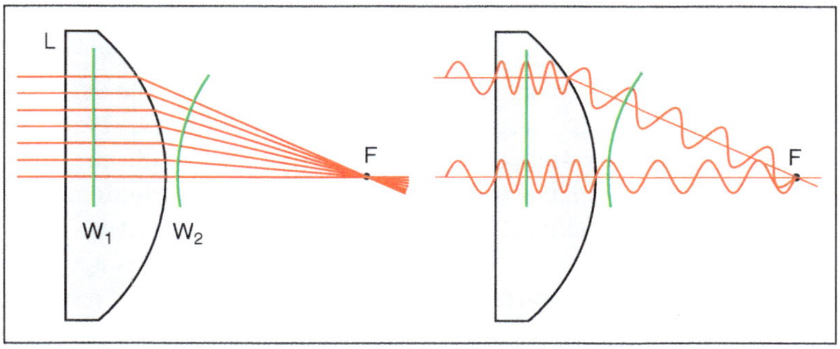

Fig. 19.6 Ideal lens; imaging of a parallel bundle of light rays into a point or, in the inverse direction, light emanating from a point. *Left*: definition of wavefronts W_1 and W_2 as surfaces perpendicular to the rays. For an ideal lens, W_1 and W_2 are plane and spherical, respectively. *Right*: at all points of a wavefront, the light has the same phase

and by the retinal cone mosaic. This chapter explains the idea and the concept of the wavefront underlying it.

19.3.1 The Concept of the Wavefront

Adaptive optics is well understood intuitively based on the concept of a wavefront. In a given optical pathway, a wavefront is a surface perpendicular to the rays. Figure 19.6 shows wavefronts of light that emerges from a point source and is converted by an ideal lens into a parallel beam of light. The wavefronts perpendicular to the rays are initially spherical and turn into planes behind the lens. Figure 19.6 shows the corresponding wave image: by definition, the wavefronts connect points of the same phase (same number of wave peaks away from the point source). In any system – not just in ideal ones – wavefronts stand perpendicular to the rays and, everywhere on a particular wavefront, the light has the same phase. In the concept of wavefronts, diffraction and other wave phenomena are not taken into account.

In less ideal systems, wavefronts deviate from ideal shapes (planes and spherical surfaces). Figure 19.7 shows wavefronts of light that emerge from an illuminated point on the retina. With ideal ocular media, optics, and adaptation, a parallel beam with a planar wavefront is produced outside the eye. In a real eye, however, apart from spherical and cylindrical aberrations, the wavefront may be influenced by local aberrations in the lens or cornea. Conversely, a parallel beam is not imaged as a sharp point on the retina.

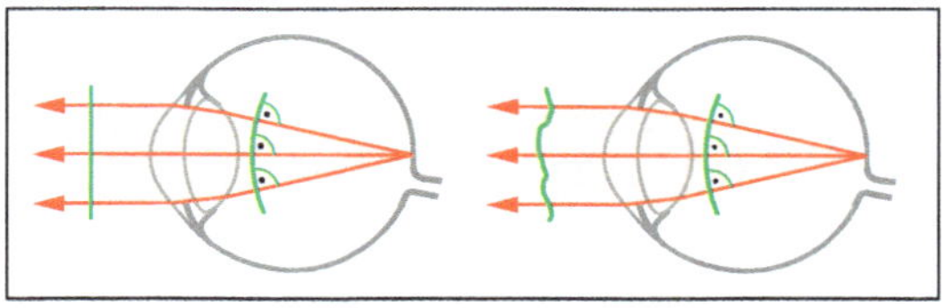

Fig. 19.7 Light originating from a point on the retina. *Left*: ideally, outside the eye, a flat wavefront is created. *Right*: aberrations cause a distorted wavefront outside the eye

19.3.2 Measuring a Wavefront

When one succeeds in measuring an individual wavefront outside the eye, created by an illuminated point on the retina (Fig. 19.7, right), precise conclusions can be drawn about the aberrations of the eye's optical apparatus or about local imperfections along the ray path. This is actually possible, e.g., with the Hartmann–Shack wavefront sensor, which is based on the fact that the wavefront is perpendicular to the light rays at every point (Fig. 19.8). In Shack's[3] implementation, the

[3]Shack's implementation goes back to the shadow masks with which the astrophysicist Johannes Hartmann checked and improved the objective lens of the large Potsdam refracting telescope in 1900.

light emerging from an illuminated point of the retina encounters a grid of small lenses. If the beam of light is completely parallel (plane wavefront), a uniform grid of points results at the sensor CCD. Points that deviate from the ideal position mean that the wavefront is locally inclined and, from this, its spatial shape can be arithmetically reconstructed.

The detected aberrations are decomposed into the Zernike circle polynomials[4] by the instrument and the respective terms are displayed in diopters.

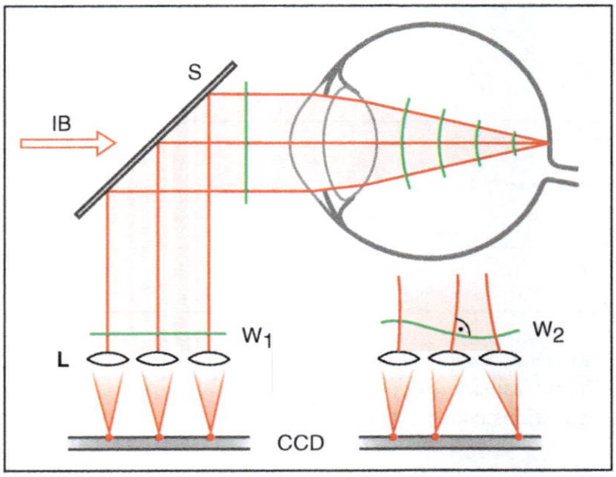

Fig. 19.8 Measuring a wavefront with the Hartmann–Shack sensor by observing the light that emerges from an illuminated point on the retina. A plane wave produces a uniform pattern of points in the focal plane of an arrangement of small lenses. Deviations from the ideal pattern allow one to make conclusions concerning the shape of the wavefront in the cross-section of the beam. *IB* beam illuminating a point on the retina, *S* semi-transparent mirror, W_1 ideal wavefront (plane), W_2 real wavefront (distorted)

Each term corresponds to a specific form of deviation from ideal imaging. The second-order terms comprise the spherical and the two astigmatic deviations (Fig. 19.9). With the higher orders, deviations with increasingly more complex dependence on location within the beam's cross-section are described.

With aberrometers, imaging errors beyond the spherical and cylindrical terms can be obtained. The success of refractive corrections can also be checked. Another type of aberrometer is displayed in Fig. 19.10 (Tscherning aberrometer). By means of an aperture mask, a grid of points is projected onto the retina. In cases of perfect imaging conditions, it should appear as a regular pattern. Deviations from this permit the recognition of aberrations.

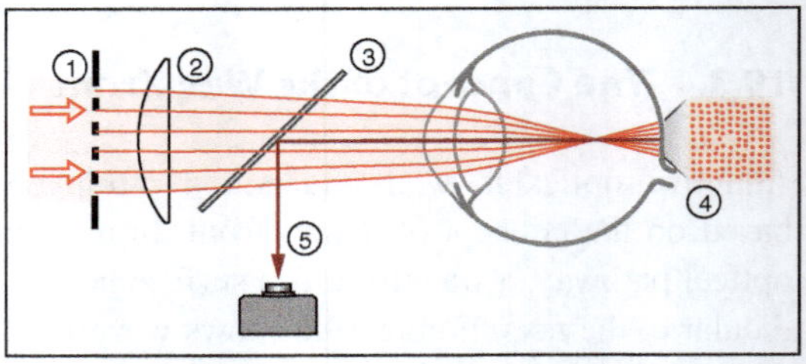

Fig. 19.10 Aberrometer according to the Tscherning principle. An aperture mask (*1*) is imaged using the objective lens (*2*) to form a pattern on the retina (*4*). Its observation (*5*) via a semitransparent mirror (*3*) reveals possible deviations from perfect imaging. The observation (*5*) takes place through a narrow aperture to avoid distortion by the imperfections of the optics of the eye

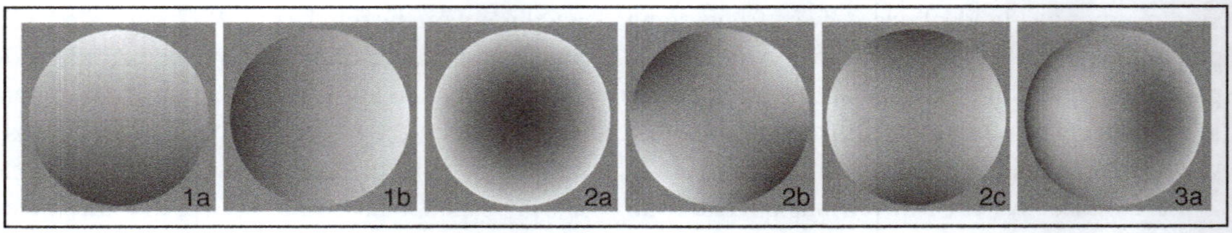

Fig. 19.9 Zernike polynomials. **1a-b** First-order terms (prismatic contributions). **2a-c** Second-order terms (spherocylindrical contributions). **3a** One of the third-order terms

[4]Frits Zernike (1888–1966), Dutch physicist, received the 1953 Nobel Prize for the development of the phase-contrast microscope. He introduced the classification according to orthogonal circle polynomials into the mathematical description of optical system aberrations.

19.4 Abbe's Limit of Resolution and the STED Microscope

Have we not always believed that object details approximately smaller than one wavelength of light cannot be resolved using a microscope? According to Abbe,[5] the smallest resolution of a diffraction-limited microscope amounts to roughly half a wavelength. The reason is the wave nature of light, similar to the Airy disk (Fig. 19.2). Fluorescence microscopy is subject to the same limitation with both the classical method (the whole field illuminated with stimulating light) and the confocal laser scanning technology. Surprisingly, Stefan Hell discovered a way out. According to his idea – now realized in the STED microscope[6] – it has been possible to achieve resolutions on the order of roughly 10 nm, compared with Abbe's limit of 200 nm with blue light (Fig. 19.11).

Fig. 19.11 Ernst Abbe

The principle can be understood in two steps. First, we call to mind laser scanning fluorescence microscopy. Interesting structures of an object are marked with fluorescent molecules. The distribution of these molecules can be scanned line by line with a fluorescence-stimulating beam focused down as small as possible. The intensity of fluorescence is continuously registered as a function of the beam position. From this information, a picture is constructed containing the distribution of the fluorescing molecules. Stimulating light and fluorescent light are separated by colored filters. However, Abbe's limit of resolution is still applicable because the diameter of the scanning beam cannot be made smaller than approximately a wavelength.

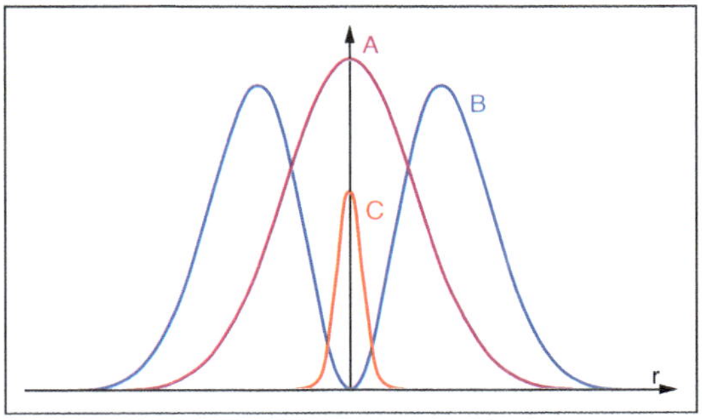

Fig. 19.12 Profile of the laser beams in the STED microscope. *A* stimulating beam, *B* de-excitation beam, *C* remaining profile of observed fluorescence. Its diameter decreases with the larger intensity of the de-excitation beam. The diameter of the beams (*A*) and (*B*) are diffraction-limited, that is, on the order of a wavelength

The decisive step is the following: Figure 19.12 shows a spot of 200 nm in diameter illuminated by a scanning beam (A). This beam overlaps with a second laser beam (B) with the same axis. Its wavelength is chosen so that the stimulated molecules are returned to a non-fluorescent state. However, its profile is special: it illuminates a ring and the intensity on the axis is zero. After this treatment, only the molecules in a small neighborhood of the axis remain in a stimulated state. A fluorescent light that is still detectable stems from this small region. Its peculiarity lies in the fact that this spot becomes smaller with the larger intensity of the de-excitation beam. With the combination of these two laser beams and guidance with high mechanical precision, the resolution can be reduced far below the Abbe limit.

[5]Ernst Abbe 1840–1905. Creator of the concepts of modern optics, such as the theory of the diffraction-limited microscope. He was one of the founders of the German optical industry, a social reformer, and the founder of the Carl-Zeiss foundation (1889).

[6]Stimulated emission depletion.

19.5 Fourier Analysis

In its intuitively accessible content, Fourier analysis is a mathematical procedure that can be explained without equations. The underlying

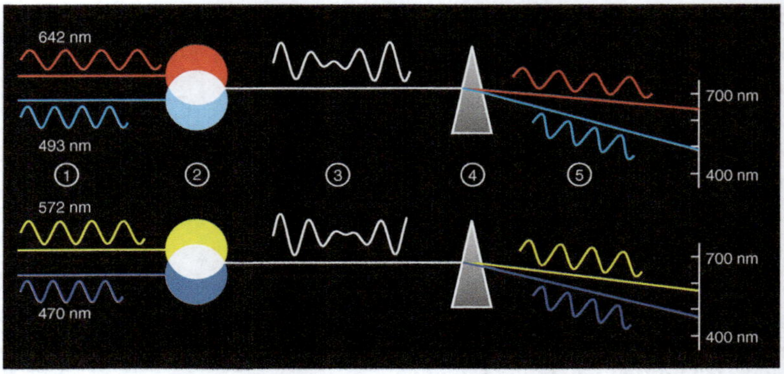

Fig. 19.13 White composing and decomposing. *Top*: the two monochromatic lights (*1*) with the 493 and 642 nm wavelengths with suitable illuminance ratios produce white light in additive mixing (*2*). The prism (*4*) decomposes the combined wave form (*3*) into the original monochromatic components (*5*). *Bottom*: a similar depiction for the wavelength pair 470/572 nm

question is: Which frequencies are hidden in a complex signal? In a female voice, higher frequencies are present than in a male voice. But what exactly does it mean to say that a certain frequency is present in a signal?

An initial example from the world of light will help to show the direction we are headed in somewhat more detail. Assume that you have various lasers available, each with a specific wavelength λ and able to project its light onto a wall. It well known that there are complementary colors (more precisely, complementary lights) that can be combined to result in white light in the right illuminance ratios. Indeed, when the 642 nm (red) and 493 nm (blue-green) wavelengths are projected onto a white wall simultaneously and the illuminances are suitably balanced, you actually see white. With the wavelength pair 470 nm/572 nm, you can also experience white. The eye is unable to distinguish between the two white light combinations involved (Fig. 19.13, left).

However, a spectrograph, consisting of a prism and a wavelength scale, can distinguish between them very well (Fig. 19.13, right): after passing through a prism, two monochromatic beams result with the associated original wavelengths. The prism decomposes the two white lights into their spectral components and reveals the oscillations out of which they are made. A mathematician would tend to say that the prism has performed a Fourier decomposition (or Fourier analysis) of the white lights. In this situation, one understands that a Fourier analysis, in the mathematical sense, means the computational decomposition of the waveforms (3) in Fig. 19.13 into monochromatic components, that is, into sinusoidal oscillations.

The decomposition of a beam of light into its monochromatic components has a very practical meaning: the exact beam path of each component can be followed individually and the refractive indices of the associated colors can be taken into account. This very successful method in optics has a much more general mathematical background: every complex oscillation (mathematically, any function of time) can be decomposed computationally by a so-called Fourier analysis into a sum of many sinusoidal oscillations. Fourier[7] found this important tool for the mathematical analysis of heat diffusion. The mentioned example illustrates that the decomposition of a complex function into monochromatic components is more than simply a mathematical curiosity. It was Fourier analysis that made the development of many modern technologies, some of which are also discussed in this book, possible. Another known example is the decomposition of a vocalized tone into the contributions of the fundamental frequency and its harmonics. In performing such analyses, one must regard the

[7]Jean B. Fourier (1768–1830), French mathematician. His turbulent life, during and after the French Revolution, is documented on Wikipedia.

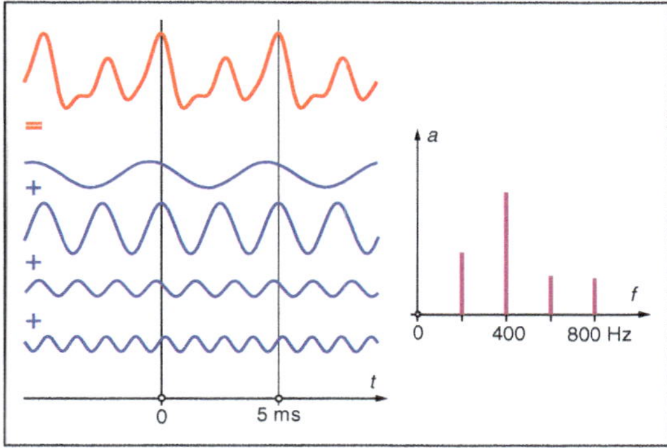

Fig. 19.14 *Left*: spectral decomposition of a vocalized vowel (*red*) into the fundamental frequency and overtones. The sum of the blue functions yields the red function. The *red signal* represents the electrical voltage beyond the microphone as a function of time in a recording of a vocalized "u." Period length: 5 ms. Repetition frequency: 200 Hz. The blue monochromatic components have frequencies of 200, 400, 600, and 800 Hz. That no higher overtones are present in this particular signal lies in the associated voice. The amplitudes and phases of the individual contributions must be chosen correctly so that the sum of the blue functions exactly reproduces the red one. *Right*: spectrum, i.e., the contributing frequencies and their amplitudes

Fourier analysis of periodic signals and of non-periodic signals separately. We turn our attention now to these two cases.

19.5.1 Fourier Decomposition of Periodic Functions

Figure 19.14 portrays a periodic function (red curve). It is said to be periodic because it always repeats itself in the same way. Here, a section from a recording of a vocalized "u" is involved; more precisely, the electrical voltage beyond the microphone is displayed as a function of time. The length of a period is 5 ms, implying that the signal has a repetition frequency of 200 Hz.

Fourier decomposition means representing the function shown in red as a sum of sinusoidal oscillations – or, put differently, decomposing it into its spectral components. Fourier's theorem indicates which frequencies appear in the decomposition in periodic functions, namely the fundamental frequency (here, 200 Hz, given by the repetition rate) and whole-number multiples of the fundamental (400, 600 Hz, ...). The amplitudes and phases of the individual components are chosen aptly. The equations of Fourier analysis indicate how, from the behavior of the function that is to be decomposed, the amplitude and phase of each component are to be calculated. This decomposition is distinct, meaning that it succeeds only with the right amplitude and phase (horizontal displacement) of each sine curve.

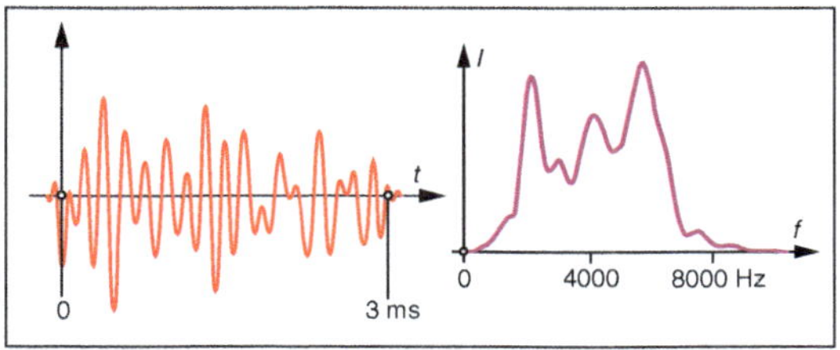

Fig. 19.15 A non-periodic function (here, a section, 3 ms in duration, from a recording of the sibilant "sh," as in "ship") exhibits a continuous spectrum. In this example, the main contributions lie in the frequencies between 1 and 8 kHz. *Left*: microphone signal (pressure amplitude) as a function of time. *t* time. *Right*: spectrum. *I* intensity (amplitude squared). *f* frequency

An essential concept in the analysis of any given oscillation is its spectrum, which indicates the frequencies hidden in the signal and their relative contribution. The spectrum of a periodic function is discrete because, in the Fourier decomposition, only very specific frequencies appear: the fundamental frequency and its whole-number multiples. In the example of Fig. 19.14, the spectrum consists of the 200, 400, 600, and 800 Hz frequencies. Often, instead of the amplitudes in a spectral plot, their squared values are shown because they correlate with the energies of the associated contributions.

19.5.2 Fourier Decomposition of Non-periodic Functions

The spectrum of a non-periodic function is continuous, meaning that every possible frequency can contribute a specific amount. For example, Fig. 19.15 shows a short section from a microphone recording of the sibilant "sh" (as in "ship").

The decomposition into monochromatic functions yields mainly contributions in the frequency range of 1–8 kHz.

19.5.3 Applications

One application in optics has already been mentioned: Fourier decomposition allows the behavior of every light frequency (or every wavelength) to be looked at individually. More precisely, before the light enters the system, we think about it in terms of its colors, study the alterations that the system makes on the individual color components, and then put the altered components back together again. Mathematically, this is possible because, in the interaction of light with lenses, prisms, absorbers, scattering media, etc., almost exclusively so-called linear systems are involved, which permit the viewing of things in terms of components.

20 Appendix: Units and Constants

20.1 Some Physical Units

The standard international (SI) units are summarized in Table 20.1. Alternate units frequently used in medicine are indicated in subsequent comments. Often, units are given with a prefix, such as 1 mJ = 1 milliJoule = 10^{-3} J. Metric prefixes are listed in Table 20.2. Photometric units are dealt with in a separate section (Sect. 20.2).

20.1.1 Length

Wavelengths are often given in μm (micrometer, 10^{-6} m), in nm (nanometer, 10^{-9} m), or in Å (angstrom, 10^{-10} m).

Table 20.1 Some standard international units

Length	meter (m)
Time	second (s)
Mass	kilogram (kg)
Force	newton (N)
Energy	joule (J)
Power	watt (W)
Pressure	pascal (Pa)
Frequency	hertz (Hz)
Temperature	kelvin (K)

20.1.2 Frequency

A frequency denotes a number of events per unit time or the number of cycles of an oscillation per unit time. Unit: 1 Hz = 1 s^{-1}.

20.1.3 Mass

1 g = 10^{-3} kg.

20.1.4 Force

1 N is the force required to accelerate 1 kg of mass at 1 m/s^2. A mass of 0.102 kg has a weight of 1 N (on earth).

20.1.5 Energy

The unit calorie (cal) is sometimes used for heat energy. Formerly, the calorie was defined as the quantity of heat required to raise the temperature of 1 g of water by 1 °C. The conversion into standard units is 1 cal = 4.18 J.

Table 20.2 Metric prefixes

peta	tera	giga	mega	kilo	deci	centi	milli	micro	nano	pico	femto	atto
P	T	G	M	k	d	c	m	μ	n	p	f	a
10^{15}	10^{12}	10^{9}	10^{6}	10^{3}	10^{-1}	10^{-2}	10^{-3}	10^{-6}	10^{-9}	10^{-12}	10^{-15}	10^{-18}

J. Flammer et al., *Basic Sciences in Ophthalmology*,
DOI 10.1007/978-3-642-32261-7_20

1 electronvolt (1 eV $=1.6\cdot10^{-19}$ J) is frequently used in atomic/molecular physics and chemistry. It is the energy required to move an electron (or any elementary electric charge) across an electric potential difference of 1 volt (1 V). Photons of light in the visible range have energies between 1.6 and 3 eV.

Energy density (or fluence) is given in units of J/m^2 or, in laser applications, often in J/cm^2. In typical applications, it is the energy delivered to an absorbing area divided by the area. Example: An energy of 1 mJ absorbed by a spot of 1 mm^2 corresponds to a fluence of 10^{-3} J/10^{-6} m^2 $=10^3$ J/m^2 $=0.1$ J/cm^2.

20.1.6 Power

Power is energy per unit time. An energy of 1 J delivered within 1 s corresponds to a power of 1 W. Very high powers can be achieved by the delivery of moderate energy within a short pulse, such as a laser pulse of energy of 1 mJ and a duration of 1 ns, which corresponds to 1 MW.

Irradiance (power density) is power per area, such as the power of a laser beam divided by its cross-section. Standard unit: 1 W/m^2. Example: A laser pointer of 1 mW power with a cross-section of 1 mm^2 has an irradiance of 10^3 W/m^2. Sunlight (outside the atmosphere) has an irradiance of 1300 W/m^2.

20.1.7 Pressure

The standard unit 1 Pa is a very small pressure, defined by a force of 1 N acting on an area of 1 m^2. Frequently used units are:

- 1 bar $=10^5$ Pa; about equal to the atmospheric pressure on the earth at sea level.
- 1 mmHg $=133.3$ Pa: the pressure exerted at the base of a column of mercury 1 mm high.
- 1 atm $=101{,}325$ Pa.
- 1 Torr $=1$ atm/760, about the same as 1 mmHg (the difference is negligible for most applications).

20.1.8 Temperature

The absolute temperature scale (given in K) starts at 0 at the absolute zero temperature, which is

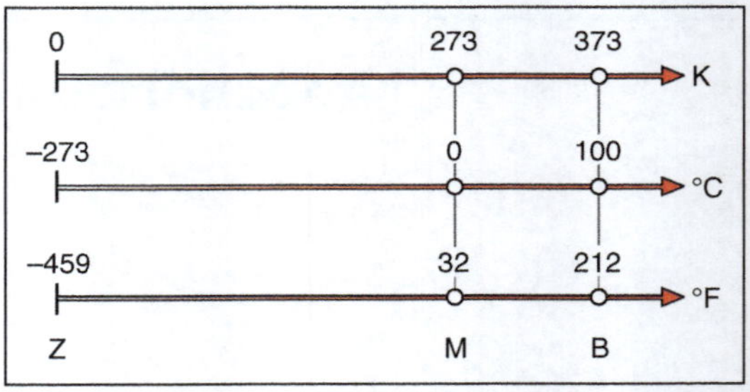

Fig. 20.1 Relationship between temperature scales

−273.16 °C on the Celsius scale and −459.67 °F on the Fahrenheit scale. The melting and boiling of water under standard pressure occur at 0 °C and 100 °C, respectively. The relations between temperature differences are 1 K $=1$ °C $=1.8$ °F. The relation between the scales (Fig. 20.1) is given by the following formulae:

Kelvin temperature = Celsius temperature + 273.16
Celsius temperature = Kelvin temperature − 273.16
Kelvin temperature = (Fahrenheit temperature + 459.67)·(5/9)
Fahrenheit temperature = (9/5)·(Kelvin temperature) − 459.67

20.1.9 Viscosity

Viscosity, a measure of internal fluid friction, is mentioned in Sect. 18.7. The standard unit of dynamic viscosity is 1 Pa·s $=1$ kg/(m s). Usually, viscosity is expressed in units of 1 mPa·s (milli-pascal second), which is about the viscosity of water at 20 °C. The dynamic viscosity is used in the Hagen–Poiseuille formula for the flux in a pipe (see Sect. 18.7).

Other units frequently used in tables include: 1 poise $=0.1$ Pa·s or 1 cp $=10^{-2}$ poise (centipoise). The viscosity of water at 20 °C is about 1 cp $=1$ mPa·s.

In some applications, kinematic viscosity is used, which is the dynamic viscosity divided by the density of the fluid. Its standard unit is 1 m^2/s. Example: from the dynamic viscosity of water ($1.0\cdot10^{-3}$ Pa·s, 20 °C) and its density (10^3 kg/m^3), the kinematic viscosity results in $1.0\cdot10^{-6}$ m^2/s. Kinematic viscosity can be expressed in stokes (1 St $=10^{-4}$ m^2/s) or in centistokes (1 cSt = 10^{-6} m²/s), which is approximately the kinematic viscosity of water.

20.1.10 Surface Tension, Interfacial Tension

Surface tension and interfacial tension are explained in Sect. 18.5. The standard unit is J/m², which is the same as N/m.

20.1.11 Room Angle

When viewed from the center of a sphere with a 1 m radius, an area of 1 m² on its surface is said to subtend a room angle of 1 sr (steradian, sterad). Our view into all directions corresponds to a room angle of $4\pi = 12.56$. The moon and sun subtend a room angle of $6 \cdot 10^{-5}$ according to their diameter of 0.5°.

20.2 Photometric Units

Light is a form of energy. The energy that a laser pointer emits in the form of light per unit time is a form of power and can be expressed in watts. The power of a beam of light is one of the so-called radiometric values. Its definition has nothing to do with the sensitivity of our eyes. A further example of a radiometric value is irradiance. It expresses the power per unit area that falls onto an illuminated desk in units of W/m².

Photometric values and units take into account the sensitivity of the eye (see Fig. 1.19). A laser pointer with a green beam of 1 mW power produces a brighter spot than a red beam of the same power. A photometric statement that accommodates this difference could be that this green beam of light has a five-times-larger luminous flux than the red one. Luminous flux is a photometric concept. We shall now proceed to define three basic quantities and associated units that are used in photometry in response to the following questions:

- How much light is emitted by a certain street lamp? (luminous flux, given in lumens). Among other factors, the answer depends on the electrical power consumption of the lamp, its efficiency in converting electric energy into light, and its spectrum.
- How well is the street illuminated by the street lamp? (illuminance, given in lux). Apart from the luminous flux of the lamp, the answer depends on the lamp's height above the ground.
- How bright does the street appear? (luminance, given in asb). The answer depends not only on the illuminance but also on the absorption of light by the street.

20.2.1 Luminous Flux

The basic photometric unit is the lumen (lm). In short, 1 lm is the luminous flux that is provided by 1.5 mW of monochromatic light with a wavelength of 0.556 μm, i.e., at the maximum of the photopic luminosity function.[1] Figure 20.2 explains the definition of the unit lumen more generally. The same luminous flux of 1 lm can be realized by 3 mW of monochromatic orange light (wavelength 0.61 μm) or with 10.5 mW of thermal light produced at a temperature of 6000 K, similar to sunlight. In all three situations, the same brightness is achieved when the light illuminates the same area of a white paper (white to reflect all colors equally). The luminous flux emitted by low-power light emitting diodes is on the order of magnitude of 1 lm. A 100-W incandescent bulb emits about 2000 lm.

20.2.2 Illuminance

A further photometric value that can be explained with the same situation is illuminance (Fig. 20.3). When light of 1 lm falls on an area of 1 m², the illuminance is 1 lm/m² = 1 lx (lux). The illuminance is a measure of the strength of the illumination, independent of the color or reflectance of the illuminated object. Let's take a simple example, where an LED designed to produce a luminous flux of 1 lm shines all of it onto a surface of 100 cm². The luminous flux amounts to 1 lm/0.01 m² = 100 lx. Table 20.3 gives some typical scene illuminances.

[1] More precisely, at 0.556 μm, a power of 1/683 W = 1.46 mW corresponds to a luminous flux of 1 lm.

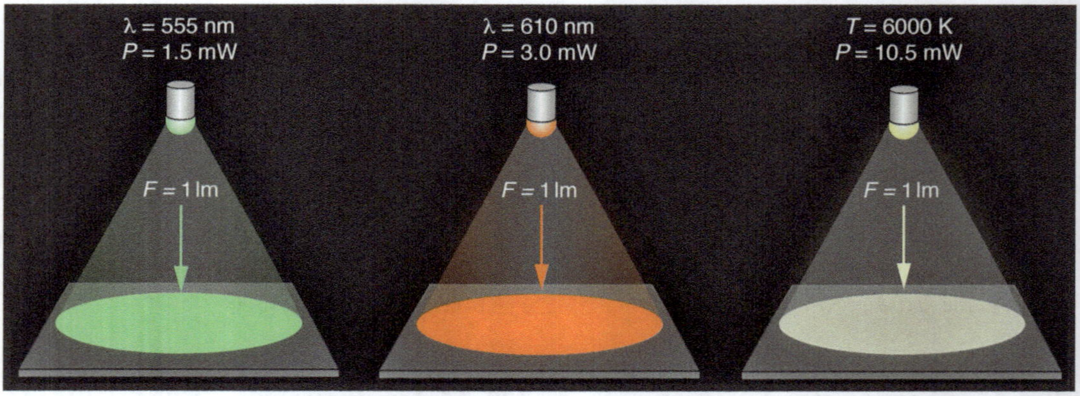

Fig. 20.2 Definition of the luminous flux F and its unit (lumen, 1 lm). The three light sources differ by their spectra (green, orange, white light). In the three situations, the luminous flux is the same (1 lm), but the physical power P differs (1.5, 3.0, and 10.5 mW). The luminous flux depends on the (physical) power carried by the light and on the spectrum. In all three situations, the same brightness is achieved when the light illuminates the same area of a piece of white paper

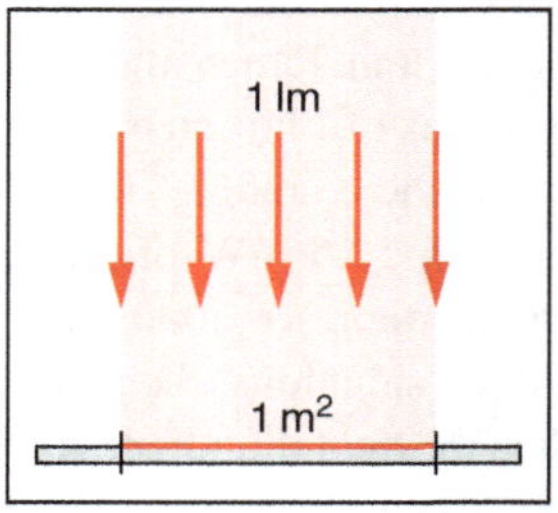

Fig. 20.3 Definition of the unit of illuminance. A luminous flux of 1 lm falling onto 1 m^2 (or 2 lm falling onto 2 m^2) produces an illuminance of 1 lx, irrespective of color and absorption of the illuminated object. The illuminance characterizes the illuminating light level, not the brightness of the object

Table 20.3 Typical scene illuminance (ground illumination by some sources). Values are orders of magnitude

Direct sunlight	10^5 lx
Full daylight	10^4 lx
Overcast day	10^3 lx
Office	$5{\cdot}10^2$ lx
Full moon	0.3 lx

20.2.3 Luminance

The concept of luminance is qualitatively simple to explain, but the exact definition is somewhat more difficult. Luminance indicates how bright an illuminated surface appears. First, we state that the brightness of a surface – say, a wall – does not change when we view it from different distances (this has a simple geometric reason: the power that enters the pupil from a given object changes by the same factor as the area on the retina). Moreover, it does not whether we concentrate on a small part of the wall or see it in its entirety.

Sometimes, a wall has the additional property such that it appears to have the same brightness when seen from any direction. In this typical case, one speaks of a Lambertian reflector. A whitewashed wall or a matte sheet of paper of any color is typical examples of Lambertian reflectors. For a start, we will discuss only this special situation (Fig. 20.4). A white Lambertian source, which absorbs no light but reflects all of it, when illuminated by an illuminance of 1 lx, has, by definition, a luminance of 1 asb (Apostilb). A gray wall, absorbing 60 % and illuminated with the same illuminance of 1 lx, has a lower luminance (0.4 asb).

An interesting equation in view of applications refers to the situation in Fig. 20.5: a camera with an objective lens of diameter D and focal length f is aimed at the wall. Assuming that one knows the luminance L of the lit wall, how large is the illuminance E_S of the sensor in the camera? The equation is provided in the figure. The distance of the camera

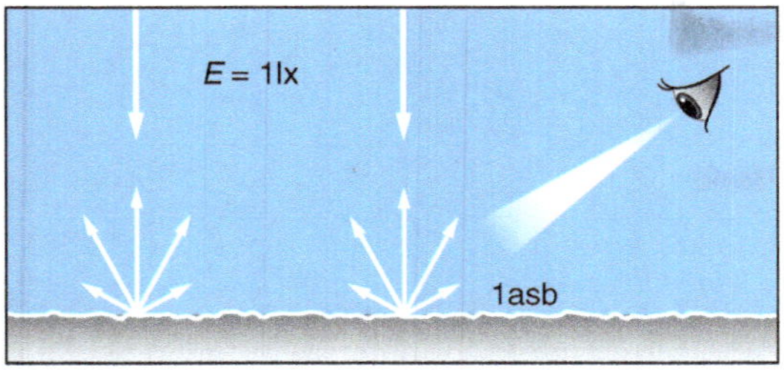

Fig. 20.4 An eye looks at a whitewashed wall that is assumed to be a Lambertian reflector. For an illuminance $E = 1$ lx, the eye sees a luminance $L = 1$ asb from any direction

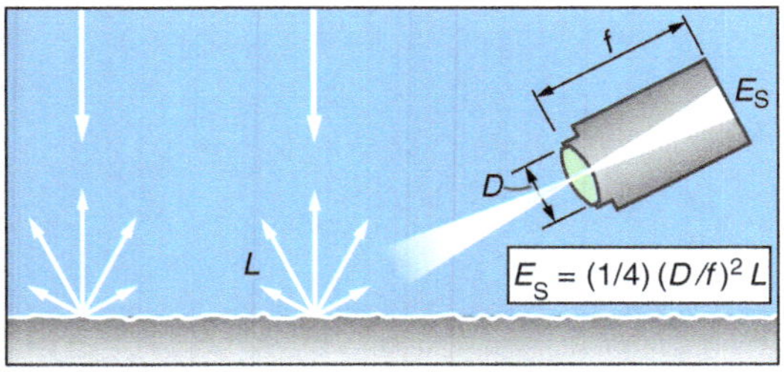

Fig. 20.5 Observing the luminance using a measuring instrument. The luminance L follows from the measured illuminance E_S in the sensor in the focus of the objective lens. D diameter of the objective lens, f focal length

from the wall plays no role; only the *f*-number (the focal length divided by the diameter of the aperture, *f*-number$=f/D$) does. Since we have assumed a Lambertian reflector, the angle from which one photographs the wall does not play a role.

The literature presents a variety of units for luminance. Here are the conversions:

1 asb (Apostilb) = 0.318 cd/m^2,

1 sb (Stilb) = 10^4 cd/m^2,

1 L (Lambert) = $(1/\pi)$ 10^4 cd/m^2,

1 fL (Foot-Lambert) = 3.426 cd/m^2.

Above, we spoke about how the luminance L for a Lambertian reflector can be calculated from its illuminance E (for a white reflector, an illuminance = 1 lx $\rightarrow$ luminance = 1 asb). We could also determine the luminance directly in accordance with Fig. 20.5, where one measures the illuminance E_S at the sensor and then calculates L using the equation. Conceivably, the result might depend on the angle of view. In this case, it is not a Lambertian reflector.

20.3 Some Physical Constants

1 Mol of a substance consists of $6.022 \cdot 10^{23}$ particles (Avogadro's constant).

The vacuum velocity of light $c = 3.0 \cdot 10^8$ m/s is exactly the same for all wavelengths of electromagnetic radiation. In a transparent medium with an index of refraction n, the speed of light is reduced to $c' = c/n$ and may depend on the wavelength (dispersion). The frequency f and wavelength λ of light in a medium with the speed of light c' are related by $f = c'/\lambda$. At the wavelength of 0.58 μm of yellow light, the frequency amounts to $5 \cdot 10^{14}$ Hz.

The Boltzmann constant $k = 1.38 \cdot 10^{-23}$ J/K can be used to estimate the order of magnitude of the mean thermal energy per atom by $k \cdot T$, where T is the absolute temperature. For noble gases, the mean thermal energy per atom is given exactly by $3\,k \cdot T/2$. At room temperature, $k \cdot T \approx 0.026$ eV.

We encountered Planck's constant $h = 6.626$ 10^{-34} J·s in the formula $E = h \cdot f = h \cdot c/\lambda$ for the energy of photons (λ = wavelength, f = frequency). For wavelengths in the visual range ($\lambda = 0.4 \ldots 0.7$ μm), $E = 2 \ldots 3$ eV. According to quantum mechanics, the formula $E = h \cdot f$ is applicable more generally to the energy of quanta of any vibration of frequency f.

Index

J. Flammer et al., *Basic Sciences in Ophthalmology*,
DOI 10.1007/978-3-642-32261-7,

GPSR Compliance

The European Union's (EU) General Product Safety Regulation (GPSR) is a set of rules that requires consumer products to be safe and our obligations to ensure this.

If you have any concerns about our products, you can contact us on ProductSafety@springernature.com

In case Publisher is established outside the EU, the EU authorized representative is:

Springer Nature Customer Service Center GmbH
Europaplatz 3
69115 Heidelberg, Germany

Batch number: 10731051

Printed by Printforce, the Netherlands